MECHANICS OF GENERALIZED CONTINUA

AF182441

Advances in Mechanics and Mathematics

VOLUME 21

Series Editors
David Y. Gao (Virginia Polytechnic Institute and State University)
Ray W. Ogden (University of Glasgow)
Romesh C. Batra (Virginia Polytechnic Institute and State University)

Advisory Board
Ivar Ekeland (University of British Columbia, Vancouver)
Tim Healey (Cornell University, USA)
Kumbakonam Rajagopal (Texas A&M University, USA)
Tudor Ratiu (École Polytechnique Fédérale, Lausanne)
David J. Steigmann (University of California, Berkeley)

Aims and Scope

Mechanics and mathematics have been complementary partners since Newton's time, and the history of science shows much evidence of the beneficial influence of these disciplines on each other. The discipline of mechanics, for this series, includes relevant physical and biological phenomena such as: electromagnetic, thermal, quantum effects, biomechanics, nanomechanics, multiscale modeling, dynamical systems, optimization and control, and computational methods.

Driven by increasingly elaborate modern technological applications, the symbiotic relationship between mathematics and mechanics is continually growing. The increasingly large number of specialist journals has generated a complementarity gap between the partners, and this gap continues to widen. *Advances in Mechanics and Mathematics* is a series dedicated to the publication of the latest developments in the interaction between mechanics and mathematics and intends to bridge the gap by providing interdisciplinary publications in the form of monographs, graduate texts, edited volumes, and a special annual book consisting of invited survey articles.

For more titles in this series, go to http://www.springer.com/series/5613

MECHANICS OF GENERALIZED CONTINUA

ONE HUNDRED YEARS AFTER THE COSSERATS

Edited By

Gérard A. Maugin
Université Pierre et Marie Curie, Paris, France

Andrei V. Metrikine
Delft University of Technology, Delft, The Netherlands

Springer

Editors:
Gérard A. Maugin
Institut Jean Le Rond d'Alembert
Université Paris VI
75252 Paris CX 5, France
gerard.maugin@upmc.fr

Andrei V. Metrikine
Faculty of Civil Engineering and Geosciences
Delft University of Technology
2628 CN Delft, The Netherlands
A.Metrikine@tudelft.nl

Series Editors:
David Y. Gao
Department of Mathematics
Virginia Polytechnic Institute
Blacksburg, VA 24061, USA
gao@vt.edu

Romesh C. Batra
Department of Mathematics
Virginia Polytechnic Institute
Blacksburg, VA 24061, USA
rbatra@vt.edu

Ray W. Ogden
Department of Mathematics
University of Glasgow
Glasgow, Scotland, UK
rwo@maths.gla.ac.uk

ISSN 1571-8689
e-ISSN 1876-9896
ISBN 978-1-4614-2574-8
e-ISBN 978-1-4419-5695-8
DOI 10.1007/978-1-4419-5695-8
Springer New York Dordrecht Heidelberg London

Mathematics Subject Classification (2010): 74-02, 74-06, 74A35, 74A60, 74E15, 74Q15, 76A15, 82D25

© Springer Science+Business Media, LLC 2010
Softcover reprint of the hardcover 1st edition 2010
All rights reserved. This work may not be translated or copied in whole or in part without the written permission of the publisher (Springer Science+Business Media, LLC, 233 Spring Street, New York, NY 10013, USA), except for brief excerpts in connection with reviews or scholarly analysis. Use in connection with any form of information storage and retrieval, electronic adaptation, computer software, or by similar or dissimilar methodology now known or hereafter developed is forbidden.
The use in this publication of trade names, trademarks, service marks, and similar terms, even if they are not identified as such, is not to be taken as an expression of opinion as to whether or not they are subject to proprietary rights.

Printed on acid-free paper

Springer is part of Springer Science+Business Media (www.springer.com)

Foreword

Welcome!

I am especially privileged and honored that Professors Maugin, Metrikine and Erofeyev, the organizers and chairmen of this meeting, the Euromech Colloquium 510 honoring the Cosserats for the 100 year anniversary of the publication of their book, have asked me to say a few words to express my welcome salute to you. Much as I would have liked to do this in person, my physical being is no longer keeping pace with my mental desires and, thus, alas, is denying me this luxury.

Sometime in the past, I remember reading an article whose author's name has slipped my memory—perhaps it was Marston Morse, Professor Emeritus at the Institute for Advanced Study, who wrote (and I paraphrase):

Discovery of new mathematical disciplines originates from two criteria:

1. **Generalization**
2. **Inversion**

Some of the earliest examples for the validity of these criteria are:

(a) The Newton–Leibniz discovery of differentiation and integration, which started calculus; and
(b) The Theory of Elasticity, which was conceived when Robert Hooke, in 1678, published an anagram: "ceiiinosssttuu", which he expressed as "*ut tensio sic vis*", meaning, the power of any material is in the same proportion within the tension thereof. Presently, this is known as "Hooke's Law".

Some 250 years later, "The modern theory of elasticity may be considered to have its birth in 1821, when Navier first gave the equations for the equilibrium and motion of elastic solids, ..." (Todhunter and Pearson).

Of course, many other scientists, Cauchy, Poisson, Stokes, and others, after 1821, improved and extended the theory to other materials, e.g., viscous fluids, and they investigated atomic and molecular foundations. This is typical—for the maturation of any discipline is the result of the contributions of many scientists and often takes a long time.

Improvements and extensions of the theory of elasticity continued in the nineteenth and early part of the twentieth century: rigorous mathematical theory of nonlinear elasticity, relativistic continuum mechanics, magneto-elasticity and other "hyphenated" sister fields, like viscoelasticity and thermoelasticity. Underlying basic postulates (e.g., frame-independence, thermodynamical restrictions, relativistic invariance) were introduced and applied in the development of field equations and admissible constitutive laws. Research in granular and porous elastic solids, composite elastic materials, polymeric materials, and statistical and molecular foundations of continua are but a few examples that still remain as active research fields.

Eugène Maurice Pierre Cosserat and his brother François Cosserat, 100 years ago, cast the seed of **Generalized Continua**, by publishing a book, in 1909, entitled *Théorie des Corps Déformables* (Hermann, Paris). The revolutionary contribution of this book is that material points of an elastic solid are considered equipped with directors, which give rise to the concept of couple stress and a new conservation law for the moment of momentum. By means of a variational principle which they called "***l'action euclidienne***", they obtained "balance laws of elasticity". The introduction of the director concept made it possible to formulate anisotropic fluids, e.g., liquid crystals, blood.

The Cosserats did not give constitutive equations. These, and the introduction of the microinertia tensor and the associated conservation law, which are crucial to the dynamic problems in solid and fluent media (e.g., liquid crystals, suspensions, etc.) were introduced later by other scientists.

Over half a century elapsed before the Cosserats' book was discovered by researchers. After 1960, independent, Cosserat-like theories were published in European countries, the USA and the USSR, under a variety of nomenclature (e.g., couple stress, polar elasticity, asymmetric elasticity, strain gradient theories, micropolar elasticity, multipolar theory, relativistic continua with directors, etc.). I recall a literature search on these subjects that was shown to me by a visiting scholar, Professor Listrov, from the USSR This book contained several hundred entries of papers published by 1970.

The next significant generalizations appear in 1964 and thereafter, in the areas of microelasticity, microfluid mechanics, micropolar continua, micromorphic electrodynamics, and others that constitute the family of micromorphic continua or microstructure theories.

The conception of these theories was based on the query, "Is it possible to construct continuum theories that can predict physical phenomena on the atomic, molecular, or nano scales?" These would require supplying additional degrees of freedom to the material point beyond a director. After all, the molecules that constitute the internal structures of the material points (particles) undergo deformations and rotations arising from the displacement and rotations of their constituent atoms. This supplies twelve degrees of freedom. A body with such an internal structure is called **Micromorphic grade 1**. Micromorphic continua of grade $N > 1$ have also been formulated.

To understand the difference between the Cosserat and the micromorphic elasticities, it is important to note that micromorphic elasticity gives rise to two different second-order strain tensors (only one of which is symmetric), and to one third-order microstrain tensor. Correspondingly, the balance laws introduce two second-order stress tensors (only one of which is symmetric), and one third-order microstress (moment-stress) tensor.

In special cases, the Micromorphic Theory leads to other special continuum theories:

Micromorphic → Microstretch → Micropolar (Cosserats) → Classical

The next important contributions are the nonlocal continuum theories that generalize constitutive equations for classical and micromorphic continuum theories, by introducing the influence of distant material points, e.g., the stress tensor is a functional of the strain tensors of all material points of a body. In this sense, micromorphic grade 1 is a nonlocal theory with a *short nonlocality* (or discrete nonlocality). Among the many important contributions of nonlocality, I mention that it eliminates the stress singularity (infinite stress) at the crack tip predicted by classical elasticity. Moreover, a natural fracture criterion was born which states that failure occurs when the maximum stress becomes or exceeds the cohesive stress.

The Present State. No doubt other generalized continuum theories are in a state of composition. But mathematical theories cannot be considered the truth without experimental verification. Unfortunately, excluding classical theories, the experimental work for all these theories is left wanting. The opportunity is here and now, for experimentalists to determine the material moduli and/or to confirm or challenge the validity of some of these theories.

A Note on the Future. Ultimately, all continuum theories must be based on the quantum field theory, or perhaps, on the quantum theory of general relativity (when unified). This offers the greatest challenges to future scientific investigators.

I am pleased to see so many interesting contributions to some of these fields included in this meeting, which are in the spirit of the Cosserats' work.

I welcome you and send my best wishes for what, I am sure, will be an inspirational and productive meeting.

Littleton, Colorado, May 2009

A. Cemal Eringen
Professor Emeritus,
Princeton University

Preface

This volume gathers in some organized and edited manner most of the contributions delivered at the EUROMECH Colloquium 510 held in Paris, May 13–16, 2009. The explicit aim of the colloquium was, on the occasion of the centennial of the publication of a celebrated book (Théorie des corps déformables) by the Cosserat brothers, to examine the evolution in time since the Cosserats, and the actuality of the notion of generalized continuum mechanics to which the Cosserats' work contributed to some important extent. Of course, the Cosserat book belongs to this collection of classics that are more often cited than read. The reason for this is twofold. First, the vocabulary and mathematical symbols have tremendously evolved since the early 1900s, and second, the Cosserat book by itself is an intrinsically difficult reading. As a matter of fact, more than introducing precisely the notion of Cosserat media (a special class of generalized continua), the Cosserats' book had a wider ambition, that of presenting a reflection on the general framework of continuum mechanics, with the notion of group permeating—not explicitly—its structure (cf. the notion of "action euclidienne"). This is reflected in many of the following contributions.

Overall, the whole landscape of contemporary generalized continuum mechanics was spanned from models to applications to structures, dynamical properties, problems with measurement of new material coefficients, numerical questions posed by the microstructure, and new possible developments (nanomaterials, fractal structures, new geometrical ideas). Remarkably absent were models and approaches using the concept of strong nonlocality (constitutive equations that are functionals over space). This is a mark of a certain evolution.

An interesting comparison can be made with the contents of the landmark *IUTAM Symposium* gathered in 1967 in Stuttgart-Freudenstadt under the chairmanship of the late E. Kröner. Most of the models presented at that meeting by luminaries such as Noll, Eringen, Rivlin, Green, Sedov, Mindlin, Nowacki, Stojanovic, and others were essentially of the Cosserat type and, still in their infancy, had a much questioned usefulness that is no longer pondered. Most of the contributions were either American or German. With the present EUROMECH we witnessed an enlargement of the classes of models with a marked interest in gradient-type theories. Also, because the political situation has drastically changed within forty tears, we realize now the importance of the Russian school. The latter was, in fact, very much ignored in the 1960s and 1970s while some Russian teams were ahead of their Western colleagues in acknowledging their debt to the Cosserats and other scientists such as Leroux, Le Corre and Laval in France. Of these heroic Soviet times, E. Aero and V. Palmov, both from St. Petersburg, who published on the subject matter in the early 1960s, were present in Paris. Professor A.C. Eringen (he also in Freundenstadt in 1967), unable to attend, kindly sent us a Welcome address that is reproduced here in the way of a Foreword.

Unfortunately, the editing of this book was saddened by the passing away of A.C. Eringen on December 06, 2010, at the age of 88, after more than sixty years of devotion to engineering science, physics and applied mathematics.

The Colloquium was financially and materially supported by the Engineering UFR of the *Université Pierre et Marie Curie* (UPMC), the STII Directorate of the French *Centre National de la Recherche Scientifique*, and the *Institut Jean Le Rond d'Alembert*, UPMC–Paris Universitas and UMR 7190 of CNRS. Members of the MPIA Team of this Institute helped much in the local organization. Ms Simona Otarasanu is to be thanked for her efficient treatment of many questions. Without the expertise of Ms Janine Indeau, the present volume would not exist.

Paris *Gérard A. Maugin*
Delft *Andrei V. Metrikine*

Contents

Contributors

Eron L. Aero Institute of Problems in Mechanical Engineering, Bolshoy av. 61, V.O., St. Petersburg 199178, Russia, 16aero@mail.ru

Elias C. Aifantis Laboratory of Mechanics and Materials, Polytechnic School, Aristotle University of Thessaloniki, Thessaloniki 54124, Greece, mom@mom.gen.auth.gr; Center for Mechanics of Material Instabilities and Manufacturing Processes, College of Engineering, Michigan Technological University, Houghton, MI 49931, USA, mom@mtu.edu

Holm Altenbach Martin-Luther-Universität Halle-Wittenberg, 06099 Halle (Saale), Germany, holm.altenbach@iw.uni-halle.de

B.R. Andrievsky Institute of Problems in Mechanical Engineering, Bolshoy av. 61, V.O., St. Petersburg 199178, Russia, bandri@yandex.ru

A. Arunachalakasi Department of Applied Mechanics, Indian Institute of Technology Madras, Chennai 600 036, India, aarajan@iitm.ac.in

Harm Askes Department of Civil and Structural Engineering, University of Sheffield, Mappin Street, Sheffield S1 3JD, UK, h.askes@sheffield.ac.uk

Petr Belov Inst. of Applied Mechanics of RAS, Russia, Moscow, Petr.Belov@boing.com

Claude Boutin DGCB, FRE CNRS 3237, Ecole Nationale des Travaux Publics de l'Etat, Université de Lyon, Lyon, France, claude.boutin@entpe.fr

A.N. Bulygin Institute of Problems in Mechanical Engineering, Bolshoy av. 61, V.O., St. Petersburg 199178, Russia, bulygin_an@mail.ru

David A. Burton Department of Physics, Lancaster University, Lancaster, UK, d.burton@lancaster.ac.uk

Youping Chen University of Florida, Gainsevilla, FL, USA, ypchen2@ufl.edu

Céline Chesnais DGCB, FRE CNRS 3237, Ecole Nationale des Travaux Publics de l'Etat, Université de Lyon, Lyon, France, celine.chesnais@entpe.fr

Stefan Diebels Chair of Applied Mechanics, Saarland University, Campus A 4.2, 66123 Saarbrü, Germany, s.diebels@mx.uni-saarland.de

Arcady V. Dyskin The University of Western Australia, 35 Stirling Highway, Crawley, Perth, WA 6009, Australia, arcady@civil.uwa.edu.au

Marcelo Epstein Department of Mechanical and Manufacturing Engineering, University of Calgary, Calgary, Alberta T2N 1N4, Canada, mepstein@ucalgary.ca

Victor A. Eremeyev South Scientific Center of RASci & South Federal University, Milchakova St. 8a, 344090 Rostov on Don, Russia, eremeyev.victor@gmail.com

Vladimir I. Erofeyev A.A. Blagonravov Mechanical Engineering Institute RAS, Nizhny Novgorod Branch, Belinskogo str., 85, Nizhny Novgorod 603024, Russia, erf04@sinn.ru

Paul Fischer University Erlangen-Nuremberg, Egerlandstrasse 5, 91058 Erlangen, Germany, paul.fischer@ltm.uni-erlangen.de

Marcel Frehner University of Vienna, Department of Geodynamics and Sedimentology, Althanstrasse 14, 1090 Vienna, Austria, marcel.frehner@univie.ac.at

Francesco Froiio Ecole Centrale de Lyon, Ecully, France, francesco.froiio@ec-lyon.fr

M.G.D. Geers Department of Mechanical Engineering, Eindhoven University of Technology, PO Box 513, 5600 MB Eindhoven, The Netherlands

H.G. Georgiadis Department of Mechanics, National Technical University of Athens, Zographou Campus, Zographou 15773, Greece, georgiad@central.ntua.gr

Inna M. Gitman Department of Mechanical Engineering, University of Sheffield, Mappin Street, Sheffield S1 3JD, UK, i.gitman@sheffield.ac.uk

P.A. Gourgiotis National Technical University of Athens, Zographou 15773, Greece

Stéphane Hans DGCB, FRE CNRS 3237, Ecole Nationale des Travaux Publics de l'Etat, Université de Lyon, Lyon, France, stephane.hans@entpe.fr

Ralf Jänicke Chair of Applied Mechanics, Saarland University, Campus A 4.2, 66123 Saarbrü, Germany, r.jaenicke@mx.uni-saarland.de

Mahmood Jabareen Faculty of Civil and Environmental Engineering, Technion-Israel Institute of Technology, Haifa 32000, Israel, cvjmah@tx.technion.ac.il

Jena Jeong École Spéciale des Travaux Publics du Bâtiment et de l'Industrie (ESTP), 28 avenue du Pré, 94234 Cachan Cedex, France, jeong@profs.estp.fr

Y. Kasyanyuk Department of Mechanical Engineering, Eindhoven University of Technology, PO Box 513, 5600 MB Eindhoven, The Netherlands

Markus Lazar Emmy Noether Research Group, Department of Physics, Darmstadt University of Technology, Hochschulstr. 6, 64289 Darmstadt, Germany, lazar@fkp.tu-darmstadt.de

Arthur Lebée Université Paris-Est, UR Navier, École des Ponts ParisTech, 6 et 8 avenue Blaise Pascal, 77455 Marne-la-Vallé, France, arthur.lebee@lami.enpc.fr

James D. Lee The George Washington University, Washington, DC, USA, jdlee@gwu.edu

Daniel Lhuillier UPMC Univ Paris 6, UMR 7190 CNRS, Institut Jean Le Rond d'Alembert, Paris, France, daniel.lhuillier@upmc.fr

Jun Li University of Illinois at Urbana-Champaign, Urbana, IL 61801, USA, junli3@illinois.edu

Sergey Lurie Inst. of Applied Mechanics of RAS, Russia, Moscow, lurie@ccas.ru

Gérard A. Maugin UMR 7190 CNRS, Institut Jean Le Rond d'Alembert, UPMC Univ Paris 6, Paris, France, gerard.maugin@upmc.fr

German A. Maximov N.N. Andreyev Acoustical Institute, Shvernika str. 4, Moscow 117036, Russia, maximov@dpt39.mephi.ru

Julia Mergheim University Erlangen-Nuremberg, Egerlandstrasse 5, 91058 Erlangen, Germany, julia.mergheim@ltm.uni-erlangen.de

Andrei V. Metrikine Faculty of Civil Engineering and Geosciences, Delft University of Technology, 2600 GA Delft, The Netherlands, A.Metrikine@tudelft.nl

Wolfgang Muschik Institut für Theoretische Physik, TU Berlin, Hardenbergstr. 36, 10623 Berlin, Germany, muschik@physik.tu-berlin.de

Ingo Münch Institut für Baustatik, Universität Karlsruhe (TH), Kaiserstrasse 12, 76131 Karlsruhe, Germany, im@bs.uka.de

Patrizio Neff Chair of Nonlinear Analysis and Modelling, Fakultät für Mathematik, Universität Duisburg-Essen, Campus Essen, Universitä, 45141 Essen, Germany, patrizio.neff@uni-due.de

Elena A. Nikitina A.A. Blagonravov Mechanical Engineering Institute RAS, Nizhny Novgorod Branch, Belinskogo str., 85, Nizhny Novgorod 603024, Russia

Martin Ostoja-Starzewski University of Illinois at Urbana-Champaign, Urbana, IL 61801, USA, martinos@illinois.edu

Vladimir Palmov St. Petersburg Polytechnical University, St. Petersburg, Russia, palmov@mail.ru

Stefanos-Aldo Papanicolopulos Department of Mechanics, National Technical University of Athens, Iroon Polytechneiou 5, 15773 Zografou, Greece, stefanos@mechan.ntua.gr

Christina Papenfuss Institut für Theoretische Physik, TU Berlin, Hardenbergstr. 36, 10623 Berlin, Germany, c.papenfuss@gmx.de

Elena Pasternak The University of Western Australia, 35 Stirling Highway, Crawley, Perth, WA 6009, Australia, elena@mech.uwa.edu.au

Robert H.J. Peerlings Department of Mechanical Engineering, Eindhoven University of Technology, PO Box 513, 5600 MB Eindhoven, The Netherlands, R.H.J.Peerlings@tue.nl

Wojciech Pietraszkiewicz Institute of Fluid-Flow Machinery of the Polish Academy of Sciences, ul. Gen. J. Fiszera 14, 80-952 Gdań, Poland, pietrasz@imp.gda.pl

Alexey V. Porubov Institute of Problems in Mechanical Engineering, Bolshoy av. 61, V.O., St. Petersburg 199178, Russia, porubov.math@mail.ioffe.ru

Julia M. Prokhorova Delft University of Technology, 2600 GA Delft, The Netherlands

Hamidréza Ramézani École Polytechnique de l'Université d'Orléans, CNRS-CRMD, 8 rue Lé, 45072 Orlé, France, hamidreza.ramezani@cnrs-orleans.fr

A. Roy Department of Mechanical Engineering, Eindhoven University of Technology, PO Box 513, 5600 MB Eindhoven, The Netherlands

Miles B. Rubin Faculty of Mechanical Engineering, Technion-Israel Institute of Technology, Haifa 32000, Israel, mbrubin@tx.technion.ac.il

Karam Sab Université Paris-Est, UR Navier, École des Ponts ParisTech, 6 et 8 avenue Blaise Pascal, 77455 Marne-la-Vallé, France, sab@enpc.fr

Carlo Sansour Division of Materials, Mechanics, and Structures, The University of Nottingham, University Park, Nottingham NG7 2RD, UK, carlo.sansour@nottingham.ac.uk

Lusine S. Sargsyan Gyumri State Pedagogical Institute, Sevak 4, Gyumri, Armenia, slusin@yahoo.com

Smuel H. Sargsyan Gyumri State Pedagogical Institute, Sevak 4, Gyumri, Armenia

Stefan Schmalholz ETH Zurich, Geological Institute, Sonneggstrasse 5, 8092 Zurich, Switzerland, schmalholz@erdw.ethz.ch

Alla V. Sharabanova A.A. Blagonravov Mechanical Engineering Institute RAS, Nizhny Novgorod Branch, Belinskogo str., 85, Nizhny Novgorod 603024, Russia

Sebastian Skatulla CERECAM, Department of Civil Engineering, The University of Cape Town, Private Bag X3, Rondebosch 7701, South Africa, sebastian.skatulla@uct.ac.za

Holger Steeb Ruhr-University Bochum, Mechanics-Continuum Mechanics, Universitätsstrasse 150, 44780 Bochum, Germany, Holger.Steeb@rub.de

Paul Steinmann University Erlangen-Nuremberg, Egerlandstrasse 5, 91058 Erlangen, Germany, paul.steinmann@ltm.uni-erlangen.de

Natalia Tuchkova Dorodnicyn Computing Centre of RAS, Russia, Moscow, tuchkova@ccas.ru

Robin W. Tucker Department of Physics, Lancaster University, Lancaster, UK, r.tucker@lancaster.ac.uk

Ioannis Vardoulakis†

Xianqiao Wang The George Washington University, Washington, DC, USA, xqwang@gwmail.gwu.edu

A. Zervos School of Civil Engineering and the Environment, University of Southampton, Southampton SO17 1BJ, UK, az@soton.ac.uk

† I. Vardoulakis (March 22nd, 1949 – September 19th, 2009), Formerly at the National Technical University of Athens, Greece.

Part I
On the Cosserat's Works

Chapter 1
Generalized Continuum Mechanics: What Do We Mean by That?

Gérard A. Maugin

Dedicated to A.C. Eringen

Abstract Discursive historical perspective on the developments and ramifications of generalized continuum mechanics from its inception by the Cosserat brothers (Théorie des corps déformables. Hermann, 1909) with their seminal work of 1909 to the most current developments and applications is presented. The point of view adopted is that generalization occurs through the successive abandonment of the basic working hypotheses of standard continuum mechanics of Cauchy, that is, the introduction of a rigidly rotating microstructure and *couple stresses* (Cosserat continua or *micropolar* bodies, nonsymmetric stresses), the introduction of a truly deformable microstructure (*micromorphic* bodies), "weak" *nonlocalization* with *gradient theories* and the notion of *hyperstresses*, and the introduction of characteristic lengths, "strong nonlocalization" with space functional constitutive equations and the loss of the Cauchy notion of stress, and finally giving up the Euclidean and even Riemannian material background. This evolution is paved by landmark papers and timely scientific gatherings (e.g., Freudenstadt, 1967; Udine, 1970, Warsaw, 1977).

Preliminary note: Over 40 years, the author has benefited from direct studies under, and lectures from, P. Germain, A.C. Eringen, E.S. Suhubi, R.D. Mindlin, W. Nowacki, V. Sokolowski, S. Stojanovic, from contacts with J.L. Ericksen, C.A. Truesdell and D.G.B. Edelen, from friendship with C.B. Kafadar, J.M. Lee, D. Rogula, H.F. Tiersten, J. Jaric, P.M. Naghdi, I.A. Kunin, L.I. Sedov, V.L. Berdichevskii, E. Kröner, and most of the authors in the present volume as co-workers or friends, all active contributors to the present subject matter. He apologizes to all these people who certainly do not receive here the fully deserved recognition for their contribution to the field.

G.A. Maugin (✉)
UMR 7190 CNRS, Institut Jean Le Rond d'Alembert, UPMC Univ Paris 6, Paris, France
e-mail: gerard.maugin@upmc.fr

G.A. Maugin, A.V. Metrikine (eds.), *Mechanics of Generalized Continua*,
Advances in Mechanics and Mathematics 21,
DOI 10.1007/978-1-4419-5695-8_1, © Springer Science+Business Media, LLC 2010

1.1 Introduction

The following question is naturally raised with the venue of EUROMECH 510 in Paris in May 2009. What do we understand by generalized continuum mechanics? Note already some ambiguity since the last expression can be alternately phrased as "generalized (continuum mechanics)" or "(generalized continuum) mechanics". We do not pursue this semantic matter. We simply acknowledge the fact that with the publication of the book of the Cosserat brothers in 1909 a true "generalized continuum mechanics" developed, first slowly and rather episodically and then with a real acceleration. That a new era was borne at the time in the field of continuum mechanics is not obvious if we remember that the Cosserats' theory was published as a supplement to the French translation (by them for some, I suppose, alimentary purpose) of Chowlson's Russian Encyclopedia of mathematics (the translation was done from the German edition). Another valid subtitle of the present contribution could be "From the classical to the less classical". But what is classical? Then the "less classical" or "generalized" will be defined by successively discarding the working hypotheses of the classical case, simple as the latter may be.

1.2 From Cauchy and the Nineteenth Century

Here we consider as a classical standard the basic model considered by engineers in solid mechanics and the theory of structures. This essentially is the theory of continua set forth by A.L. Cauchy in the early nineteenth century for isotropic homogeneous elastic solids in small strains. The theory of continua respecting Cauchy's axioms and simple working hypotheses is such that the following holds true:

1. *Cauchy's postulate*. The traction exerted on a facet cut in the solid depends on the geometry of that facet only at the *first order* (the local unit normal); it will be linear in that normal. From this the notion of a *stress tensor* follows, the so-called stress being the only "internal force" in the theory.
2. *It is understood* that both physical space (of Newton) and material manifold (the set of material particles constituting the body) are Euclidean and connected, whence the notion of displacement is well defined.
3. *Working hypothesis (i)*. There are no applied couples in both volume and surface.
4. *Working hypothesis (ii)*. There exists no "microstructure" described by additional internal degrees of freedom.

According to Points 3 and 4, the Cauchy stress tensor is symmetric. This results from the application of the balance of angular momentum. Isotropy, homogeneity, and small strains are further hypotheses but they are not so central to our argument. Then generalizations of various degrees consist in relaxing more or less these different points above, hence the notion of *generalized continuum*. This notion of generalization depends also on the culture and physical insight of the scientists. For instance, the following generalizations are "weak" ones:

- "Generalized" Hooke's law (linear, homogeneous, but *anisotropic* medium);
- Hooke–Duhamel law in thermoelasticity;
- Linear homogeneous piezoelectricity in obviously anisotropic media (no center of symmetry).

These are "weak" generalizations because they do not alter the main mathematical properties of the system. Of course, thermoelasticity and linear piezoelectricity require adding new independent variables (e.g., temperature θ or scalar electric potential φ). In some sense, the problem becomes four-dimensional for the basic field (elastic displacement and temperature in one case, elastic displacement and electric potential in the other). The latter holds in this mere simplicity under the hypothesis of weak electric fields, from which there follows the neglect of the so-called ponderomotive forces and couples, e.g., the couple $\mathbf{P} \times \mathbf{E}$ when electric field and polarization are not necessarily aligned; see Eringen and Maugin [24]. Such theories, just like standard elasticity, do not involve a *length scale*. But classical linear *inhomogeneous* elasticity presents a higher degree of generalization because a characteristic length intervenes necessarily.

From here on, we envisage three true (in our view) generalizations.

1.2.1 The Cauchy Stress Tensor Becomes Nonsymmetric for Various Reasons

This may be due to

(i) The existence of body couples (e.g., in electromagnetism: $\mathbf{P}\times\mathbf{E}$ or/and $\mathbf{M}\times\mathbf{H}$; the case of intense EM fields or linearization about intense bias fields);
(ii) The existence of surface couples (the introduction of "internal forces" of a new type of the so-called *couple stresses*); the medium possesses internal degrees of freedom that modify the balance of angular momentum;
(iii) The existence of internal degrees of freedom of a nonmechanical nature in origin, e.g., polarization inertia in ferroelectrics, intrinsic spin in ferromagnetics (see Maugin's book [57]);
(iv) The existence of internal degrees of freedom of "mechanical" nature.

This is where the Cosserats' model comes into the picture.

The first example in this class pertains to a *rigid microstructure* (three additional degrees of freedom corresponding to an additional rotation at each material point, independently of the vorticity). Examples of media of this type go back to the early search for a continuum having the capability to transmit transverse waves (as compared to acoustics in a pure fluid), i.e., in relation to optics. The works of McCullagh [64] and Lord Kelvin must be singled out (cf. Whittaker [85]). Pierre Duhem [9] proposes to introduce a triad of three rigidly connected directors (unit vectors) to represent this rotation. In modern physics, there are other tools for this, including Euler's angles (not very convenient), quaternions and spinors. It is indeed the Cosserats, among other studies in elasticity, who really introduced internal degrees of freedom of the rotational type (these are *micropolar continua* in the sense of Eringen) and the dual concept of *couple stress*. Hellinger [36], in a brilliant essay, recognized at once the new potentialities offered by this generalization but did not elaborate on these. A modern rebirth of the field had to await works in France by crystallographs (Laval [45–47]; Le Corre [49]), in Russia by Aero and Kuvshinskii [1], and Palmov [71], in Germany by Schaeffer [77], Günther [34], Neuber [67], and in Italy by Grioli [33] and Capriz—see Capriz's book of 1989 [3]. But the best formulations are those obtained by considering a field of orthogonal transformations (rotations) and not the directors themselves, see Eringen [19–21], Kafadar and Eringen [37], Nowacki [70], although we note some obvious success of the "director" representation, e.g., in *liquid crystals* (Ericksen [17]; Leslie [52]) and the kinematics of the deformation of slender bodies (Ericksen, Truesdell, Naghdi)—in this volume see the contribution of Lhuillier. But there was in the mid 1960s a complete revival of continuum mechanics (cf. Truesdell and Noll [82]) which, by paying more attention to the basics, favored the simultaneous formulation of many more or less equivalent theories of generalized continua in the line of thought of the Cosserats (works by Mindlin and Tiersten [66], Mindlin and Eshel [65], Green and Rivlin [32], and Green and Naghdi [31], Toupin [80, 81], Truesdell and Toupin [83], and Eringen and Suhubi [25, 26], etc.).

More precisely, in the case of a *deformable microstructure* at each material point, the vector triad of directors of Duhem–Cosserats becomes deformable and the additional degree of freedom at each point, or micro-deformation, is akin to a general linear transformation (nine degrees of freedom). These are *micromorphic continua* in Eringen's classification. A particular case is that of continua with microstretch. A truly new notion here is that of the existence of a conservation law of *micro-inertia* (Eringen [18], Stokes [79]). In the present volume, this is illustrated by several contributions. A striking example is due to Drouot and Maugin [8] dealing with fluid solutions of macromolecules, while Pouget and Maugin [73] have provided a fine example of truly micromorphic solids with the case of piezoelectric powders treated as continua.

Remark 1.1. Historical moments in the development of this avenue of generalization have been the IUTAM symposium organized by E. Kröner in Freudenstadt in 1967 (see Kröner [41]) and the CISM Udine summer course of 1970 (Mindlin, Eringen, Nowacki, Stojanovic, Sokolowski, Maugin, Jaric, Micunovic, etc. were present).

Remark 1.2. Strong scientific initial motivations for the studies of generalized media at the time (1960–1970s) were (i) the expected elimination of field singularities in many problems with standard continuum mechanics, (ii) the continuum description of *real* existing materials such as granular materials, suspensions, blood flow, etc. But further progress was hindered by a notorious lack of knowledge of new (and too numerous) material coefficients despite trials of estimating such coefficients, e.g., by Gauthier and Jashman [28] at the Colorado School of Mines by building artificially microstructured solids.

Remark 1.3. Very few French works were concluded in the 1960–1970s if we note the exceptional work of Duvaut [10, 11] on finite strains after a short stay in the

USA, the variational principle for micromorphic bodies by Maugin [53] from the USA, and those on micropolar fluids by C. Hartmann [35] under the influence of R. Berker (who had been the teacher of Eringen in Istanbul).

Remark 1.4. The intervening of a rotating microstructure allows for the introduction of wave modes of rotation of the "optical" type with an obvious application to many solid crystals (e.g., crystals equipped with a polar group such as $NaNO_2$, cf. Pouget and Maugin [74]).

1.2.2 The Loss of Validity of the Cauchy Postulate

Then the geometry of a cut intervenes at a higher order than one (variation of the unit normal, role of the curvature, edges, apices and thus capillarity effects). We may consider two different cases referred to as the *weakly nonlocal theory* and the *strongly nonlocal theory* (distinction introduced by the author at the Warsaw meeting of 1977, cf. Maugin [55]). Only the first type does correspond to the exact definition concerning a cut and the geometry of the cut surface. This is better referred to as *gradient theories of the nth order*; it is understood that the standard Cauchy theory is, in fact, a *theory of the first gradient* (by this we mean the first gradient of the displacement or the theory involving just the strain and no gradient of it in the constitutive equations).

1.2.2.1 Gradient Theories

Now, to tell the truth, gradient theories abound in physics, starting practically with all continuum theories in the nineteenth century. Thus, Maxwell's electromagnetism is a first-gradient theory (of the electromagnetic potentials); the Korteweg [39] theory of fluids is a theory of the first gradient of density (equivalent to a second-gradient theory of displacement in elasticity); Einstein's [12] (also [13]) theory of gravitation (general relativity, 1916) is a second-gradient theory of the metric of curved space–time, and Le Roux [50] (also [51]) seems to be the first public exhibition of a second-gradient theory of (displacement) elasticity in small strains (using a variational formulation). There was a renewal of such theories in the 1960s with the works of Casal [4] on capillarity, and of Toupin [80], Mindlin and Tiersten [66], Mindlin and Eshel [65], and Grioli [33] in elasticity.

However, it is with a neat formulation basing on the *principle of virtual power* that some order was imposed in these formulations with an unambiguous deduction of the (sometimes tedious) boundary conditions and a clear introduction of the notion of *internal forces* of higher order, i.e., *hyperstresses* of various orders (see, Germain [29, 30], Maugin [56]). Phenomenological theories involving gradients of other physical fields than displacement or density, coupled to deformation, were envisaged consistently by the author in his Princeton PhD thesis (Maugin [54]) dealing with typical ferroïc electromagnetic materials. This is justified by a microscopic approach, i.e., the continuum approximation of a crystal lattice with medium-range interactions; with distributed magnetic spins or permanent electric dipoles. This also applies to the pure mechanical case (see, for instance, the Boussinesq paradigm in Christov et al. [5]).

Very interesting features of these models are:

F1. Inevitable introduction of characteristic lengths;
F2. Appearance of the so-called capillarity effects (surface tension) due to the explicit intervening of curvature of surfaces;
F3. Correlative boundary layers effects;
F4. Dispersion of waves with a possible competition and balance between nonlinearity and dispersion, and the existence of solitonic structures (see Maugin [60], Maugin and Christov [63]);
F5. Intimate relationship with the Ginzburg–Landau theory of phase transitions and, for fluids, van der Waals' theory.

Truly sophisticated examples of the application of these theories are found in

(i) The coupling of a gradient theory (of the carrier fluid) and consideration of a microstructure in the study of the inhomogeneous diffusion of microstructures in polymeric solutions (Drouot and Maugin [8]);
(ii) The elimination of singularities in the study of structural defects (dislocations, disclinations) in elasticity combining higher-order gradients and polar microstructure (cf. Lazar and Maugin [48]).

Most recent works consider the application of the notion of gradient theory in elastoplasticity for nonuniform plastic strain fields (works by Aifantis, Fleck, Hutchinson, and many others)—but see the thermodynamical formulation in Maugin [58]. In the present volume, this trend is exemplified by the first-hand synthesis contribution of E.C. Aifantis.

Insofar as general mathematical principles at the basis of the notion of gradient theory are concerned, we note the fundamental works of Noll and Virga [69] and Dell'Isola and Seppecher [7], the latter with a remarkable economy of thought.

1.2.2.2 Strongly Nonlocal Theory (Spatial Functionals)

Initial concepts in this framework were established by Kröner and Datta [42], Kunin [43, 44], Rogula [76], Eringen and Edelen [23]. As a matter of fact, the Cauchy construct does *not* apply anymore. In principle, only the case of *infinite bodies* should be considered as any cut would destroy the prevailing long-range ordering. Constitutive equations become integral expressions over space, perhaps with a more or less rapid attenuation with distance of the spatial kernel. This, of course, inherits from the action-at-a-distance dear to the Newtonians, while adapting the disguise of a continuous framework. This view is justified by the approximation of an infinite crystal lattice; the relevant kernels can be justified through this discrete approach. Of course, this raises the matter of solving integro-differential equations instead of

PDEs. What about boundary conditions that are in essence foreign to this representation of matter-matter interaction? There remains a possibility of the existence of a "weak-nonlocal" limit by the approximation by gradient models.

The historical moment in the recognition of the usefulness of strongly nonlocal theories was the EUROMECH colloquium on nonlocality organized by D. Rogula in Warsaw (cf. Maugin [55]). A now standard reference is Eringen's book [22], also Kunin [44]. A recent much publicized application of the concept of nonlocality is that to *damage* by Pijaudier–Cabot and Bazant [72].

Note in conclusion to this point that any field theory can be generalized to a nonlocal one while saving the notions of linearity and anisotropy; but loosing the usual notion of flux. Also, it is of interest to pay attention to the works of Lazar and Maugin [48] for a comparison of field singularities in the neighborhood of structural defects in different "generalized" theories of elasticity (micropolar, gradient-like, strongly nonlocal or combining these). In this respect, see Lazar's contribution in this volume.

1.2.3 Loss of the Euclidean Nature of the Material Manifold

Indeed, the basic relevant problem emerges as follows. How can we represent *geometrically* the fields of structural defects (such as *dislocations* associated with a loss of continuity of the elastic displacement, or *disclinations* associated with such a loss for rotations). A similar question is raised for *vacancies and point* defects. One possible answer stems from the consideration of a *non-Euclidean material manifold*, e.g., a manifold without curvature but with an affine connection, or an Einstein–Cartan space with *both* torsion and curvature, etc. With this, one enters a true "geometrization" of continuum mechanics of which conceptual difficulties compare favorably with those met in modern theories of gravitation. Pioneers in the field in the years 1950–1970 were K. Kondo [38] in Japan, E. Kröner [40] in Germany, Bilby in the UK, Stojanovic [78] in what was then Yugoslavia, W. Noll [68] and C.C. Wang [84] in the USA. Modern developments are due to, among others, M. Epstein and the author [14, 15], M. Elzanowski and S. Preston (see the theory of material inhomogeneities by Maugin [59]). Main properties of this type of approach are (i) the relationship to the multiple decomposition of finite strains (Bilby, Kroener, Lee) and (ii) the generalization of theories such as the theory of volumetric growth (Epstein and Maugin [16]) or the theory of phase transitions within the general *theory of local structural rearrangements* (local evolution of reference; see Maugin [62], examining Kröner's inheritance and also the fact that true *material inhomogeneities* (dependence of material properties on the material point) are then seen as *pseudo-plastic effects* [61]. All local structural rearrangements and other physical effects (e.g., related to the diffusion of a dissipative process) are reciprocally seen as pseudo material inhomogeneities (Maugin [62]). An original geometric solution is presented in the book of Rakotomanana [75] which offers a representation of a material manifold that is everywhere dislocated. Introduction of the notion of fractal sets opens new horizons (cf. Ostoja-Starzewski's contribution in this volume). An antiquated forerunner work of all this may be guessed in Burton [2], but only with obvious good will by a perspicacious reader.

1.3 Conclusion

Since the seminal work of the Cosserats, three more or mess successful paths haven been taken towards the generalization of continuum mechanics. These were recalled above. They are also fully illustrated in the various contributions that follow. An essential difference between the bygone times of the pioneers and now is that artificial materials can be man-made that are indeed generalized continua. In addition, mathematical methods have been developed (homogenization techniques) that allow one to show that generalized continua are deduced as macroscopic continuum limits of some structured materials. This is illustrated by the book of Forest [27].

References

1. Aero, E.L., Kuvshinskii, E.V.: Fundamental equations of the theory of elastic media with rotationally interacting particles. Sov. Phys. Solid State **2**, 1272–1281 (1961). Engl. Transl.; in Russian (1960)
2. Burton, C.V.: Theory concerning the constitution of matter. Philos. Mag. **33**(201), 191–204 (1891)
3. Capriz, G.: Continua with Microstructure. Springer, New York (1989)
4. Casal, P.: Capillarité interne en mécanique. C. R. Acad. Sci. Paris **256**, 3820–3822 (1963)
5. Christov, C.I., Maugin, G.A., Porubov, A.S.: On Boussinesq's paradigm on nonlinear wave propagation. C. R. Méc. **335**(9–10), 521–535 (2007). Acad. Sci. Paris, Special Issue on Boussinesq
6. Cosserat, E., Cosserat, F.: Théorie des corps déformables. Hermann, Paris (1909). Reprint, Gabay, Paris (2008)
7. Dell'Isola, F., Seppecher, P.: The relationship between edge contact forces, double forces and interstitial working allowed by the principle of virtual power. C. R. Acad. Sci. Paris IIb **321**, 303–308 (1995)
8. Drouot, R., Maugin, G.A.: Phenomenological theory for polymer diffusion in non-homogeneous velocity gradient flows. Rheolog. Acta **22**(4), 336–347 (1983)
9. Duhem, P.: Le potentiel thermodynamique et la pression hydrostatique. Ann. Ecol. Norm. **10**, 187–230 (1893)
10. Duvaut, G.: Application du principe de l'indifférence matérielle à un milieu élastique matériellement polarisé. C. R. Acad. Sci. Paris **258**, 3631–3634 (1964)
11. Duvaut, G.: Lois de comportement pour un milieu isotrope matériellement polarisé de degré 2. C. R. Acad. Sci. Paris **259**, 3178–3179 (1964)
12. Einstein, A.: Die Grundlage der allgemeinen Relativitätstheorie. Ann. Phys. **49**, 769–822 (1916)
13. Einstein, A.: The Meaning of Relativity. Princeton University Press, Princeton (1956)
14. Epstein, M., Maugin, G.A.: The energy-momentum tensor and material uniformity in finite elasticity. Acta Mech. **83**(3–4), 127–133 (1990)

15. Epstein, M., Maugin, G.A.: Notions of material uniformity and homogeneity. In: Tatsumi, T., Watanabe, E., Kambe, T. (eds.) Theoretical and Applied Mechanics, Opening Lecture of MS1, ICTAM, Kyoto, 1996, pp. 201–215. Elsevier, Amsterdam (1997)
16. Epstein, M., Maugin, G.A.: Thermomechanics of volumetric growth in uniform bodies. Int. J. Plast. **16**(7–8), 951–978 (2000)
17. Ericksen, J.L.: Anisotropic fluids. Arch. Ration. Mech. Anal. **4**, 231–237 (1960)
18. Eringen, A.C.: Theory of micropolar fluids. J. Math. Mech. **16**, 1–18 (1966)
19. Eringen, A.C.: Theory of micropolar elasticity. In: Liebowitz, H. (ed.) Fracture: A Treatise, vol. II, pp. 621–729. Academic Press, New York (1968)
20. Eringen, A.C.: Microcontinuum Field Theories, I—Foundations and Solids. Springer, New York (1999)
21. Eringen, A.C.: Microcontinuum Field Theories, II—Fluent Media. Springer, New York (2001)
22. Eringen, A.C.: Nonlocal continuum field theories. Springer, New York (2002)
23. Eringen, A.C., Edelen, D.G.B.: On nonlocal elasticity. Int. J. Eng. Sci. **10**(3), 233–248 (1972)
24. Eringen, A.C., Maugin, G.A.: Electrodynamics of Continua, 2 vols. Springer, New York (1990)
25. Eringen, A.C., Suhubi, E.S.: Nonlinear theory of simple microelastic solids I. Int. J. Eng. Sci. **2**(2), 189–203 (1964)
26. Eringen, A.C., Suhubi, E.S.: Nonlinear theory of simple microelastic solids II. Int. J. Eng. Sci. **2**(4), 389–404 (1964)
27. Forest, S.: Milieux continus généralisés et matériaux hétérogènes. Presses de l'Ecole des Mines, Paris (2006)
28. Gauthier, R.D., Jashman, W.E.: A quest for micropolar elastic constants. J. Appl. Mech., Trans. ASME Ser. E **42**(2), 369–374 (1975)
29. Germain, P.: La méthode des puissances virtuelles en mécanique des milieux continus, Première partie: théorie du second gradient. J. Méc. (Paris) **12**, 235–274 (1973)
30. Germain, P.: The method of virtual power in continuum mechanics—II: Microstructure. SIAM J. Appl. Math. **25**(3), 556–575 (1973)
31. Green, A.E., Naghdi, P.M.: Micropolar and director theories of plates. Q. J. Mech. Appl. Math. **20**, 183–199 (1967)
32. Green, A.E., Rivlin, R.S.: Multipolar continuum mechanics. Arch. Ration. Mech. Anal. **17**, 113–147 (1964)
33. Grioli, G.: Elasticità asimmetrica. Ann. Mat. Pura Appl., Ser. IV **50**, 389–417 (1960)
34. Günther, W.: Zur Statik und Kinematik des Cosseratschen Kontinuums. Abh. Braunschweig. Wiss. Ges. **10**, 195 (1958)
35. Hartmann, C.: Ecoulement d'un fluide micropolaire autour de la sphère à faible nombre de Reynolds. J. Méc. (Paris) **12**(1), 97–120 (1973)
36. Hellinger, E.: Die allgemeinen Ansätze der Mechanik der Kontinua. In: Klein, F., Wagner, K. (eds.) Enz. Math. Wiss, vol. 4, pp. 602–694. Springer, Berlin (1914)
37. Kafadar, C.B., Eringen, A.C.: Micropolar media—I—The classical theory. Int. J. Eng. Sci. **9**(3), 271–308 (1971)
38. Kondo, K.: Non-Riemannian geometry of imperfect crystals from a macroscopic viewpoint. In: Kondo, K. (ed.) RAAG Memoirs of the Unifying Study of Basic Problems in Engineering and Physical Sciences by Means of Geometry, vol. 1, pp. 459–480. Gakujutsu Bunken Fukyukai, Tokyo (1955)
39. Korteweg, D.J.: Sur la forme que prennent les équations du mouvement des fluides si l'on tient compte des forces capillaires causées par des variations de densité considérables mais continues et sur la théorie de la capillarité dans l'hypothèse d'une variation de la densité. Arch. Néer. Sci. Exactes Nat., Sér. II **6**, 1–24 (1901)
40. Kröner, E.: Kontinuumstheorie der Versetzungen und Eigenspannungen. Springer, Berlin (1958)
41. Kröner, E. (ed.): Generalized Continua. Proc. IUTAM Symp. Freudenstadt. Springer, Berlin (1968)

42. Kröner, E., Datta, B.K.: Nichtlokal Elastostatik: Ableitung aus der Gittertheorie. Z. Phys. **196**(3), 203–211 (1966)
43. Kunin, I.A.: Model of elastic medium with simple structure and space dispersion. Prikl. Mat. Mekh. **30**, 542–550 (1966)
44. Kunin, I.A.: Elastic Media with Microstructure I & II. Springer, Berlin (1982). Translated from the 1975 Russian edition
45. Laval, J.: L'élasticité du milieu cristallin. J. Phys. Radium **18**(4), 247–259 (1957)
46. Laval, J.: L'élasticité du milieu cristallin II. J. Phys. Radium **18**(5), 289–296 (1957)
47. Laval, J.: L'élasticité du milieu cristallin III. J. Phys. Radium **18**(6), 369–379 (1957)
48. Lazar, M., Maugin, G.A.: On microcontinuum field theories: the Eshelby stress tensor and incompatibility conditions. Philos. Mag. **87**, 3853–3870 (2007)
49. Le Corre, Y.: La dissymétrie du tenseur des efforts et ses conséquences. J. Phys. Radium **17**(11), 934–939 (1956)
50. Le Roux, J.: Etude géométrique de la torsion et de la flexion, dans les déformations infinitésimales d'un milieu continu. Ann. Ecole Norm. Super. **28**, 523–579 (1911)
51. Le Roux, J.: Recherches sur la géométrie des déformations finies. Ann. Ecole Norm. Super. **30**, 193–245 (1913)
52. Leslie, F.M.: Constitutive equations for liquid crystals. Arch. Ration. Mech. Anal. **28**, 265–283 (1968)
53. Maugin, G.A.: Un principe variationnel pour des milieux micromorphiques non dissipatifs. C. R. Acad. Sci. Paris A **271**, 807–810 (1970)
54. Maugin, G.A.: Micromagnetism and polar media. PhD thesis, Princeton University (1971)
55. Maugin, G.A.: Nonlocal theories or gradient-type theories: a matter of convenience? Arch. Mech. **31**(1), 15–26 (1979). PL, Proc. Euromech Coll. on Nonlocal Theories, Warsaw (1977)
56. Maugin, G.A.: Method of virtual power in continuum-mechanics: application to coupled fields. Acta Mech. **35**(1–2), 1–70 (1980)
57. Maugin, G.A.: Continuum Mechanics of Electromagnetic Solids. North-Holland, Amsterdam (1988)
58. Maugin, G.A.: Internal variables and dissipative structures. J. Non-Equilib. Thermodyn. **15**(2), 173–192 (1990)
59. Maugin, G.A.: Material Inhomogeneities in Elasticity. Chapman & Hall, London (1993)
60. Maugin, G.A.: Nonlinear Waves in Elastic Crystals. Oxford University Press, Oxford (1999)
61. Maugin, G.A.: Pseudo-plasticity and pseudo-inhomogeneity effects in materials mechanics. J. Elast. **71**(1–3), 81–103 (2003)
62. Maugin, G.A.: Geometry and thermomechanics of structural rearrangements: Ekkehart Kroener's Legacy. Z. Angew. Math. Mech. **83**(2), 75–84 (2003). GAMM'2002, Kröener's Lecture, Augsbug (2002)
63. Maugin, G.A., Christov, C.I.: Nonlinear waves and conservation laws (nonlinear duality between elastic waves and quasi-particles). In: Christov, C.I., Guran, A. (eds.) Topics in Nonlinear Wave Mechanics, p. 40. Birkhäuser, Boston (2002)
64. McCullagh, J.: An essay towards a dynamical theory of crystalline reflexion and refraction. Trans. R. Irish Acad. Sci. **21**, 17–50 (1839)
65. Mindlin, R.D., Eshel, N.N.: On first strain-gradient theories in linear elasticity. Int. J. Solids Struct. **4**(1), 109–124 (1968)
66. Mindlin, R.D., Tiersten, H.F.: Effects of couple stresses in linear elasticity. Arch. Ration. Mech. Anal. **11**, 415–448 (1962)
67. Neuber, H.: On the general solution of linear elastic problems in isotropic and anisotropic Cosserat continua. In: Contribution to 11th International Conference of Applied Mechanics, München, 1964
68. Noll, W.: Materially uniform simple bodies with inhomogeneities. Arch. Ration. Mech. Anal. **27**, 1–32 (1967)
69. Noll, W., Virga, E.G.: On edge interactions and surface tension. Arch. Ration. Mech. Anal. **111**, 1–31 (1990)
70. Nowacki, W.: Theory of Asymmetric Elasticity. Pergamon, Oxford (1986)

71. Palmov, A.: Fundamental equations of the theory of asymmetric elasticity. Prokl. Mat. Mekh. **28**, 401–408 (1964)
72. Pijaudier-Cabot, G., Bazant, Z.P.: Nonlocal damage theory. J. Eng. Mech. ASCE **113**(10), 1512–1533 (1987)
73. Pouget, J., Maugin, G.A.: Non-linear electroacoustic equations for piezoelectric powders. J. Acoust. Soc. Am. **74**(3), 925–940 (1983)
74. Pouget, J., Maugin, G.A.: Nonlinear dynamics of oriented elastic solids—I—Basic equations. J. Elast. **22**(2–3), 135–155 (1989)
75. Rakotomanana, L.R.: A Geometric Approach to Thermomechanics of Dissipating Continua. Birkhäuser, Boston (2003)
76. Rogula, D.: Influence of spatial acoustic dispersion on dynamical properties of dislocations. Bull. Acad. Pol. Sci., Sér. Si. Tech. **13**, 337–385 (1965)
77. Schaefer, H.: Das Cosserat-Kontinuum. Z. Angew. Math. Mech. **47**, 34 (1967)
78. Stojanovic, R.: Mechanics of Polar Continua. CISM, Udine (1969)
79. Stokes, V.K.: Theories of Fluids with Microstructure. Springer, Berlin (1984)
80. Toupin, R.A.: Elastic materials with couple stress. Arch. Ration. Mech. Anal. **11**, 395–414 (1962)
81. Toupin, R.A.: Theories of elasticity with couple-stress. Arch. Ration. Mech. Anal. **17**, 85–112 (1964)
82. Truesdell, C.A., Noll, W.: Nonlinear field theories of mechanics. In: Flügge, S. (ed.) Handbuch der Physik, Bd. III/3. Springer, Berlin (1965)
83. Truesdell, C.A., Toupin, R.A.: The classical theory of fields. In: Flügge, S. (ed.) Handbuch der Physik, Bd. III/1. Springer, Berlin (1960)
84. Wang, C.C.: On the geometric structure of simple bodies, a mathematical foundation for the theory of continuous distributions of dislocations. Arch. Ration. Mech. Anal. **27**, 33–94 (1967)
85. Whittaker, E.T.: A History of the Theories of Aether and Elasticity, vols. 1 and 2. Thomas Nelson, New York (1951). Reprint in one volume, Dover, New York (1953)

Chapter 2
On Semi-Holonomic Cosserat Media

Marcelo Epstein

Dedicated to the memory of the Cosserat brothers on the centenary of the publication of their magnum opus [1]

Abstract The notions of semi-holonomic and quasi-holonomic Cosserat media are introduced and their differences outlined. Contrary to the classical holonomic and non-holonomic counterparts, the definition of semi- and quasi-holonomic media is not kinematic but constitutive. Possible applications include granular media embedded in a rigid matrix and colloidal suspensions in an ideal incompressible fluid.

2.1 Cosserat Bodies

In Continuum Mechanics, a *material body* $\mathscr{B}$ is defined as a 3-dimensional differentiable manifold that can be covered with a single coordinate chart. A *configuration* κ is defined as an embedding of $\mathscr{B}$ into the 3-dimensional Euclidean space $\mathbb{E}^3$:

$$\kappa : \mathscr{B} \longrightarrow \mathbb{E}^3. \tag{2.1}$$

In terms of coordinate charts X^I $(I = 1,2,3)$ and x^i $(i = 1,2,3)$ in the body and in space, respectively, the configuration κ is given by three smooth functions:

$$X^I \mapsto x^i = \kappa^i(X^I). \tag{2.2}$$

To convey the presence of extra kinematic degrees of freedom, however, these definitions need to be expanded so that the differential geometry can properly reflect the existence of the microstructure and its possible deformability. We recall that,

M. Epstein (✉)
Department of Mechanical and Manufacturing Engineering, University of Calgary, Calgary, Alberta T2N 1N4, Canada
e-mail: mepstein@ucalgary.ca

G.A. Maugin, A.V. Metrikine (eds.), *Mechanics of Generalized Continua*, Advances in Mechanics and Mathematics 21,
DOI 10.1007/978-1-4419-5695-8_2, © Springer Science+Business Media, LLC 2010

given an m-dimensional differentiable manifold $\mathscr{M}$, its *principal frame bundle* $F\mathscr{M}$ is obtained by adjoining at each point $x \in \mathscr{M}$ the collection $F_x\mathscr{M}$ of all the possible bases of its tangent space $T_x\mathscr{M}$. The set thus obtained has a canonical structure of a differentiable manifold of dimension $m + m^2$. It is endowed with the *natural projection* map:

$$\pi_M : F\mathscr{M} \longrightarrow \mathscr{M} \tag{2.3}$$

that assigns to each point $p \in F\mathscr{M}$ the point $\pi(p) \in \mathscr{M}$ to which it is attached. If x^i $(i = 1, \ldots, m)$ is a coordinate chart on $U \subset \mathscr{M}$ with natural basis $\mathbf{e}_i = \partial/\partial x^i$, we can construct an associated chart in $F\mathscr{M}$ by assigning to each point $p \in \pi_M^{-1}(U)$ the numbers $\{x^i, p^i{}_j\}$, where $p^i{}_j$ is the ith component of the jth vector of the frame p in the natural basis $\{\mathbf{e}_1, \ldots, \mathbf{e}_m\}$. Expressed in terms of coordinates, the natural projection is given by:

$$x^i, p^i_j \mapsto x^i. \tag{2.4}$$

We define a *Cosserat body* as the principal frame bundle $F\mathscr{B}$ of an ordinary material body $\mathscr{B}$. The physical intent is that, while the underlying body $\mathscr{B}$ represents the *macro-medium*, each *fibre* $F_x\mathscr{B}$ represents the *micro-particle* or *grain* at $x \in \mathscr{B}$.

Concomitantly with the enlargement of the scope of material bodies, we need to envisage a more general definition of the notion of configuration. To this end, we consider *fibre-preserving* maps:

$$K : F\mathscr{B} \longrightarrow F\mathbb{E}^3 \tag{2.5}$$

such that K is a *principal fibre-bundle morphism* between $F\mathscr{B}$ and its image. By fibre preservation, we mean the commutativity of the diagram:

$$\begin{array}{ccc} F\mathscr{B} & \xrightarrow{\;K\;} & F\mathbb{E}^3 \\ \downarrow{\scriptstyle \pi_B} & & \downarrow{\scriptstyle \pi_E} \\ \mathscr{B} & \xrightarrow{\;\kappa\;} & \mathbb{E}^3 \end{array} \tag{2.6}$$

where κ is a well-defined map between the base manifolds. Thus, a *Cosserat deformation* K automatically implies the existence of an ordinary deformation κ, representing the deformation of the macro-structure. By fibre-bundle morphism we imply that, fibre by fibre, each of the restrictions $K|_X$ $(X \in \mathscr{B})$ commutes with the multiplicative right-action of the general linear group $\mathrm{GL}(3; \mathbb{R})$. In terms of coordinates, this means that there exists an X-dependent matrix K^i_I such that any Cosserat configuration is completely defined by twelve smooth functions:

$$x^i = \kappa^i(X^J), \tag{2.7}$$

and

$$K^i_I = K^i_I(X^J). \tag{2.8}$$

The physical meaning of these assumptions is that each grain can undergo only homogeneous deformations, as represented by the local matrix K^i_I. In other words, each grain behaves as a pseudo-rigid body. A more detailed treatment can be found in [2–4].

Remark 2.1. As already pointed out, the original formulation by the Cosserat brothers considered the case in which K^i_I is orthogonal. In the terminology of [5], this case corresponds to the micropolar continuum. The more general case in which K^i_I is an arbitrary non-singular matrix corresponds to the micromorphic continuum of [5]. We use the terminology "Cosserat body" in this more general sense.

We can see that in a Cosserat body there exist two, in principle independent, mechanisms for dragging vectors by means of a deformation: The first mechanism is the ordinary dragging of vectors by means of the deformation gradient of the macro-medium, represented by the matrix with entries $F^i_I = x^i_{,I}$. The second mechanism is the one associated with the deformation of the micro-particle or grain, and is represented by the matrix with entries K^i_I. Note that in a second-grade body these two mechanisms are identified with each other, thus suggesting that different kinds of Cosserat media may be obtained by either kinematic restrictions of this kind or by constitutive restrictions. In fact, the Cosserat brothers themselves already advanced these possibilities and introduced the outmoded terminology of "trièdre caché" (hidden triad) and "W caché" (hidden strain-energy function) to refer, respectively, to these kinematic or constitutive restrictions. We will follow in their steps.

2.2 Various Jets

Given two smooth manifolds, $\mathscr{M}$ and $\mathscr{N}$, of dimensions m and n, respectively, we say that two maps $f, g : \mathscr{M} \longrightarrow \mathscr{N}$ have the same k-jet at a point $X \in \mathscr{M}$ if: (i) $f(X) = g(X)$; (ii) in a coordinate chart in $\mathscr{M}$ containing X and a coordinate chart in $\mathscr{N}$ containing the image $f(X)$, all the partial derivatives of f and g up to and including the order k are respectively equal.

Although the above definition is formulated in terms of charts, it is not difficult to show by direct computation that the property of having the same derivatives up to and including order k is, in fact, independent of the coordinate systems used in either manifold. Notice that, in order for this to work, it is imperative to equate *all* the lower-order derivatives. If, for example, we were to equate just the second derivatives, without regard to the first, the equality of the second derivatives would not be preserved under arbitrary coordinate transformations.

The property of having the same k-jet at a point is, clearly, an equivalence relation. The corresponding equivalence classes are called *k-jets at X*. Any function in a given k-jet is then called a *representative* of the k-jet. The k-jet at X of which a given function $f : \mathscr{M} \longrightarrow \mathscr{N}$ is a representative is denoted by $j^k_X f$. The collection of all k-jets at $X \in \mathscr{M}$ is denoted by $J^k_X(\mathscr{M}, \mathscr{N})$. The point X is called the *source* of $j^k_X f$ and the image point $f(X)$ is called its *target*.

Let a smooth map $f : \mathscr{M} \longrightarrow \mathscr{N}$ be given in terms of coordinates X^I ($I = 1,\ldots,m$) and x^i ($i = 1,\ldots,n$) in $\mathscr{M}$ and $\mathscr{N}$, respectively, by the functions:

$$x^i = x^i(X^1,\ldots,X^m), \quad i = 1,\ldots,n. \tag{2.9}$$

The jet $j^2_X f$, for example, is then given by the following coordinate expressions:

$$x^i(X^1,\ldots,X^m), \qquad \left[\frac{\partial x^i}{\partial X^I}\right]_X, \qquad \left[\frac{\partial^2 x^i}{\partial X^J \partial X^I}\right]_X, \tag{2.10}$$

a total of $n + mn + m^2 n$ numbers.

We are particularly interested in the case of 1-jets. Let us evaluate, accordingly, the coordinate expression of $j^1_X K$, where K is a Cosserat configuration, as defined in coordinates by (2.7) and (2.8). Notice that the dimension of both the source and the target manifolds in this case is 12. Following the definition, we conclude that the $j^1_X K$ consists of the following elements:

$$x^i, \qquad K^i_I, \qquad \left[\frac{\partial x^i}{\partial X^I}\right]_X, \qquad \left[\frac{\partial K^i_I}{\partial X^J}\right]_X, \tag{2.11}$$

which we can abbreviate as:

$$x^i, \qquad K^i_I, \qquad F^i_I = x^i_{,I}, \qquad K^i_{I,J}. \tag{2.12}$$

If no further restrictions are imposed on K, we speak of the components (2.12) as the representatives of a *non-holonomic* 1-jet at $X \in \mathscr{B}$. It is possible, however, to demand in an intrinsic manner, independent of the coordinates, that the functions K under consideration satisfy the following compatibility requirement in a neighborhood of X:

$$K^i_I \equiv x^i_{,I}. \tag{2.13}$$

In this case, the collection of 1-jets obtained is smaller. Not only the second and third entries in (2.12) are the same, but also, by virtue of the identical satisfaction of (2.13) in a neighborhood of X, we must have:

$$K^i_{I,J} = x^i_{,IJ} = K^i_{J,I}. \tag{2.14}$$

In other words, the last element of the jet is symmetric with respect to its lower indices. We will indicate the coordinate expression of these *holonomic jets* as follows:

$$x^i, \qquad F^i_I, \qquad K^i_{I,J} = K^i_{J,I}. \tag{2.15}$$

Finally, there exists a third type of jet, somewhat intermediate between the two extremes just presented. It is obtained when the potential representatives K are restricted to satisfy the condition:

$$K^i_I(X) = x^i_{,I}(X). \tag{2.16}$$

In other words, we demand the satisfaction of (2.13) not identically in a neighborhood of X, but just at the point X itself. The 1-jets thus obtained are known as *semi-holonomic jets*. The coordinate expression of a semi-holonomic jet is:

$$x^i, \qquad K^i_I, \qquad K^i_{I,J}. \tag{2.17}$$

Notice that the last entry is no longer necessarily symmetric.

Remark 2.2. Given an actual arbitrary configuration K, it will give rise automatically to point-wise non-holonomic jets. If the configuration is restricted so that Condition (2.13) is satisfied over the whole base manifold $\mathscr{B}$, it will give rise to everywhere holonomic jets. In this sense, it is possible to speak of non-holonomic or holonomic configurations, respectively. On the other hand, it is not possible to define semi-holonomic configurations. Indeed, if Condition (2.16) were to be imposed at each point, we would immediately revert to Condition (2.13), thus obtaining a holonomic configuration.

2.3 Semi-Holonomic Cosserat Media

The last section ended in a definitely pessimistic note. Indeed, if semi-holonomic configurations cannot be properly defined, there seems to be no point in attempting a definition of semi-holonomic media. This kinematic impasse, however, can perhaps be resolved by means of a constitutive statement. We could say, for example, that a non-holonomic Cosserat medium is semi-holonomic if its constitutive equation involves only the semi-holonomic part of the 1-jet of the configuration. Physically, this would correspond to a response that is in some sense oblivious of the presence of the macro-medium. In this section, we look into this and other possibilities with some care.

Since we are contemplating a particular case of non-holonomic Cosserat media, it will be useful to record the law governing the change of constitutive law of such a medium under a change of reference configuration. For specificity, we will limit ourselves to a single scalar constitutive law, such as the free-energy density per unit mass ψ. Let the constitutive law with respect to a reference configuration K_0 be given in a coordinate system X^I by the expression:

$$\psi = \psi_0\left(K^i_I, F^i_I, K^i_{I,J}; X^I\right), \tag{2.18}$$

and let the counterpart for a reference configuration K_1 with coordinates Y^A be given by:

$$\psi = \psi_1\left(K^i_A, F^i_A, K^i_{A,B}; Y^A\right), \tag{2.19}$$

with an obvious notational scheme. The deformation from K_0 to K_1 is given by twelve quantities denoted as:

$$Y^A = Y^A\left(X^I\right), \qquad K^A_I = K^A_I\left(X^J\right). \tag{2.20}$$

By the law of composition of jets (or derivatives), we obtain the following relation between the constitutive expressions:

$$\psi_1\big(K^i_A, F^i_A, K^i_{A,B}; Y^A(X^J)\big)$$
$$= \psi_0\big(K^i_A K^A_I, F^i_A F^A_I, K^i_{A,B} K^A_I F^B_J + K^i_A K^A_{I,J}; X^J\big), \qquad (2.21)$$

where $F^A_I = Y^A_{,I}$.

The point of bringing this transformation to bear is the proof of the following:

Proposition 2.1. *If the constitutive law* (2.18), *in the reference configuration* K_0, *is independent of the second argument* (F^i_I), *so is the expression of the same constitutive law in any other reference configuration* K_1 *independent of the second argument* (F^i_A).

Proof. The proof is an immediate consequence of the transformation law (2.21). □

As a direct corollary of this proposition, we can propose the following definition.

Definition 2.1. A non-holonomic Cosserat medium is said to be *semi-holonomic* at X if its constitutive law at X is independent of the deformation gradient of the macro-medium.

From the mathematical standpoint, it is necessary to note that this definition does not imply the existence of a *canonical* projection of a non-holonomic jet onto a semi-holonomic part. In fact, such a canonical projection does not exist. What the definition implies is that, once a non-canonical choice is effected in one particular reference configuration, this choice can be convected to all other configurations by means of the correct application of the transformation equation (2.21). In particular, this convection involves the gradient of the change of reference configuration (F^A_I). Another way to state the choice of a particular "projection" is to say that a particular parallelism (whose physical meaning may, for example, be related to the existence of some particular stress-free configuration) must be chosen as part and parcel of the constitutive law of a semi-holonomic Cosserat medium.

From the physical point of view, a semi-holonomic Cosserat medium may be said to consist of an incoherent matrix upon which a coherent micro-medium has been installed. The interaction between the grains may "remember" the existence of a particular configuration of the macro-medium whereby the constitutive law takes a particularly simple form. It is interesting to remark that, since it plays no other role, the macro-medium of a semi-holonomic medium may be, in a possible application, assumed to be rigid.

The converse of the above statement is not true: a non-holonomic Cosserat medium with a rigid matrix is not automatically semi-holonomic. Indeed, by a direct application of the principle of frame indifference, the constitutive law (2.18) can be reduced to the form:

$$\psi = \psi\big(\mathbf{R}^T\mathbf{K}, \mathbf{U}, \mathbf{R}^T\nabla\mathbf{K}; X\big), \qquad (2.22)$$

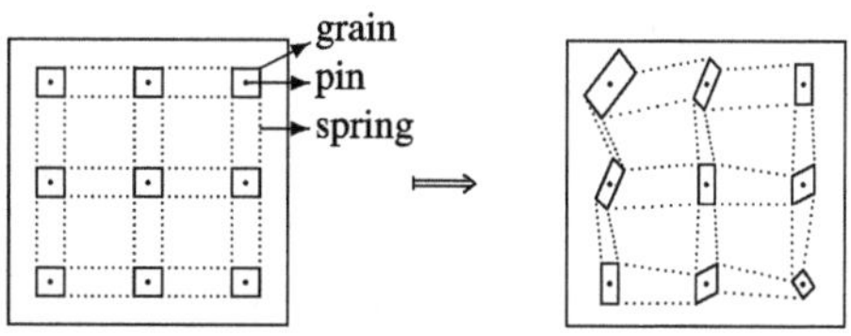

Fig. 2.1 A rigid-matrix semi-holonomic Cosserat medium

where the polar decomposition $\mathbf{F} = \mathbf{RU}$ has been exploited and where block letters stand for the collections of homonymous indexed quantities used in previous formulas. Using now the polar decomposition:

$$\mathbf{K} = \mathbf{R}'\mathbf{U}', \tag{2.23}$$

we may write (2.22) as:

$$\psi = \psi\big(\mathbf{R}\mathbf{U}', \mathbf{U}, \mathbf{r}\mathbf{R}'^T\nabla\mathbf{K}; X\big), \tag{2.24}$$

where:

$$\mathbf{R} = \mathbf{R}^T\mathbf{R}' \tag{2.25}$$

is the (referential) *relative rotation* of the grain with respect to the macro-medium. If the macro-medium is rigid, we must have necessarily $\mathbf{U} = \mathbf{I}$. But for a semi-holonomic body the constitutive law must be independent of *both* components $\mathbf{U}$ and $\mathbf{R}$ of the polar decomposition of $\mathbf{F}$. It follows, therefore, that rigidity alone does not imply semi-holonomy. If, on the other hand, the constitutive law of a rigid-matrix Cosserat medium is independent of the rotation $\mathbf{R}$, we may choose $\mathbf{R} = \mathbf{R}'$ (or, equivalently, $\mathbf{R} = \mathbf{I}$), thereby leading to the following reduced equation of a semi-holonomic Cosserat body:

$$\psi = \psi\big(\mathbf{U}', \mathbf{R}'^T\nabla\mathbf{K}; X\big). \tag{2.26}$$

In the physical interpretation, we may say that the grains are attached to the rigid macro-medium by means of ideal frictionless pins, so that there is no energetic cost to produce a relative rotation between them. In the admittedly imperfect pictorial representation of Fig. 2.1, the grains in the reference configuration are depicted as squares pin-jointed at their centers to the rigid matrix and connected to their neighbors by means of springs (represented by broken lines) designed to detect differential stretches and rotations between contiguous grains. The grains themselves behave as *pseudo-rigid bodies*, so that their deformed versions are represented by parallelograms.

The reduced form (2.26) of the constitutive law of a semi-holonomic Cosserat material applies whether or not the matrix is rigid, since in either case the response is independent of both $\mathbf{U}$ and $\mathbf{R}$.

2.4 Quasi-Holonomic Cosserat Media

As defined, a semi-holonomic Cosserat medium may not necessarily have any material symmetries. We want to contrast the above definition with the following one that, by demanding the maximum possible symmetry of the macro-medium, appears to carry the same physical meaning.

Definition 2.2. A non-holonomic Cosserat medium is said to be *quasi-holonomic* at X if, for some (local) reference configuration, its symmetry group $\mathscr{H}$ at X contains the subgroup given by:

$$\mathscr{G} = \{\{I, G, 0\} \mid G \in \mathrm{GL}(3;\mathbb{R})\}, \tag{2.27}$$

where I is the unit of $\mathrm{GL}(3;\mathbb{R})$.

The reason to suspect that this definition might be equivalent to the previous one is that, due to the assumed arbitrariness of G, it seems to imply that the deformation of the macro-medium plays no role in the constitutive response. A direct application of the definition of a non-holonomic symmetry, however, leads to the conclusion that a quasi-holonomic medium must have a constitutive law of the form:

$$\psi = \psi(K^i_I, K^i_{I,J} F^{-J}_j; X^I), \tag{2.28}$$

in the special reference configuration used in the definition.[1]

Physically, this means that the price to pay for this large symmetry group is, surprisingly, the reappearance of the deformation gradient of the macro-medium in the last argument of the constitutive law so as to permit the interaction between the grains to take into account their relative spatial locations (rather than those pulled back to some putative, perhaps unstressed, reference configuration).

The purpose of the following simple example is to shed light on the subtle difference between semi-holonomic and quasi-holonomic media, as conceived in Definitions 2.1 and 2.2, respectively. To this end, we consider the successive application of two deformations, the first of which can be regarded as a change of reference configuration so as to bring the notation in line with that of the previous section. The (Cartesian) coordinate systems X^I, Y^A, x^i are assumed to coincide with each other. The first deformation is a uniaxial contraction along the X^1-axis, namely:

$$Y^1 = 0.8X^1, \qquad Y^2 = X^2, \qquad Y^3 = X^3, \qquad K^A_I = \delta^A_I. \tag{2.29}$$

The second deformation is a micro-rotation about the Y^3 axis that increases linearly with Y^1. Specifically:

---

[1] In any other reference configuration, the symmetry group will contain a conjugate of the group $\mathscr{G}$ and the form of the constitutive law will be, accordingly, somewhat more involved.

$$x^1 = Y^1, \qquad x^2 = Y^2, \qquad x^3 = Y^3,$$
$$\{K^i_A\} = \begin{bmatrix} \cos\left(\frac{\pi}{3}Y^1\right) & -\sin\left(\frac{\pi}{3}Y^1\right) & 0 \\ \sin\left(\frac{\pi}{3}Y^1\right) & \cos\left(\frac{\pi}{3}Y^1\right) & 0 \\ 0 & 0 & 1 \end{bmatrix}. \tag{2.30}$$

The effect of each of the two deformations on a unit-width strip in the X^1, X^2 and Y^1, Y^2 planes, respectively, is shown in Figs. 2.2 and 2.3, while Fig. 2.4 shows the composition. Notice that, at the moment of composition, it is the *already contracted* strip that encounters the values of the rotation field already in place (as dictated by the second deformation), thus resulting in a maximum value for the rotation of the grain in the deformed strip of 48° rather than 60°, which was the value at the right-hand end of the strip as far as the second deformation alone was concerned. If the Cosserat body is semi-holonomic, the gradient of the rotation would be obtained by dividing 48° by the original unit width. On the other hand, if the Cosserat body is quasi-holonomic, it is the width measured in the final deformed configuration that matters in the calculation of the gradient. Since this width is of 0.8, we verify that the rotation gradient in the composite deformation turns out to be identical to the gradient in the second deformation. In other words, the pre-application of the first deformation (in this case a contraction of the macro-medium) is irrelevant for a quasi-holonomic medium.

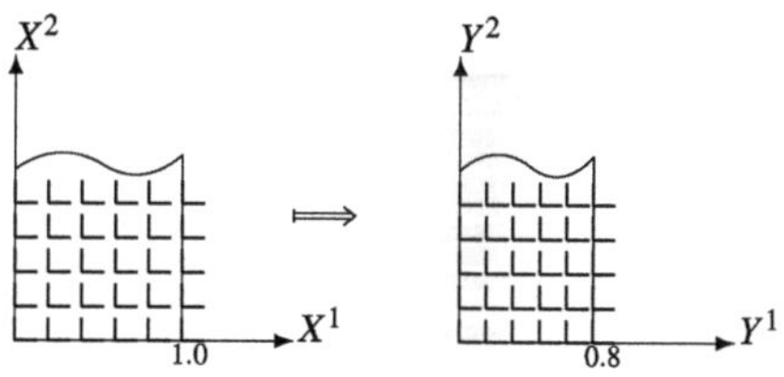

Fig. 2.2 First deformation

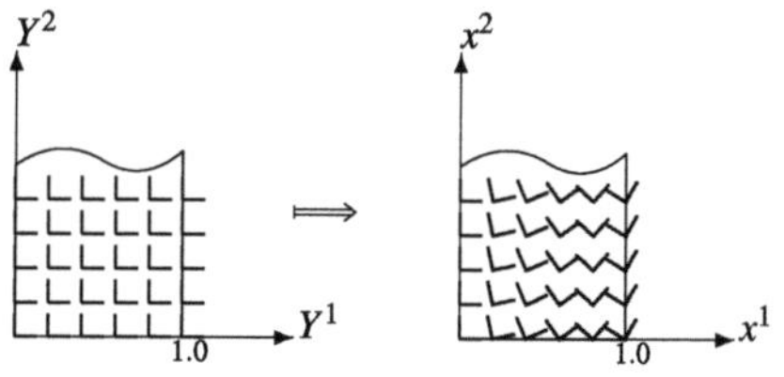

Fig. 2.3 Second deformation

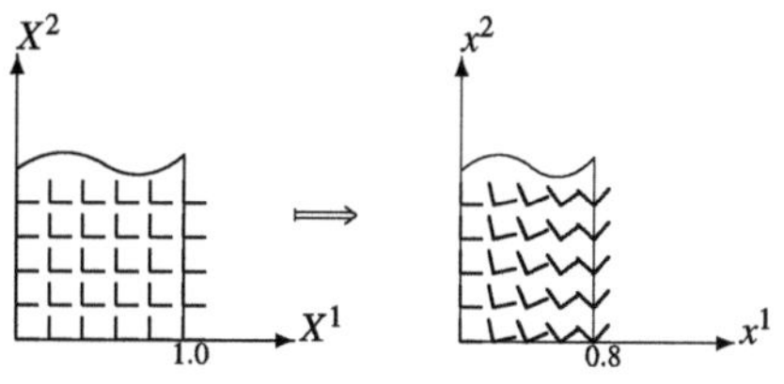

Fig. 2.4 Composition

Among various possible physical applications of both semi-holonomic and quasi-holonomic Cosserat media, beyond those with a rigid matrix, we mention the modeling of aggregates [7], such as colloidal suspensions [6], when the underlying continuum upon which the interacting particles dwell is, say, an ideal incompressible fluid. The choice of model depends on the physical nature of the interactions between the dispersed particles.

Acknowledgements This work has been partially supported by the Natural Sciences and Engineering Research Council of Canada.

References

1. Cosserat, E., Cosserat, F.: Théorie des corps déformables. Hermann et Fils, Paris (1909)
2. Epstein, M., de León, M.: Homogeneity conditions for generalized Cosserat media. J. Elast. **43**, 189–201 (1996)
3. Epstein, M., de León, M.: Geometrical theory of uniform Cosserat media. J. Geom. Phys **26**, 127–170 (1998)
4. Epstein, M., Elżanowski, M.: Material Inhomogeneities and Their Evolution. Springer, Berlin (2007)
5. Eringen, A.C.: Microcontinuum Field Theories. Springer, Berlin (1999)
6. Moosaie, A., Atefi, G.: Analysis of concentrated suspension flow by utilizing a Cosserat-type continuum theory. J. Dispers. Sci. Technol. **28**, 901–906 (2007)
7. Zhang, X., Jeffrey, R.G., Mai, Y.-W.: A micromechanics-based Cosserat-type model for dense particulate solids. Z. Angew. Math. Phys. **57**, 682–707 (2006)

Part II
Cosserat Media
(Rigidly Rotating Microstructure)

Chapter 3
On the Theories of Plates Based on the Cosserat Approach

Holm Altenbach and Victor A. Eremeyev

Abstract The classical isotropic linear elastic material behavior is presented by two material parameters, e.g., the Young's modulus and the Poisson's ratio, while the Cosserat continuum is given by six material parameters. The latter continuum model can be the starting point for the deduction of the governing equations of the Cosserat plate theory via a through-the-thickness integration. In contrast, the basic equations of the Cosserat plate theory can be established applying the direct approach. It can be shown that both systems of equations are similar in the main terms. The assumed identity of both systems results in consistent stiffness parameters identification for the two-dimensional theory based on the direct approach and, in addition, in some constraints. Using the experimental results of Lakes, one can show in which cases the additional material properties coming from the tree-dimensional Cosserat material model have a significant influence on the stiffness parameters.

3.1 Introduction

The continuum model of the Cosserat brothers [5] is founded on the *a priori* introduction of the independence of both the translations and rotations. From this it follows that one symmetric stress tensor is not enough to represent the response of the continuum on the external loadings. The base of such a new continuum model generalizing the Cauchy's model was known since Euler because he introduced two

H. Altenbach (✉)
Martin-Luther-Universität Halle-Wittenberg, 06099 Halle (Saale), Germany
e-mail: holm.altenbach@iw.uni-halle.de

V.A. Eremeyev
South Scientific Center of RASci & South Federal University, Milchakova St. 8a, 344090 Rostov on Don, Russia
e-mail: eremeyev.victor@gmail.com

G.A. Maugin, A.V. Metrikine (eds.), *Mechanics of Generalized Continua*,
Advances in Mechanics and Mathematics 21,
DOI 10.1007/978-1-4419-5695-8_3, © Springer Science+Business Media, LLC 2010

independent laws of motion: the balance of momentum and the balance of moment of momentum (see [18] among others).

Applications of the Cosserat continuum could not be established for a long time. Only in the 1950s, the Cosserat continuum was recognized as a starting point for various constructions of generalized continuum models. Let us note that some of them are presented for the three-dimensional case to describe complex behavior of solids and fluids. Let us mention here only the pioneering works summarized in the proceedings [11]. In addition, the Cosserat approach was used in establishing less than three-dimensional continuum theories to model shells, plates and rods.

Below we limit our discussion to the plate theory. Any set of governing equations for plates can be deduced using the conventional three-dimensional continuum equations together with the some engineering hypotheses or mathematical techniques. Finally, the manipulated equations are integrated over the thickness. Another possibility is the *a priori* introduction of the two-dimensional field equations for a so-called deformable surface. The latter is an elegant and a more natural way to formulate the plate equations, but the identification of the effective properties is a non-trivial problem. Let us note only some fundamental publications presenting the basic items of the direct approach [9, 16, 17, 8], and the reviews of the Cosserat approach in the shell and plate theories [1, 3].

Here we firstly present two sets of plate equations of the Cosserat type. The first set is introduced by the direct approach, while the second is based on the three-dimensional equations of the Cosserat continuum and a through-the-thickness integration. Secondly, both sets will be compared and analyzed. This way one gets the equivalent terms in both sets showing which terms in the equations introduced by the direct approach correspond to which terms in the integrated three-dimensional equations. As in any dimension-reduction problem, some constraints can be obtained. Last but not least, using the experimental data of Lakes [12, 13] for an open-cell and a closed-cell foam, the influence of the additional material parameters is discussed. It can be shown that in some cases the classical continuum model does not allow the description of the material behavior with a sufficient accuracy, and so a generalized continuum model must be applied.

3.2 A Priori Two-Dimensional Governing Equations

Let us introduce the basic equations for micropolar plates based on the direct approach. We consider a deformable plane surface, see Fig. 3.1. Each material point of this surface is an infinitesimal rigid body with 3 translational and 3 rotational degrees of freedom. The balances of momentum and moment of momentum are formulated as follows

$$\begin{aligned} \mathbf{F}_s^* &\equiv \int_{\mathscr{M}_*} \mathbf{q}\,\mathrm{d}A + \int_{\mathscr{C}_*} \mathbf{t}_s\,\mathrm{d}s = \mathbf{0}, \\ \mathbf{M}_s^* &\equiv \int_{\mathscr{M}_*} (\mathbf{x}\times\mathbf{q}+\mathbf{c})\,\mathrm{d}A + \int_{\mathscr{C}_*} (\mathbf{x}\times\mathbf{t}_s+\mathbf{m}_s)\,\mathrm{d}s = \mathbf{0}, \end{aligned} \tag{3.1}$$

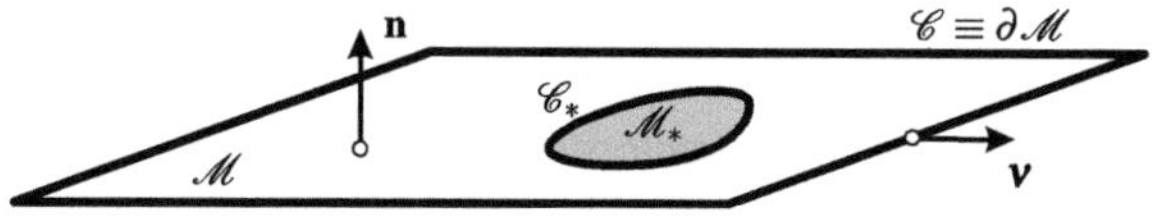

Fig. 3.1 Deformable plane surface

where $\mathbf{q}$ and $\mathbf{c}$ are the vectors of surface loads (forces and moments), $\mathbf{x}$ is the position vector of the plane surface $\mathscr{M}$, $\times$ denotes the cross-product, $\mathbf{t}_s$ and $\mathbf{m}_s$ are the surface analogues of the stress vector and the couple stress vector, respectively. Here the direct (index-free) notation in the sense of [15] is used. From (3.1) we obtain the local form of the balances of momentum and moment of momentum as well as the static boundary conditions. In the case of sufficiently smooth fields, the local equilibrium equations can be stated as follows

$$\nabla_s \cdot \mathsf{T} + \mathbf{q} = \mathbf{0}, \qquad \nabla_s \cdot \mathsf{M} + \mathsf{T}_\times + \mathbf{c} = \mathbf{0}. \tag{3.2}$$

Here the surface (plane) nabla operator ∇_s is given as $\nabla_s = \mathbf{i}_\alpha(\partial/\partial x_\alpha)$ with the Cartesian coordinates x_α, and $\mathbf{i}_\alpha$ being the unit base vectors. The Greek indices take the values 1 and 2. The tensors T and M denote the surface stress and couple stress tensors, respectively. They relate to $\mathbf{t}_s$ and $\mathbf{m}_s$ by the equations $\boldsymbol{\nu} \cdot \mathsf{T} = \mathbf{t}_s$ and $\boldsymbol{\nu} \cdot \mathsf{M} = \mathbf{m}_s$, where $\boldsymbol{\nu}$ is the outward normal vector to $\mathscr{C}$. T and M take the form

$$\mathsf{T} = T_{\alpha\beta}\mathbf{i}_\alpha\mathbf{i}_\beta + T_{\alpha 3}\mathbf{i}_\alpha\mathbf{n}, \qquad \mathsf{M} = M_{\alpha\beta}\mathbf{i}_\alpha\mathbf{i}_\beta + M_{\alpha 3}\mathbf{i}_\alpha\mathbf{n} \quad (\alpha, \beta = 1, 2). \tag{3.3}$$

Let us introduce the linear strain measures

$$\mathsf{e} = \nabla_s\mathbf{v} + \mathsf{A} \times \boldsymbol{\theta}, \qquad \mathsf{k} = \nabla_s\boldsymbol{\theta}, \tag{3.4}$$

where $\mathbf{A} \equiv \mathbf{I} - \mathbf{n} \otimes \mathbf{n}$ is the first metric tensor, and $\mathbf{n}$ the unit normal to $\mathscr{M}$. Applying the methodology presented in [4, 14], one can show that the linear strain measures are work-conjugated to the stress measures T and M.

For the isotropic plate, the surface strain energy W can be introduced by [4, 6, 8]

$$\begin{aligned} 2W &= \alpha_1 \operatorname{tr}^2 \mathsf{e}_S + \alpha_2 \operatorname{tr} \mathsf{e}_S^2 + \alpha_3 \operatorname{tr}\left(\mathsf{e}_S \cdot \mathsf{e}_S^T\right) + \alpha_4 \mathbf{n} \cdot \mathsf{e}^T \cdot \mathsf{e} \cdot \mathbf{n} \\ &\quad + \beta_1 \operatorname{tr}^2 \mathsf{k}_S + \beta_2 \operatorname{tr} \mathsf{k}_S^2 + \beta_3 \operatorname{tr}\left(\mathsf{k}_S \cdot \mathsf{k}_S^T\right) + \beta_4 \mathbf{n} \cdot \mathsf{k}^T \cdot \mathsf{k} \cdot \mathbf{n}. \end{aligned} \tag{3.5}$$

Here $\mathsf{e}_S = \mathsf{e} \cdot \mathsf{A}$, $\mathsf{k}_S = \mathsf{k} \cdot \mathsf{A}$, and α_i, β_i are the elastic stiffness parameters, $i = 1, 2, 3, 4$. The constitutive equations are $\mathsf{T} \equiv \partial W/\partial \mathsf{e}$ and $\mathsf{M} \equiv \partial W/\partial \mathsf{k}$.

The surface strain energy W must be positive definite, from which it follows [7]

$$\begin{aligned} &2\alpha_1 + \alpha_2 + \alpha_3 > 0, \qquad \alpha_2 + \alpha_3 > 0, \qquad \alpha_3 - \alpha_2 > 0, \qquad \alpha_4 > 0, \\ &2\beta_1 + \beta_2 + \beta_3 > 0, \qquad \beta_2 + \beta_3 > 0, \qquad \beta_3 - \beta_2 > 0, \qquad \beta_4 > 0. \end{aligned} \tag{3.6}$$

Note that for an isotropic three-dimensional micropolar solid we have only 6 elastic moduli, while the micropolar plate theory contains 8 elastic stiffness parameters. The increase in the number of parameters can be explained by at least two reasons:

- *Reduced symmetry of the constitutive equations.*
  In the case of two-dimensional equations, a smaller number of symmetry groups in comparison with the three-dimensional theory can be established. This fact is well-known also for other plate theories, see, e.g., [2, 6].
- *Reduction of three-dimensional equations to two-dimensional.*
  In the isotropic elasticity, material behavior is presented by two material parameters (e.g., Lamé's moduli). The number in the full Kirchhoff plate theory including both the in-plane and out-of-plane behavior is greater. Similar conclusions can be given for other plate theories, too.

3.3 Reduction of the Three-Dimensional Micropolar Elasticity Equations by the Through-the-Thickness Integration

Let us introduce the balance equations of the micropolar elasticity [10]. The equilibrium conditions of any part of a micropolar body occupying the volume $V_* \subset V$ consist of the following relations

$$\begin{aligned} \mathbf{F}^* &\equiv \int_{V_*} \rho \mathbf{f} \, \mathrm{d}V + \int_{S_*} \mathbf{t} \, \mathrm{d}A = \mathbf{0}, \\ \mathbf{M}^* &\equiv \int_{V_*} \rho(\mathbf{r} \times \mathbf{f} + \mathbf{l}) \, \mathrm{d}V + \int_{S_*} (\mathbf{r} \times \mathbf{t} + \mathbf{m}) \, \mathrm{d}A = \mathbf{0}, \end{aligned} \tag{3.7}$$

where $\mathbf{f}$ and $\mathbf{l}$ are the mass forces and mass couples vectors, respectively, ρ is the density, $\mathbf{r}$ is the position vector, $S_* = \partial V_*$, $\mathbf{t}$ and $\mathbf{m}$ are the stress and couple stress vectors, respectively. $\mathbf{F}^*$ and $\mathbf{M}^*$ are the total force and the total couple acting on V_*, respectively. Hence, for any part of the micropolar body, $(3.7)_1$ states that the vector of total force is zero, while $(3.7)_2$ states that the vector of total moment is zero. With the relations $\mathbf{n} \cdot \boldsymbol{\sigma} = \mathbf{t}$, $\mathbf{n} \cdot \boldsymbol{\mu} = \mathbf{m}$, the local equilibrium equations are

$$\nabla \cdot \boldsymbol{\sigma} + \rho \mathbf{f} = \mathbf{0}, \qquad \nabla \cdot \boldsymbol{\mu} + \boldsymbol{\sigma}_\times + \rho \mathbf{l} = \mathbf{0}. \tag{3.8}$$

Now the nabla operator ∇ is a three-dimensional operator, ρ is the density, and $\boldsymbol{\sigma}$ and $\boldsymbol{\mu}$ are the stress and couple stress tensors, respectively. $\boldsymbol{\sigma}_\times$ is the vectorial invariant of a second-order tensor $\boldsymbol{\sigma}$.

The small strains of the micropolar continuum are usually presented by the displacement vector $\mathbf{u}$ and the vector of microrotation $\boldsymbol{\vartheta}$. The linear strain measures, i.e., the stretch tensor E and the wryness tensor K, are given by the relations

$$\mathsf{E} = \nabla \mathbf{u} + \boldsymbol{\vartheta} \times \mathsf{I}, \qquad \mathsf{K} = \nabla \boldsymbol{\vartheta}. \tag{3.9}$$

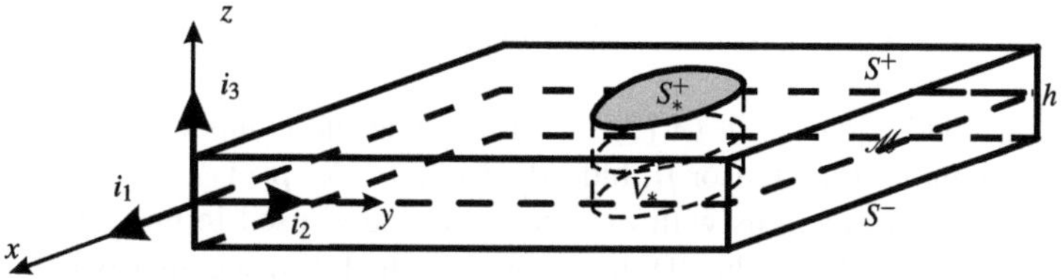

Fig. 3.2 Plate-like body

The classical isotropic elasticity can be deduced by setting $\kappa = \alpha = \beta = \gamma = 0$. In this case, the stress tensor $\boldsymbol{\sigma}$ will be a symmetric tensor. In addition, $\boldsymbol{\mu}$ and $\mathbf{l}$ vanish.

Finally, the isotropic elastic constitutive equations are

$$\boldsymbol{\sigma} = \lambda \mathsf{I} \operatorname{tr} \mathsf{E} + \mu \mathsf{E}^T + (\mu + \kappa)\mathsf{K}, \quad \boldsymbol{\mu} = \alpha \mathsf{I} \operatorname{tr} \mathsf{K} + \beta \mathsf{K}^T + \gamma \mathsf{K}. \tag{3.10}$$

The integration procedure is performed as follows. Let us assume that our plate-like body occupies a volume with one dimension which is significantly smaller in comparison with the other two. The coordinate z denotes this special direction and h is the plate thickness; z takes the values $-h/2 \leq z \leq h/2$ (Fig. 3.2). The boundary conditions of the upper $(+)$ and lower $(-)$ plate surfaces can be given by

$$\mathbf{n}^{\pm} \cdot \boldsymbol{\sigma} = \mathbf{t}^{\pm}, \qquad \mathbf{n}^{\pm} \cdot \boldsymbol{\mu} = \mathbf{m}^{\pm}, \tag{3.11}$$

where $\mathbf{t}^{\pm}$ and $\mathbf{m}^{\pm}$ are the surface loads (forces and moments) and $\mathbf{n}^{\pm} = \mathbf{i}_3$.

The main idea of the reduction procedure is the application of the 3D equilibrium conditions (3.7) to any volume V_* of the plate-like body and the transformation of the results to the 2D case as in (3.1). Following [1], we transform (3.7) into the relations

$$\begin{aligned}
\mathbf{F}^* &= \int_{\mathscr{M}_*} \mathbf{q}\, \mathrm{d}A + \int_{\mathscr{C}_*} \boldsymbol{\nu} \cdot \langle \boldsymbol{\sigma} \rangle\, \mathrm{d}s = \mathbf{0}, \qquad \langle (\ldots) \rangle = \int_{-h/2}^{h/2} (\ldots)\, \mathrm{d}z, \\
\mathbf{M}^* &= \int_{\mathscr{M}_*} [\mathbf{x} \times \mathbf{q} + \mathbf{c}]\, \mathrm{d}A \\
&\quad + \int_{\mathscr{C}_*} \left[\boldsymbol{\nu} \cdot \langle \boldsymbol{\mu} \rangle - \boldsymbol{\nu} \cdot \langle z \boldsymbol{\sigma} \times \mathbf{i}_3 \rangle - \boldsymbol{\nu} \cdot \langle \boldsymbol{\sigma} \rangle \times \mathbf{x}\right] \mathrm{d}A = \mathbf{0},
\end{aligned} \tag{3.12}$$

where the following notations are introduced

$$\mathbf{q} = \langle \rho \mathbf{f} \rangle + \mathbf{t}^+ + \mathbf{t}^-,$$
$$\mathbf{c} = \langle \rho \mathbf{l} \rangle + \mathbf{m}^+ + \mathbf{m}^- + \mathbf{i}_3 \times \langle \rho z \mathbf{f} \rangle + \tfrac{h}{2} \mathbf{i}_3 \times (\mathbf{t}^+ - \mathbf{t}^-).$$

The comparison of (3.12) and (3.1) leads to the determination of the stress resultant and stress couple tensors by the following relations

$$\mathsf{T} = \langle \mathsf{A} \cdot \boldsymbol{\sigma} \rangle, \qquad \mathsf{M} = \langle \mathsf{A} \cdot \boldsymbol{\mu} \rangle - \langle \mathsf{A} \cdot z\boldsymbol{\sigma} \times \mathbf{i}_3 \rangle. \tag{3.13}$$

From the second equation in (3.13), it follows that the components $M_{\alpha 3}$ depend only on the couple stress tensor $\boldsymbol{\mu}$. Indeed, $\mathsf{M} \cdot \mathbf{i}_3 = \langle \mathbf{A} \cdot \boldsymbol{\mu} \cdot \mathbf{i}_3 \rangle$.

To establish the relations with the vectors $\mathbf{u}$ and $\boldsymbol{\vartheta}$ used in the 3D theory and their analogues $\mathbf{v}$ and $\boldsymbol{\theta}$ in the 2D theory, we use the following approximation of $\mathbf{u}$ and $\boldsymbol{\vartheta}$, see [1] for details,

$$\begin{aligned} &\mathbf{u}(x, y, z) = \mathbf{v}(x, y) - z\boldsymbol{\phi}(x, y), \\ &\boldsymbol{\vartheta} = \boldsymbol{\phi}(x, y) \times \mathbf{i}_3 + \vartheta_3(x, y)\mathbf{i}_3 = \boldsymbol{\theta}, \quad \boldsymbol{\phi} \cdot \mathbf{i}_3 = 0. \end{aligned} \tag{3.14}$$

This means that the couple stress tensor $\boldsymbol{\mu}$ does not depend on z, while the stress tensor $\boldsymbol{\sigma}$ depends on z linearly as in [10]. But the approximation (3.14) is more restrictive than the one applied by Eringen [10]. As a result, the effective stiffness parameters can be estimated as

$$\alpha_1 = \tilde{\lambda} h, \quad \tilde{\lambda} \equiv \frac{\lambda(2\mu + \kappa)}{\lambda + 2\mu + \kappa},$$

$$\alpha_2 = \mu h, \qquad \alpha_3 = (\mu + \kappa) h, \qquad \alpha_4 = (\mu + \kappa) h,$$

$$\beta_1 = \alpha h - \mu \frac{h^3}{12}, \qquad \beta_2 = \beta h - \lambda \frac{h^3}{12},$$

$$\beta_3 = \gamma h + (2\mu + \kappa + \tilde{\lambda}) \frac{h^3}{12}, \qquad \beta_4 = \gamma h.$$

The in-plane stiffness parameters α_1, α_3, α_3, and the transverse shear stiffness α_4 depend linearly on h. The dependence of β_i, $i = 1, \ldots, 4$, on h is more complicated. The parameters β_i have linear asymptotes when h tends to zero, i.e., $\beta_i \sim h$. The considered case differs from the cases of Kirchhoff's and Reissner's plates when $\beta_i \sim h^3$.

Introducing the engineering constants

$$G = \frac{2\mu + \kappa}{2}, \qquad \nu = \frac{\lambda}{2\lambda + 2\mu + \kappa}, \qquad l_b^2 = \frac{\gamma}{2(2\mu + \kappa)},$$

where G is the shear modulus, ν the Poisson ratio, l_b the characteristic length under bending, see [10], we obtain the expression

$$D = \frac{Gh^3}{12(1 - \nu)} \left[1 + 2\frac{l_b^2}{h^2} \right]. \tag{3.15}$$

The dependence $D/\widetilde{D}$ on h is given in Fig. 3.3, where $\widetilde{D} = Gh^3/[12(1 - \nu)]$ is the value of the bending stiffness used by Eringen [10]. From Fig. 3.3 it can be seen that the micropolar properties are inessential if $h > 2l_b$. The values of the elastic stiffness parameters for two porous materials are given in Table 3.1. Here we have used the experimental data presented by Lakes [12, 13]. The index $*$ denotes the case of the material without the micropolar properties, i.e., if $\kappa = \alpha = \beta = \gamma = 0$.

To calculate the elastic stiffness we have used the approximation (3.14). Let us note that in the literature on the shell theory there are many different approaches to derive the shell equations from the 3D elasticity which lead to different values of the stiffness parameters, see [4, 14] among others. For the Cosserat continuum the derivation of the shell theories are given in many publications, see the reviews in [1, 3] and [19]. This means that the elastic stiffness parameters presented above can be considered as some estimates. In other words, these parameters show the influence of the micropolar properties. In particular, (3.15) demonstrates the size-effect which is well-known in the micropolar elasticity, see [10, 12, 13]. The analysis of (3.15) shows that the influence of the micropolar properties is essential if the thickness of the plate h has the same value as the characteristic length of the microstructure of the material.

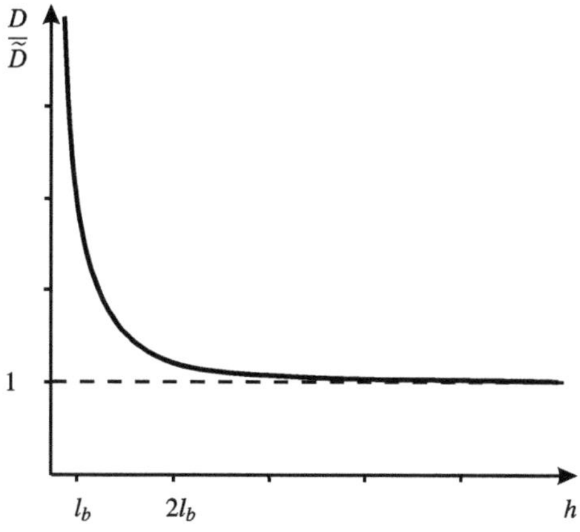

Fig. 3.3 Dimensionless bending stiffness $D/\widetilde{D}$ vs. the dimensionless thickness h/l_b

Table 3.1 Effective stiffness of a plate made of different foams (h has dimension m)

Elastic constants	Foam, PU	Foam, PU *	Foam, PS	Foam, PS *
α_1, N/m 10^6	$0.165h$	$0.165h$	$138.67h$	$138.67h$
α_2, N/m 10^6	$1.001h$	$1.1h$	$99.84h$	$104h$
α_3, N/m 10^6	$1.199h$	$1.1h$	$108.16h$	$104h$
α_4, N/m 10^6	$1.199h$	$1.1h$	$108.16h$	$104h$
β_1, N·m 10^6	$-2.6 \times 10^{-6}h - 0.083h^3$	$-0.092h^3$	$-6.7 \times 10^{-6}h - 8.3h^3$	$-8.67h^3$
β_2, N·m 10^6	$-10^{-4}h - 0.014h^3$	$-0.014h^3$	$-2.5 \times 10^{-5}h + 11.6h^3$	$-11.5h^3$
β_3, N·m 10^6	$1.1 \times 10^{-4}h + 0.197h^3$	$0.197h^3$	$4.5 \times 10^{-5}h + 28.9h^3$	$28.8h^3$
β_4, N·m 10^6	$1.1 \times 10^{-4}h$	0	$4.5 \times 10^{-5}h$	0

3.4 Conclusion

In this paper, we present the general six-parametric or micropolar linear plate linear theory with two vector fields of the translations and rotations as the independent kinematic variables. Within the proposed theory one may take into account an external surface drilling moment. We discuss the relations between the direct approach and the through-the-thickness integration procedure to derive the plate equilibrium equations and the constitutive equations. The influence of the micropolar properties on the stiffness parameters of the plate is illustrated for two foams.

Acknowledgements The second author was supported by the RFBR with the grant No. 09-01-00459 and by the DFG with the grant No. AL 341/31-1.

References

1. Altenbach, H., Eremeyev, V.A.: On the linear theory of micropolar plates. Z. Angew. Math. Mech. **4**(89), 242–256 (2009)
2. Altenbach, H., Zhilin, P.A.: A general theory of elastic simple shells. Usp. Mekh. **11**(4) (1988). In Russian
3. Altenbach, J., Altenbach, H., Eremeyev, V.A.: On generalized Cosserat-type theories of plates and shells. A short review and bibliography. Arch. Appl. Mech. **80**, 73–92 (2010)
4. Chróścielewski, J., Makowski, J., Pietraszkiewicz, W.: Statyka i dynamika powłok wielopłatowych. Nieliniowa teoria i metoda elementów skończonych. Wydawnictwo IPPT PAN, Warszawa (2004)
5. Cosserat, E., Cosserat, F.: Théorie des corps déformables. Herman et Fils, Paris (1909)
6. Eremeyev, V.A., Pietraszkiewicz, W.: Local symmetry group in the general theory of elastic shells. J. Elast. **85**(2), 125–152 (2006)
7. Eremeyev, V.A., Zubov, L.M.: On constitutive inequalities in nonlinear theory of elastic shells. Z. Angew. Math. Mech. **87**(2), 94–101 (2007)
8. Eremeyev, V.A., Zubov, L.M.: Mechanics of Elastic Shells. Nauka, Moscow (2008). In Russian
9. Ericksen, J.L., Truesdell, C.: Exact theory of stress and strain in rods and shells. Arch. Ration. Mech. Anal. **1**(1), 295–323 (1958)
10. Eringen, A.C.: Microcontinuum Field Theory. I. Foundations and Solids. Springer, New York (1999)
11. Kröner, E. (ed.): Mechanics of Generalized Continua. Proceedings of the IUTAM-Symposium on the Generalized Cosserat Continuum and the Continuum Theory of Dislocations with Applications, Freudenstadt and Stuttgart (Germany), 1967, vol. 16. Springer, Berlin (1968)
12. Lakes, R.S.: Experimental microelasticity of two porous solids. Int. J. Solids Struct. **22**, 55–63 (1986)
13. Lakes, R.S.: Experimental micro mechanics methods for conventional and negative Poisson's ratio cellular solids as Cosserat continua. Trans. ASME. J. Eng. Mat. Technol. **113**, 148–155 (1991)
14. Libai, A., Simmonds, J.G.: The Nonlinear Theory of Elastic Shells, 2nd edn. Cambridge University Press, Cambridge (1998)
15. Lurie, A.I.: Theory of Elasticity. Foundations of Engineering Mechanics. Springer, Berlin (2005)

16. Naghdi, P.: The theory of plates and shells. In: Flügge, S. (ed.) Handbuch der Physik, vol. VIa/2, pp. 425–640. Springer, Heidelberg (1972)
17. Rubin, M.B.: Cosserat Theories: Shells, Rods and Points. Kluwer, Dordrecht (2000)
18. Truesdell, C.: Die Entwicklung des Drallsatzes. Z. Angew. Math. Mech. **44**(4/5), 149–158 (1964)
19. Zubov, L.M.: Micropolar-shell equilibrium equations. Dokl. Phys. **54**(6), 290–293 (2009)

Chapter 4
Cracks in Cosserat Continuum—Macroscopic Modeling

Arcady V. Dyskin and Elena Pasternak

Abstract Modeling of particulate and layered materials (e.g., concrete and rocks, rock masses) by Cosserat continua involves characteristic internal lengths which can be commensurate with the microstructural size (the particle size or layer thickness). When fracture propagation in such materials is considered, the criterion of their growth is traditionally based on the parameters of the crack-tip stress singularities referring to the distances to the crack tip smaller than the characteristic lengths and hence smaller than the microstructural size. This contradicts the very notion of continuum modeling which refers to the scales higher than the microstructural size. We propose a resolution of this contradiction by considering an intermediate asymptotics corresponding to the distances from the crack tip larger than the microstructural sizes (the internal Cosserat lengths) but yet smaller than the crack length. The approach is demonstrated using examples of shear crack in particulate and bending crack in layer materials.

4.1 Introduction

In a variety of materials with microstructure, there are two important types whose adequate continuum modeling requires the consideration of rotational degrees of freedom—the particulate and layered materials with week bonding. Such materials represent a wide class of natural and structural materials including rocks and rock masses, soils, concrete and composites. As the continuum modeling is based on a homogenization procedure which introduces the averaging scale (e.g., the volume element size), the distances smaller than the averaging scale cannot be addressed

A.V. Dyskin (✉) · E. Pasternak
The University of Western Australia, 35 Stirling Highway, Crawley, Perth, WA 6009, Australia
e-mail: arcady@civil.uwa.edu.au

E. Pasternak
e-mail: elena@mech.uwa.edu.au

G.A. Maugin, A.V. Metrikine (eds.), *Mechanics of Generalized Continua*, Advances in Mechanics and Mathematics 21,
DOI 10.1007/978-1-4419-5695-8_4, © Springer Science+Business Media, LLC 2010

in such a modeling even if all distances are equally accessible in the continuum. When the higher order continua such as Cosserat continuum are employed, they bring their own characteristic lengths. Then the interplay between the characteristic lengths of the continuum and the scale of averaging becomes important and in some cases can control the results of the modeling. This paper investigates crack propagation in the Cosserat continua whose characteristic lengths commensurate with the microstructural length of the material (the particle size or the layer thickness).

4.2 The Length Scales in Continuum Description

Continuum modeling is based on the introduction of the volume element of size H satisfying the following double inequality under the hypothesis of separation of scales:

$$l_m \ll H \ll L, \tag{4.1}$$

where l_m is the characteristic size of the material microstructure (e.g., the grain or defect size, the distance between the microstructural elements, etc.), L is the characteristic size of the area under consideration, the wave length, etc. The first inequality in (4.1) is required for the volume element to be representative. The second inequality stipulates that the characteristic length l_g of redistribution of the fields involved in the description be considerably greater than H since the fields in the continuum description are approximated by uniform within H. In other words, the variations of the field over distance H should be negligible. This can be quantified as follows.

Consider, for the sake of simplicity, a scalar function $f(x)$ continuously differentiable in a vicinity of 0, which signifies the origin of a suitable coordinate frame. If the variations of f are negligible over H then

$$\left|f(H) - f(0)\right| \ll \left|f(0)\right|.$$

Keeping only the linear term in the Taylor expansion, this condition gives

$$l_g = \left|\frac{f(0)}{f'(0)}\right| \gg H. \tag{4.2}$$

If the function is regular at the origin then $l_g \sim L$ and (4.2) are equivalent to the right-hand inequality (4.1). If, however, the field is singular at the origin then we shall write (4.2) at a point $x \neq 0$ and then allow x tend to zero. In particular, when the singularity is a power law, $|f(x)| \sim x^a$, as in the cases of corner singularities, Condition (4.2) is replaced by $x/a \gg H$. This corresponds to the asymptotics

$$x/H \to \infty \tag{4.3}$$

which is the outer asymptotics of the singularity. Thus in dealing with singularities, one first has to obtain the asymptotics (4.3) by fixing x and tending H (and subsequently l_m) to zero and then tending L to infinity. Upon rescaling this leads to the inner asymptotics $x \to 0$.

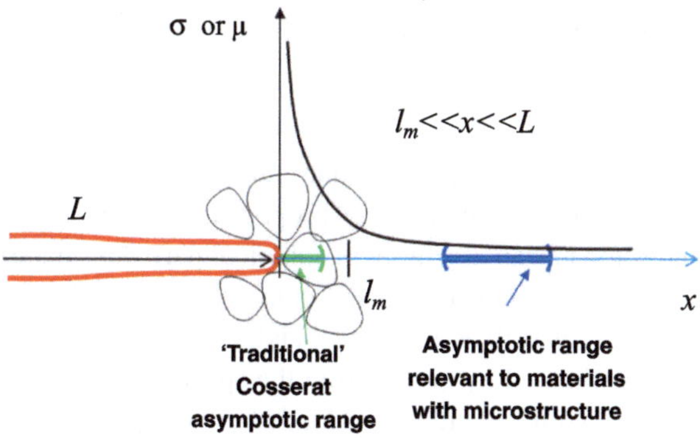

Fig. 4.1 In particulate material, the traditional asymptotic range falls within the microstructural elements. The proposed asymptotics works on the boundary between the microstructural scale and the scale of the crack

Standard continua do not inherit a microstructural length, hence this asymptotics becomes redundant. For non-standard continua possessing characteristic lengths, the asymptotics (4.3) can be put to use to achieve considerable simplification. The practical way of doing this is to neglect the terms of the type of $Hf'(x)$ as compared to $f(x)$. By continuing the expansion in the Taylor series, one can obtain that in this asymptotics the terms of the type of $H^2f''(x),\ldots$ should also be neglected.

In essence, a continuum constructed to represent a heterogeneous material cannot address the distances shorter than the volume element size. This fact manifests itself in the vicinity of a stress singularity: The stress distribution computed in the continuum may misrepresent the stress state of the original homogeneous material in the vicinity of the crack tip. Instead, the inner region of the outer asymptotics should be used. Therefore, instead of the traditional modeling based just on the $x \to 0$ singularity, one needs to obtain the outer asymptotics first and then use it to determine the singularity, see Fig. 4.1.

Of course, if the continuum is a classical one without a characteristic length then the complete solution and the outer asymptotics coincide, so the traditional approach does not change. The difference only comes for continua with internal crack lengths, such as Cosserat continua.

The characteristic lengths, l, are related to the microstructural length of the material through the values of the material constants. In principle, the characteristic lengths can be considerably higher than the microstructural length such that if the latter is beyond the resolution of the continuum the formers are not. We, however, distinguish one important case when $l \sim H$ such that the Cosserat characteristic lengths are also beyond the resolution of the continuum. We call it the *small scale Cosserat continuum*. The following sections will provide two examples of small scale continua and consider the simplest cases of crack propagation in them.

4.3 Small Scale Isotropic Cosserat Continuum as a Model of Particulate Material

We start with an isotropic Cosserat continuum model of particulate materials (such as rock, concrete, ceramics). A model was proposed in [1] whereby the particulate material was represented as an assembly of spherical particles connected at random points. Each connection was considered as a combination of 6 elastic springs: 1 tensile and 2 shear springs, as well as 1 torsional and 2 bending springs. The stiffnesses of the springs were estimated by considering the springs as elastic cylinders of diameter commensurate with the grain size loaded either by uniform stress (for the normal and shear stiffnesses) or linearly distributed stress with vanishing average (for the torsional and bending stiffnesses). After applying homogenization by differential expansions (e.g., [7]) an isotropic Cosserat continuum was obtained with the characteristic lengths of the order of the grain size. Thus $l \sim H$, i.e., the Cosserat characteristic lengths are also beyond the resolution of the continuum, and hence the asymptotics of $l \to 0$ discussed in the previous section applies.

Consider, for the sake of simplicity, a two dimensional Cosserat continuum in the plane (x_1, x_2). The Lamé equations read (e.g., [5, 4])

$$\begin{cases} (\lambda+2\mu)[\frac{\partial^2 u_1}{\partial x_1^2}+\frac{\partial^2 u_2}{\partial x_1 \partial x_2}]+(\mu+\alpha)[\frac{\partial^2 u_1}{\partial x_2^2}-\frac{\partial^2 u_2}{\partial x_1 \partial x_2}]+2\alpha\frac{\partial \varphi_3}{\partial x_2}=0, \\ (\lambda+2\mu)[\frac{\partial^2 u_1}{\partial x_1 \partial x_2}+\frac{\partial^2 u_2}{\partial x_2^2}]+(\mu+\alpha)[\frac{\partial^2 u_2}{\partial x_1^2}-\frac{\partial^2 u_1}{\partial x_1 \partial x_2}]-2\alpha\frac{\partial \varphi_3}{\partial x_1}=0, \\ l_2^2[\frac{\partial^2 \varphi_3}{\partial x_1^2}+\frac{\partial^2 \varphi_3}{\partial x_2^2}]-\varphi_3+\frac{1}{2}[\frac{\partial u_2}{\partial x_1}-\frac{\partial u_1}{\partial x_2}]=0. \end{cases} \quad (4.4)$$

The only characteristic length present here is the length l_2 of the following group of lengths which characterize the isotropic Cosserat continuum:

$$l_1=\sqrt{B/(4\mu)}; \qquad l_2=\sqrt{B/(4\alpha)}; \qquad l=\sqrt{l_1^2+l_2^2}, \quad (4.5)$$

where λ and μ are the Lamé constants (μ is the shear modulus), α is the Cosserat shear modulus, B is the bending stiffness.

As l_2 is the only length parameter explicit in (4.4), the asymptotics of small volume elements requires that $l_2 \to 0$, which reduces the third equation of (4.4) to

$$\varphi_3=\frac{1}{2}\left(\frac{\partial u_2}{\partial x_1}-\frac{\partial u_1}{\partial x_2}\right), \quad (4.6)$$

implying that the asymptotics of small characteristic size formally leads to the *Cosserat continuum with constrained rotations*. We can use the corresponding solutions to obtain the required asymptotics.

4.4 Shear Crack in Small Scale Isotropic Cosserat Continuum: A Mechanism of In-Plane Growth

Traditional treatment of cracks in Cosserat continuum [4, 2] is based on extracting of asymptotics $r \ll l$, where r is the distance to the crack tip. Subsequently, the Cosserat continuum with constrained rotations had to be employed to avoid insurmountable technical difficulties. Here we use the above intermediate asymptotics and obtain the Cosserat continuum with constrained rotations as a consequence of the asymptotics rather than a palliative.

Consider a shear crack of length $2L$ under uniform loading. The boundary conditions for this problem are

$$\sigma_{21} = \tau_0, \qquad \sigma_{22} = 0, \qquad \mu_{23} = 0 \quad \text{for} \ -L \le x_1 \le L,\ x_2 = 0, \tag{4.7}$$

while the stresses and displacements are continuous outside $[-L, L]$.

We represent the crack as a continuous distribution of dislocations (displacement discontinuities) and disclinations (rotation discontinuities in the case of Cosserat continuum) of unknown density in the otherwise continuous material and then equating the stress produced by all dislocations at the points of the crack contour to the boundary conditions. However, due to symmetry of boundary conditions (4.7), only dislocations with Burgers vector parallel to x_1 need to be taken into account [1].

A solution for this dislocation in the Cosserat continuum with constrained rotations (as required by the small scale asymptotics) can be found in [2], from where the main terms of asymptotics $r \gg l$ are:

$$\begin{aligned}
\frac{\sigma_{11}}{\mu b} &= \frac{1}{r}\left[-\frac{3\sin\theta + \sin 3\theta}{4\pi(1-\nu)} + O\left(\frac{l^2}{r^2}\right)\right],\\
\frac{\sigma_{22}}{\mu b} &= \frac{1}{r}\left[\frac{\sin 3\theta - \sin\theta}{4\pi(1-\nu)} + O\left(\frac{l^2}{r^2}\right)\right],\\
\frac{\sigma_{12}}{\mu b} &= \frac{1}{r}\left[\frac{\cos\theta + \cos 3\theta}{4\pi(1-\nu)} + O\left(\frac{l^2}{r^2}\right)\right],\\
\frac{\sigma_{21}}{\mu b} &= \frac{1}{r}\left[\frac{\cos\theta + \cos 3\theta}{4\pi(1-\nu)} + O\left(\frac{l^2}{r^2}\right)\right],\\
\mu_{13} &= 2\frac{\mu b l^2}{\pi r^2}\cos 2\theta + O\left(\frac{l^2}{r^2}\right),\\
\mu_{23} &= 2\frac{\mu b l^2}{\pi r^2}\sin 2\theta + O\left(\frac{l^2}{r^2}\right).
\end{aligned} \tag{4.8}$$

Here b is the value of the Burgers vector, $r^2 = x_1^2 + x_2^2$, θ is the polar angle.

The main asymptotic terms for stress (4.8) coincide with the solution for a conventional dislocation in the classical isotropic elastic continuum [3]. Furthermore, moment stress μ_{23} created by the dislocation on the crack line ($\theta = 0, \pm\pi$) is equal to zero. This means that the dislocation does not contribute to the moment stress on the crack line. Therefore, the boundary conditions (4.7) can be satisfied by the appropriate distribution of the dislocations without disclinations. (There is no coupling between the dislocations and disclination for shear cracks.) Subsequently, the problem reduces to the conventional one for a shear crack whose solution for a uniform load is known. Furthermore, since we are only interested in the stress concentration at the crack tip, we can obtain a general result using the expression for the displacement discontinuity at the crack tip. We use the expression for the distribution of the relevant displacement component u_1 in a vicinity of a Mode II crack tip [8] as well as its discontinuity across the crack line and its derivative, which is the dislocation density, ρ_1:

$$u_1 = \frac{K_{\mathrm{II}}}{\mu}\sqrt{\frac{r}{2\pi}}\sin\frac{\theta}{2}\left[2(1-\nu)+\cos^2\frac{\theta}{2}\right],$$
$$\rho_1(r) = \frac{2K_{\mathrm{II}}(1-\nu)}{\mu\sqrt{2\pi r}}. \tag{4.9}$$

Here (r,θ) is the polar coordinate frame with the origin placed at the crack tip, K_{II} is the conventional Mode II stress intensity factor, ν is the Poisson's ratio.

Using the expression for the dislocation-generated moment stresses (4.8) one obtains the moment stress at the crack tip on the line of crack continuation ($\theta = 0$) whose leading asymptotic term (as $r \to 0$) reads

$$\mu_{13} = \frac{K_{\mathrm{II}}(1-\nu)}{\sqrt{2\pi}r^{3/2}}l^2, \qquad \mu_{23} = 0. \tag{4.10}$$

This is a stronger singularity than the classical one. However, it does not contribute to the energy release rate because of the absence of reciprocal rotation discontinuity in Mode II. In the spirit of the philosophy proposed, the way to treat it is to compare the values of $\mu_{13}(l)$, which are finite. In particular, the criterion of crack propagation can be formulated in terms of the critical value of the moment stress needed to break the bonds and make the particles rotate. It is important that the maximum moment stress is acting at the continuation of the crack, which forms a mechanism of often observed in-plane shear crack propagation.

4.5 Anisotropic Cosserat Continuum for Layered Material with Sliding Layers

In 2D layered materials, the role of rotation is played by the bending angle (the derivative of layer deflection). For the corresponding Cosserat continuum, in the case of freely sliding layers the displacement and rotation fields, in the absence of body forces and moments, are governed by the following set of equations [6]:

$$\frac{\partial \sigma_{11}}{\partial x_1} + \frac{\partial \sigma_{21}}{\partial x_2} = 0, \qquad \frac{\partial \sigma_{12}}{\partial x_1} + \frac{\partial \sigma_{22}}{\partial x_2} = 0, \qquad \frac{\partial \mu_{31}}{\partial x_1} + \sigma_{12} - \sigma_{21} = 0,$$
$$\sigma_{11} = A_{11}\gamma_{11} + A_{12}\gamma_{22}, \qquad \sigma_{22} = A_{12}\gamma_{11} + A_{22}\gamma_{22}, \tag{4.11}$$
$$\sigma_{21} = 0, \qquad \sigma_{12} = G\gamma_{12}, \qquad \mu_{13} = B\kappa_{13},$$

where

$$\gamma_{11} = \frac{\partial u_1}{\partial x_1}, \qquad \gamma_{21} = \frac{\partial u_1}{\partial x_2} + \varphi_3, \qquad \gamma_{22} = \frac{\partial u_2}{\partial x_2},$$
$$\gamma_{12} = \frac{\partial u_2}{\partial x_1} - \varphi_3, \qquad \kappa_{13} = \frac{\partial \varphi_3}{\partial x_1}, \tag{4.12}$$

E, ν are the layer's Young's modulus, and Poisson's ratio, $G = E/(2(1+\nu))$, b is the layer thickness,

$$A_{11} = A_{22} = \frac{(1-\nu)E}{(1+\nu)(1-2\nu)},$$
$$A_{12} = \frac{\nu E}{(1+\nu)(1-2\nu)}, \qquad B = \frac{Eb^2}{12(1-\nu^2)}. \tag{4.13}$$

4.6 Bending Crack

We consider now a special type of the crack—a bending crack which is oriented normal to the layering and is represented as a continuous distribution of disclinations, which are discontinuities in rotations, see Fig. 4.2. The leading terms of the outer asymptotics of stress and moment stress are [6]:

$$\sigma_{22}(0, x_2) = -\frac{Eb^{1/2}}{4(1-\nu^2)\sqrt{2\pi}3^{1/4}} \frac{1}{\sqrt{|x_2|}},$$
$$\mu_{13}(0, x_2) = -\frac{Eb^{3/2}3^{1/4}}{24(1-\nu^2)\sqrt{2\pi}} \frac{\operatorname{sgn}(x_2)}{\sqrt{|x_2|}}. \tag{4.14}$$

The crack can be modeled as a distribution of disclinations sitting on the x_2 axis, as shown in Fig. 4.2, then the only stress component the disclinations produce there is μ_{13}. This leads to an integral equation whose solution for a semi-infinite crack loaded by a couple of concentrated moments, m, applied at a distance a from its tip is [6]:

$$\mu_{13} \sim \frac{\mathrm{M}_3}{(x_2)^{1/4}}, \qquad \mathrm{M}_3 = -\frac{m\sqrt{2}}{2\pi a^{3/4}}. \tag{4.15}$$

The stress singularity here is much weaker than in the Mode II crack (Sect. 4.4). Similarly to the case of a crack in particulate materials, the criterion of crack prop-

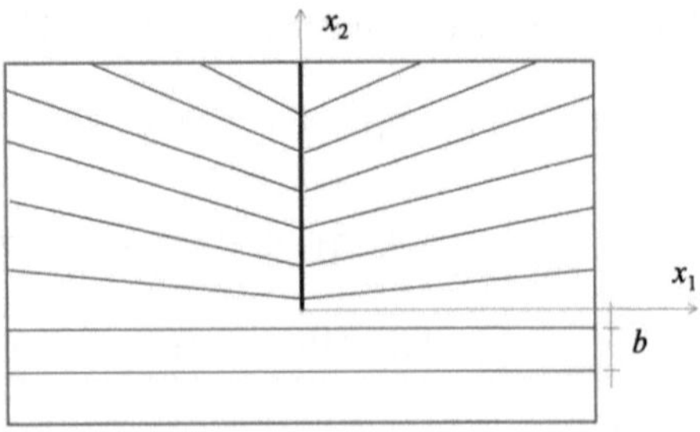

Fig. 4.2 Bending crack in layered material

agation can be formulated in terms of $\mu_{13}(b)$ compared with the bending moment per unit area required to break the layers.

4.7 Conclusions

Modeling particulate materials with the small-scale Cosserat continuum—a continuum with characteristic scales of the order of the microstructural size—justifies the asymptotics of small characteristic lengths. In 2D isotropic Cosserat continuum, this asymptotics formally produces a Cosserat continuum with constraint rotations, which is the first simplification of the proposed modeling. Furthermore, the stress singularities should be considered at the distances greater than the characteristic lengths, as opposed to the traditional approach.

The Mode II crack in a small scale isotropic Cosserat continuum produces moment stress with a strong ($3/2$ power) singularity controlled by the conventional Mode II stress intensity factor. The bending crack (discontinuity in rotations) in a small scale Cosserat model of layered material with sliding layers produces moment stress with a weak ($1/4$ power) singularity. In both cases, the singularity can only be interpreted up to the distances to the crack tip exceeding the microstructural length of the material. Subsequently, the moment stress values at these distances are to be used in the (force) criteria of crack propagation.

Acknowledgements We acknowledge financial support from the Australian Research Council via Discovery Grants DP0773839 (EP) and DP0559737 (AVD). We also wish to thank Greg Sevel for his help with the paper formatting.

References

1. Dyskin, A.V., Pasternak, E.: Rotational mechanism of in-plane shear crack growth in rocks under compression. In: Potvin, Y., Carter, J., Dyskin, A., Jeffrey, R. (eds.) 1st Southern Hemisphere International Rock Mechanics Symposium SHIRMS 2008, vol. 2, pp. 111–120. Australian Centre for Geomechanics, Crawley (2008)

2. Gourgiotis, P., Georgiadis, H.: Distributed dislocation approach for cracks in couple-stress elasticity: shear modes. Int. J. Fract. **147**, 83–102 (2007)
3. Landau, L., Lifshitz, E.: Theory of Elasticity. Braunschweig, Oxford (1959)
4. Morozov, N.: Mathematical Issues of the Crack Theory. Nauka, Moscow (1984). In Russian
5. Nowacki, W.: The linear theory of micropolar elasticity. In: Nowacki, W., Olszak, W. (eds.) Micropolar Elasticity, pp. 1–43. Springer, Berlin (1974)
6. Pasternak, E., Dyskin, A., Mhlhaus, H.B.: Cracks of higher modes in Cosserat continua. Int. J. Fract. **140**, 189–199 (2006)
7. Pasternak, E., Mhlhaus, H., Dyskin, A.: On the possibility of elastic strain localisation in a fault. Pure Appl. Geophys. **161**, 2309–2326 (2004)
8. Tada, H., Paris, P., Irwin, G.: The Stress Analysis of Cracks. Handbook, 3rd edn. ASME Press, New York (1985)

Chapter 5
Micropolar Fluids: From Nematic Liquid Crystals to Liquid-Like Granular Media

Daniel Lhuillier

Abstract The micropolar model is certainly the best continuum-mechanical model to describe the collective behavior of molecules or rigid particles interacting via short-range forces and couples. We look at the necessary modifications of the original model for it to describe two unusual materials: nematic liquid crystals on the one hand and liquid-like granular materials on the other hand.

5.1 Continuum-Mechanical Description

The continuum-mechanical description of matter with rotational degrees of freedom is now well-known. In the present contribution, we focus on liquid-like materials only, known by the generic name of micropolar fluids. Encompassing the seminal works of Aero [1], Cowin [2], Eringen [5] and many others, the description of micropolar fluids involves the mass and momentum conservation together with balance equations for the intrinsic moment of momentum and the internal energy

$$\frac{\mathrm{d}\rho}{\mathrm{d}t} = -\rho \nabla \cdot \mathbf{V}, \tag{5.1}$$

$$\rho \frac{\mathrm{d}\mathbf{V}}{\mathrm{d}t} = \nabla \cdot \sigma + \mathbf{F}^{\mathrm{ext}}, \tag{5.2}$$

$$\rho \frac{\mathrm{d}\mathbf{m}}{\mathrm{d}t} = \nabla \cdot \mu - \varepsilon \cdot \sigma + \mathbf{L}^{\mathrm{ext}}, \tag{5.3}$$

$$\rho \frac{\mathrm{d}e}{\mathrm{d}t} = \sigma^T : (\nabla \mathbf{V} + \varepsilon \cdot \boldsymbol{\omega}) + \mu^T : \nabla \boldsymbol{\omega} - \nabla \cdot \mathbf{q}. \tag{5.4}$$

Here ρ is the mass density of the material, $\rho\mathbf{V}$ and $\rho(\mathbf{m} + r \times \mathbf{V})$ are the momentum and moment of momentum per unit volume, $\mathbf{F}^{\mathrm{ext}}$ and $\mathbf{L}^{\mathrm{ext}}$ are external forces

D. Lhuillier (✉)
UPMC Univ Paris 6, UMR 7190 CNRS, Institut Jean Le Rond d'Alembert, Paris, France
e-mail: daniel.lhuillier@upmc.fr

G.A. Maugin, A.V. Metrikine (eds.), *Mechanics of Generalized Continua*,
Advances in Mechanics and Mathematics 21,
DOI 10.1007/978-1-4419-5695-8_5, © Springer Science+Business Media, LLC 2010

and couples, σ and μ are the stress and couple-stress tensors, and $\mathbf{q}$ is the heat flux. Moreover, $\boldsymbol{\omega}$ is the rotating velocity of the material (independent of the bulk angular velocity represented by $(1/2)\nabla \times \mathbf{V}$), while ρe is the internal energy per unit volume, defined as the total energy minus the kinetic energy $(1/2)\rho\mathbf{V}^2 + (1/2)\rho\mathbf{m}\cdot\boldsymbol{\omega}$. Lastly, ε is the alternating third-order tensor, a superscript T on a tensor means its transpose and $\mathrm{d}/\mathrm{d}t = \partial/\partial t + \mathbf{V}\cdot\nabla$ is the material time-derivative. Our purpose is to apply that well-known set of equations to the description of rather unusual materials such as nematic liquid crystals and liquid-like granular media. The former material offers an opportunity to define an "ordered" micropolar fluid and the latter will provide us with an example of "agitated" micropolar fluid for which the velocity fluctuations play an important role.

5.2 Ordered Micropolar Fluids

Nematic liquid crystals are made of small rod-like molecules, almost perfectly aligned along some direction. Since these small molecules interact through short-range forces and couples, the micropolar framework is relevant. And, in fact, Eringen [6] and Lee and Eringen [10] did propose an extended version of the micropolar fluid model to describe the dynamics of nematic liquid crystals. However, their model was progressively superseded by the Ericksen–Leslie model which is now considered as the standard model for that material [3]. What is the reason that led to abandon the Lee–Eringen model? It must be stressed that the two models do introduce a unit vector to describe the microstructure. But while Ericksen and Leslie considered their unit vector $\mathbf{n}$ as giving the mean direction of the molecular axis with a simple transport equation like

$$\frac{\mathrm{d}\mathbf{n}}{\mathrm{d}t} = \boldsymbol{\omega} \times \mathbf{n}, \tag{5.5}$$

Lee and Eringen introduced a micro-rotation tensor depending on a scalar ϕ and a unit vector $\mathbf{n}^E$ which evolves in time according to

$$\sin\phi \frac{\mathrm{d}\mathbf{n}^E}{\mathrm{d}t} \times \mathbf{n}^E + (1-\cos\phi)\frac{\mathrm{d}\mathbf{n}^E}{\mathrm{d}t} = \boldsymbol{\omega} \times \mathbf{n}^E.$$

It is clear that no value of the parameter ϕ is compatible with (5.5). This means that the unit vector $\mathbf{n}^E$ cannot be considered as the "director", i.e., a body-fixed unit vector aligned with the molecular axis. The use of $\mathbf{n}^E$ is thus not recommended and consequently a second feature that needs to be modified as compared to the original Lee–Eringen model is the description of the elastic energy that appears whenever the director field departs from uniformity. The most convenient expression is certainly the Franck–Oseen distortion energy [3] which depends explicitly on the director and its gradients and leads to writing the internal energy of the nematic liquid crystal as

$$e = e(\rho, s, \mathbf{n}, \nabla\mathbf{n}). \tag{5.6}$$

One thus ends with the picture of an ordered micropolar fluid for which the internal energy not only depends on the usual variables (the mass density ρ and the specific entropy s) but also on the director which plays the role of an order parameter. With Alejandro Rey (Mac-Gill University) we revisited the micropolar model with due account for (5.5) and (5.6). The complete set of results is presented in [13], while below we focus on the main steps and findings. The starting point is the Gibbs relation which is the differential form of (5.6)

$$\mathrm{d}e = T\,\mathrm{d}s - p\,\mathrm{d}\left(\frac{1}{\rho}\right) + \frac{\partial e}{\partial \mathbf{n}} \cdot \mathrm{d}\mathbf{n} + \frac{\partial e}{\partial (\nabla \mathbf{n})} : \mathrm{d}(\nabla \mathbf{n}), \tag{5.7}$$

where p and T are the pressure and the temperature of the medium. At equilibrium the entropy of the system is a maximum for fixed values of its mass, momentum, moment of momentum, and total energy. With the help of the Gibbs relation that maximization of entropy is obtained with the equilibrium conditions

$$\begin{aligned} &\nabla T = 0, \qquad \nabla \boldsymbol{\omega} = 0, \qquad \nabla \mathbf{V} + \nabla \mathbf{V}^T = 0, \\ &\boldsymbol{\omega} - \frac{1}{2} \nabla \times \mathbf{V} = 0, \qquad \mathbf{n} \times \mathbf{h} = 0, \end{aligned} \tag{5.8}$$

where $\mathbf{h}$ is the molecular field defined by

$$\mathbf{h} = -\rho \frac{\partial e}{\partial \mathbf{n}} + \rho \nabla \cdot \left(\frac{\partial e}{\partial \nabla \mathbf{n}}\right). \tag{5.9}$$

The Gibbs relation (5.7) is useful not only at equilibrium but also out of equilibrium. From it one can derive the entropy balance in the form

$$\begin{aligned} T\left[\rho \frac{\mathrm{d}s}{\mathrm{d}t} + \nabla \cdot \left(\frac{\mathbf{q}}{T}\right)\right] &= -\frac{\mathbf{q}}{T} \cdot \nabla T + (\sigma + pI - \Sigma)^T : (\nabla \mathbf{V} + \varepsilon \cdot \boldsymbol{\omega}) \\ &\quad + (\boldsymbol{\mu} - \mathsf{M})^T : \nabla \boldsymbol{\omega} + \boldsymbol{\omega} \cdot [\mathbf{n} \times \mathbf{h} + \varepsilon : \Sigma - \nabla \cdot \mathsf{M}], \end{aligned} \tag{5.10}$$

where

$$\Sigma_{ij} = -\rho \frac{\partial e}{\partial (\nabla_i n_k)} \nabla_j n_k, \tag{5.11}$$

$$\mathsf{M}_{ij} = \rho \varepsilon_{jlk} n_l \frac{\partial e}{\partial (\nabla_i n_k)} \tag{5.12}$$

are the elastic stress and elastic couple stress, respectively. The right-hand side of (5.10) represents the entropy production rate and must vanish at equilibrium. Taking the equilibrium conditions (5.8) into account, one can check that all terms vanish at equilibrium except the last one because at equilibrium the molecular rotation is uniform but has no reason to vanish. To ensure that the entropy production rate does vanish at equilibrium one is obliged to require

$$\mathbf{n} \times \mathbf{h} + \varepsilon : \Sigma - \nabla \cdot \mathsf{M} = 0. \tag{5.13}$$

That imposed relation between the elastic stresses and couples can also be obtained as a necessary condition for the internal energy to be a scalar function of n and $\nabla \mathbf{n}$, and that relation was coined "rotational identity" in [3]. Here we showed that the rotational identity is necessary to avoid spurious terms in the entropy production. The next step is to define the viscous stress σ^V and the viscous couple stress μ^V by

$$\sigma = -pI + \Sigma + \sigma^V, \tag{5.14}$$

$$\mu = \mathsf{M} + \mu^V. \tag{5.15}$$

The viscous stress is not symmetric. Its symmetric part will be noted σ^{VS} and its antisymmetric part will be associated with the viscous torque

$$\mathbf{L}^V = -\varepsilon : \sigma^V. \tag{5.16}$$

The entropy balance is now rewritten as

$$T\left[\rho \frac{\mathrm{d}s}{\mathrm{d}t} + \nabla \cdot \left(\frac{\mathbf{q}}{T}\right)\right] = -\frac{\mathbf{q}}{T} \cdot \nabla T + \sigma^{VS} : \mathsf{D} + \mathbf{L}^V \cdot (\Omega - \omega) + \left(\mu^V\right)^T : \nabla \omega, \tag{5.17}$$

where $\mathsf{D} = (1/2)(\nabla \mathbf{V} + \nabla \mathbf{V}^T)$ is the strain rate and $\Omega = (1/2)\nabla \times \mathbf{V}$ is the bulk rotation rate. The right-hand side of (5.17) displays the four types of dissipative phenomena that are specific to micropolar fluids. It is worthy to notice that, with the rotational identity (5.13) and the definitions (5.14) to (5.16), the angular momentum balance (5.3) finally appears as

$$\rho \frac{\mathrm{d}\mathbf{m}}{\mathrm{d}t} = \nabla \cdot \mu^V + \mathbf{n} \times \mathbf{h} + \mathbf{L}^V + \mathbf{L}^{\text{ext}}. \tag{5.18}$$

There is a last intriguing difference between the micropolar model and the Ericksen–Leslie model. It concerns the entropy production rate which is written [3] as

$$-\frac{\mathbf{q}}{T} \cdot \nabla T + \sigma^{VS} : \mathsf{D} + \mathbf{h} \cdot \left(\frac{\mathrm{d}\mathbf{n}}{\mathrm{d}t} - \Omega \times \mathbf{n}\right) \tag{5.19}$$

and looks rather different from the right-hand side of (5.17). In fact, there is no contradiction between the dissipation rates of the two models. Suppose that we may neglect the rotary inertia, the viscous couple stress and the external torques acting on the rod-like molecules. As a consequence of these assumptions, the moment of momentum balance simplifies to

$$0 \approx \mathbf{n} \times \mathbf{h} + \mathbf{L}^V \tag{5.20}$$

so that with (5.5)

$$\mathbf{L}^V \cdot (\Omega - \boldsymbol{\omega}) \approx \mathbf{h} \cdot \left(\frac{\mathrm{d}\mathbf{n}}{\mathrm{d}t} - \Omega \times \mathbf{n} \right). \tag{5.21}$$

We conclude that Lee and Eringen [10] gave a description of micro-rotation and elasticity in nematics which was rigorous but apparently not compatible with the existence and role of a body-fixed director. It was shown above that the micropolar fluid model, completed with (5.5) and (5.6), offers a natural extension to the Ericksen–Leslie model. The main difference between a simple micropolar fluid and an ordered one is the presence of three elastic (non-dissipative) quantities $\mathbf{h}$, Σ and M. The micropolar model is able to reveal all the assumptions implicitly made to get the Ericksen–Leslie model, as witnessed, for example, by the simplified moment of momentum balance written in (5.20). The reader is referred to [13] for more details on the comparison between the two models.

5.3 Agitated Micropolar Fluids

Granular materials are media made from particles of macroscopic (millimeter) size. They behave like a gas when the particles are dilute but they have a more complex dynamic behavior at high particle concentrations, when the grains display more or less permanent contacts with their immediate neighbors. To give an order of magnitude, for spherical particles all of the same size, that complex behavior is observed for a particle volume fraction ϕ between $\phi \approx 0.5$ and $\phi \approx 0.65$, that is to say, between the random loose packing and the random close packing. Those dense granular materials have a solid-like behavior when submitted to relatively small shear stresses but above some critical shear they display a liquid-like behavior, much like the sand at the surface of a dune which is either motionless or is flowing as an avalanche depending on the shear exerted by gravity. We are interested in these dense granular liquids and will consider small loads only, so that the grains can be considered as rigid. Moreover, the grains will be considered as so massive that the interstitial gas will have no influence on their dynamics. Since our granular liquid is a collection of rigid particles interacting through contact (or collision) forces and contact (or collisions) couples mainly (hence short-range forces and couples), it is expected that the micropolar framework is quite relevant to describe its dynamics. Our purpose in what follows is to prove that this is indeed the case. To simplify the issue, we suppose that all the particles are rigid spheres with the same radius a, the same mass m and the same moment of inertia $J = 2ma^2/5$. A given grain is labeled by α. At contact, or when colliding, particle β exerts on particle α a force $\mathbf{F}^{\alpha\beta}$ and a couple $\mathbf{L}^{\alpha\beta}$. The trajectory of particle α is given by the position $\mathbf{R}^\alpha(t)$ of its center, and its motion is depicted by its velocity $\mathbf{v}^\alpha(t) = \mathrm{d}\mathbf{R}^\alpha/\mathrm{d}t$ and its angular velocity $\boldsymbol{\omega}^\alpha(t)$. The equations of motion of particle α are

$$m\frac{\mathrm{d}\mathbf{v}^\alpha}{\mathrm{d}t} = \sum_{\beta \neq \alpha} \mathbf{F}^{\alpha\beta} + \mathbf{f}^{\text{ext}}, \tag{5.22}$$

$$J\frac{\mathrm{d}\omega^{\alpha}}{\mathrm{d}t} = \sum_{\beta\neq\alpha} \frac{\mathbf{R}^{\beta}-\mathbf{R}^{\alpha}}{2} \times \mathbf{F}^{\alpha\beta} + \sum_{\beta\neq\alpha} \mathbf{L}^{\alpha\beta} + \mathbf{l}^{\text{ext}}. \tag{5.23}$$

In these equations, $\mathbf{f}^{\text{ext}}$ and $\mathbf{l}^{\text{ext}}$ are the external force and couple acting on each particle. We want to check that the continuum-mechanical version of those equations are given by the micropolar equations (5.1) to (5.4) and to provide some rigorous definitions for the quantities involved. To achieve that goal, we introduce a Dirac-like function δ_α which is located on the instantaneous position of the center of particle α

$$\delta_\alpha(\mathbf{x},t) = \delta\big(\mathbf{x}-\mathbf{R}^{\alpha}(t)\big). \tag{5.24}$$

Since δ_α follows the trajectory of particle α, it obeys the evolution equation

$$\frac{\partial\delta_\alpha}{\partial t} + \nabla\cdot\big(\delta_\alpha\mathbf{v}^{\alpha}\big) = 0. \tag{5.25}$$

To get the Eulerian equations of motion is now straightforward. After multiplying the equations of motion of particle α by δ_α, one takes the sum over all particles, takes (5.25) into account and performs a statistical average denoted by $\langle\cdots\rangle$. All the details are given in [12], and we give only the main results below. Mass conservation is expressed as in (5.1) with definitions

$$\rho = \Big\langle \sum_\alpha \delta_\alpha m \Big\rangle = n_p m, \qquad \rho\mathbf{V} = \Big\langle \sum_\alpha \delta_\alpha m \mathbf{v}^{\alpha} \Big\rangle, \tag{5.26}$$

where $n_p = \langle \sum_\alpha \delta_\alpha \rangle$ is the number density of the particles which, according to (5.25), obeys the conservation equation $\partial n_p/\partial t + \nabla\cdot(n_p\mathbf{V}) = 0$. Momentum conservation and moment of momentum balance for the collection of grains stem from the equations of motion (5.22) and (5.23) for individual grains. They can be presented like in (5.2) and (5.3) with the definitions $\mathbf{F}^{\text{ext}} = n_p\mathbf{f}^{\text{ext}}$, $\mathbf{L}^{\text{ext}} = n_p\mathbf{l}^{\text{ext}}$, $\rho\mathbf{m} = \langle \sum_\alpha \delta_\alpha J\omega_\alpha \rangle = n_p J\omega$ and

$$\sigma_{ij} = \Big\langle (1/2)\sum_\alpha\sum_{\beta\neq\alpha} \delta_\alpha \big(R^{\beta}-R^{\alpha}\big)_i F_j^{\alpha\beta} \Big\rangle - \Big\langle \sum_\alpha \delta_\alpha m v_i'^{\alpha} v_j'^{\alpha} \Big\rangle, \tag{5.27}$$

$$\mu_{ij} = \Big\langle (1/2)\sum_\alpha\sum_{\beta\neq\alpha} \delta_\alpha \big(R^{\beta}-R^{\alpha}\big)_i L_j^{\alpha\beta} \Big\rangle - \Big\langle \sum_\alpha \delta_\alpha J v_i'^{\alpha} \omega_j'^{\alpha} \Big\rangle. \tag{5.28}$$

Note the similarities between the definitions of the stress and the couple stress and the presence of a kinetic contribution linked to the velocity fluctuations $\mathbf{v}'^{\alpha} = \mathbf{v}^{\alpha} - \mathbf{V}$ and the angular velocity fluctuations $\omega'^{\alpha} = \omega^{\alpha} - \omega$. The last issue is to check whether we can obtain an equation for the internal energy like (5.4). It is now widely accepted that velocity fluctuations play an important role in granular materials where the concept of a "granular temperature" is widely used. Two quantities that proved to be important in that context are the mean fluctuational kinetic energies k_v and k_ω defined as

$$\rho k_v = \left\langle \sum_\alpha \delta_\alpha m \frac{(v'^\alpha)^2}{2} \right\rangle, \qquad \rho k_\omega = \left\langle \sum_\alpha \delta_\alpha J \frac{(\omega'^\alpha)^2}{2} \right\rangle. \tag{5.29}$$

One can manipulate the equations of motion (5.22) and (5.23) of particle α to deduce the transport equations for the total fluctuational kinetic energy

$$\rho \frac{\mathrm{d}}{\mathrm{d}t}(k_v + k_\omega) = \sigma^T : (\nabla \mathbf{V} + \varepsilon \cdot \boldsymbol{\omega}) + \mu^T : \nabla \boldsymbol{\omega} - \nabla \cdot \mathbf{q}^k - \gamma, \tag{5.30}$$

where γ is the dissipation rate of the so-called "granular temperature" $k_v + k_\omega$

$$\gamma = \left\langle \frac{1}{2} \sum_\alpha \sum_{\beta \neq \alpha} \delta_\alpha \big(\boldsymbol{\omega}^\beta - \boldsymbol{\omega}^\alpha\big) \cdot \mathbf{L}^{\alpha\beta} \right\rangle + \left\langle \frac{1}{2} \sum_\alpha \sum_{\beta \neq \alpha} \delta_\alpha \mathbf{V}^{\alpha\beta} \cdot \mathbf{F}^{\alpha\beta} \right\rangle, \tag{5.31}$$

where $\mathbf{V}^{\alpha\beta}$ is the relative velocity at the contact between particles α and β

$$\mathbf{V}^{\alpha\beta} = \mathbf{v}^\beta - \mathbf{v}^\alpha - (1/2)\big(\boldsymbol{\omega}^\beta + \boldsymbol{\omega}^\alpha\big) \times \big(\mathbf{R}^\beta - \mathbf{R}^\alpha\big). \tag{5.32}$$

Because of the dissipation rate γ, we do not have a transport equation looking like (5.4), yet. But we remark that the dissipated energy is absorbed by the grains (resulting in an increase of their "true" temperature) and is then diffused by collisions or contacts. One can model these two processes with a transport equation for the internal energy e_0 per unit mass

$$\rho \frac{\mathrm{d}e_0}{\mathrm{d}t} = \gamma - \nabla \cdot \mathbf{q}^e. \tag{5.33}$$

It is now clear that if we define the effective internal energy of the medium as

$$e = e_0 + k_v + k_\omega \tag{5.34}$$

then e will obey (5.4) with a total energy flux $\mathbf{q} = \mathbf{q}^e + \mathbf{q}^k$. Hence granular matter can be described by the micropolar model provided the internal energy is understood as the sum of the true internal energy and the total fluctuational kinetic energy. In other words, (5.34) is the main characteristic feature of an "agitated" micropolar fluid. A second specific property of granular liquids is the unusual expression of their stress tensor, itself a consequence of the unusual features of the contact forces between grains. A very suggestive way to obtain the expression of the stress tensor was proposed by Kanatani [9]. As usual, the stress tensor is split into a pressure stress and a dissipative stress $\sigma = -pI + \tau$. Neglecting the role of the inter-grain couples, the dissipation rate is approximately $\tau : \mathsf{D}$, and considering result (5.31) Kanatani writes

$$\tau : \mathsf{D} \approx n_p V^T F^T, \tag{5.35}$$

where V^T is the mean value of the tangential velocity at contact and F^T the mean value of the tangential force at contact. F^T is a solid friction force proportional to the normal component F^N itself proportional to the granular pressure, hence $F^T \approx \mu(p/n_p a)$ where μ is the solid friction coefficient and a is the mean grain size.

The relative velocity at contact is approximately $V^T \approx a\sqrt{\mathsf{D}:\mathsf{D}}$. As a consequence, the order of magnitude of the dissipation is $\tau : \mathsf{D} \approx \mu p\sqrt{\mathsf{D}:\mathsf{D}}$ and the dissipative stress is

$$\tau \approx \mu p \frac{\mathsf{D}}{\sqrt{\mathsf{D}:\mathsf{D}}}. \tag{5.36}$$

The dissipative stress in granular liquids is thus very different from the usual viscous stress proportional to the strain rate D. It is proportional to the pressure and it depends on the direction of the strain rate and not on its magnitude. That unusual expression was extended in different ways to take collisions into account [8, 7].

5.4 Conclusions

We showed that it was possible to achieve the rheological description of two unusual materials within the micropolar framework. The versatility of the micropolar model is thus demonstrated. A generalization to the case of deformable molecules or deformable particles is offered by the micromorphic fluid model and was already used in the past to describe polymer solutions [4] and bubbly fluids [11].

References

1. Aero, E.L., Bulygin, A.N., Kuvshinskii, E.V.: Asymmetric hydromechanics. J. Appl. Math. Mech. **29**, 333–346 (1965). PMM **29**, 297–308 (1965)
2. Cowin, C.S.: Polar fluids. Phys. Fluids **11**, 1919–1932 (1968)
3. de Gennes, P.G., Prost, J.: The Physics of Liquid Crystals, 2nd edn. Oxford University Press, Oxford (1993)
4. Drouot, R., Maugin, G.A.: Internal variables and the thermodynamics of macromolecule solutions. Int. J. Eng. Sci. **21**, 705–724 (1983)
5. Eringen, A.C.: Theory of micropolar fluids. J. Math. Mech. **16**, 1–18 (1966)
6. Eringen, A.C.: Micropolar theory of liquid crystals. In: Liquid Crystals and Ordered Fluids, vol. 3, pp. 443–473. Plenum, New York (1978)
7. Jop, P., Forterre, Y., Pouliquen, O.: A constitutive law for dense granular flows. Nature **441**, 727–730 (2006)
8. Josserand, C., Lagrée, P.-Y., Lhuillier, D.: Stationary shear flows of dry granular materials: a tentative continuum modeling. Eur. Phys. J. **14**, 127–135 (2004)
9. Kanatani, K.: A micropolar continuum theory for the flow of granular materials. Int. J. Eng. Sci. **17**, 419–432 (1979)
10. Lee, J.D., Eringen, A.C.: Wave propagation in nematic liquid crystals. J. Chem. Phys. **54**, 5027–5034 (1971)
11. Lhuillier, D.: The equations of motion of a one-velocity, one-temperature bubbly fluid (revisited). J. Phys. **46**, 1325–1333 (1985)
12. Lhuillier, D.: Constitutive relations for steady flows of dense granular liquids. Physica A **383**, 267–275 (2007)
13. Lhuillier, D., Rey, A.D.: Nematic liquid crystals and ordered micropolar fluids. J. Non-Newtonian Fluid Mech. **120**, 169–174 (2004)

Chapter 6
Linear Cosserat Elasticity, Conformal Curvature and Bounded Stiffness

Patrizio Neff, Jena Jeong, Ingo Münch, and Hamidréza Ramézani

Abstract We describe a principle of bounded stiffness and show that bounded stiffness in torsion and bending implies a reduction of the curvature energy in linear isotropic Cosserat models leading to the so called conformal curvature case $\frac{1}{2}\mu L_c^2\|\operatorname{dev}\operatorname{sym}\nabla\operatorname{axl}\overline{A}\|^2$ where $\overline{A} \in \mathfrak{so}(3)$ is the Cosserat microrotation. Imposing bounded stiffness greatly facilitates the Cosserat parameter identification and allows a well-posed, stable determination of the one remaining length scale parameter L_c and the Cosserat couple modulus μ_c.

6.1 Introduction

Non-classical size-effects are becoming increasingly important for materials at the micro- and nanoscale regime. There are many possibilities in order to include size-effects on the continuum scale. One such prominent model is the Cosserat model. In its simplest isotropic linear version, the Cosserat model introduces six material

P. Neff (✉)
Chair of Nonlinear Analysis and Modelling, Fakultät für Mathematik, Universität Duisburg-Essen, Campus Essen, Universitätsstr. 2, 45141 Essen, Germany
e-mail: patrizio.neff@uni-due.de

J. Jeong
École Spéciale des Travaux Publics du Bâtiment et de l'Industrie (ESTP), 28 avenue du Président Wilson, 94234 Cachan Cedex, France
e-mail: jeong@profs.estp.fr

I. Münch
Institut für Baustatik, Universität Karlsruhe (TH), Kaiserstrasse 12, 76131 Karlsruhe, Germany
e-mail: im@bs.uka.de

H. Ramézani
École Polytechnique de l'Université d'Orléans, CNRS-CRMD, 8 rue Léonard de Vinci, 45072 Orléans, France
e-mail: hamidreza.ramezani@cnrs-orleans.fr

G.A. Maugin, A.V. Metrikine (eds.), *Mechanics of Generalized Continua*, Advances in Mechanics and Mathematics 21,
DOI 10.1007/978-1-4419-5695-8_6, © Springer Science+Business Media, LLC 2010

parameters. However, parameter identification for Cosserat solids remains a difficult and challenging issue.

We motivate how to a priori reduce the number of curvature parameters in the linear, isotropic, centro-symmetric Cosserat model by requiring what we identify with *bounded stiffness*. First, we recall the Cosserat model and we motivate bounded stiffness in general. Imposing bounded stiffness reduces the curvature energy to the conformally invariant case, which is the weakest possible requirement for well-posedness of the linear isotropic Cosserat model [1].

It is well-known that a Cosserat solid displays size-effects. These size effects refer to a non-classical dependence of rigidity of an object upon one or more of its dimensions. In classical linear elasticity for a circular cylinder with radius a, the rigidity in tension is proportional to a^2, and the rigidity in torsion and bending is proportional to a^4. For the Cosserat solid, the ratio of rigidity to its classical value is increased: thinner samples of the same material respond stiffer. For certain parameter ranges of the Cosserat solid, this effect may be dramatic. For example, the rigidity in torsion could become proportional to a^2 so that the normalized torsional rigidity (normalized against the classical value) has a singularity proportional to $1/a^2$. However, Lakes [3] already notes: "...infinite stiffening effects are unphysical."

Our principle of bounded stiffness requires simply that the stiffness increase for thinner and thinner samples (normalized against the classical stiffness) should be bounded independently of the wire radius a, i.e., a singularity free response. In bending and torsion, we can directly read off the corresponding requirement. It leads in a straightforward way to what we term the conformal curvature case $\frac{1}{2}\mu L_c^2\|\operatorname{dev}\operatorname{sym}\nabla\operatorname{axl}\overline{A}\|^2$. In separate contributions [10, 2, 7, 9, 1, 8], we have investigated, in more detail, this novel conformal curvature case from alternative perspectives.

6.2 The Linear Elastic Cosserat Model

For the *displacement* $u : \Omega \subset \mathbb{R}^3 \mapsto \mathbb{R}^3$ and the *skew-symmetric infinitesimal microrotation* $\overline{A} : \Omega \subset \mathbb{R}^3 \mapsto \mathfrak{so}(3)$, we consider the *two-field* minimization problem

$$I(u,\overline{A}) = \int_\Omega W_{\mathrm{mp}}(\overline{\varepsilon}) + W_{\mathrm{curv}}(\nabla\operatorname{axl}\overline{A}) - \langle f,u\rangle\,\mathrm{d}V \\ - \int_{\partial\Omega}\langle f_s,u\rangle - \langle M_s,u\rangle\,\mathrm{d}S \mapsto \min \quad \text{w.r.t.} \quad (u,\overline{A}), \qquad (6.1)$$

under the following constitutive requirements and boundary conditions:

$$\overline{\varepsilon} = \nabla u - \overline{A}, \quad \textit{first Cosserat stretch tensor}$$

$$u_{|\Gamma} = u_{\mathrm{d}}, \quad \textit{essential displacement boundary conditions}$$

$$W_{\mathrm{mp}}(\overline{\varepsilon}) = \mu\|\operatorname{sym}\overline{\varepsilon}\|^2 + \mu_c\|\operatorname{skew}\overline{\varepsilon}\|^2 + \frac{\lambda}{2}\operatorname{tr}[\operatorname{sym}\overline{\varepsilon}]^2, \quad \textit{strain energy}$$

$$\begin{aligned}
&\phi := \operatorname{axl} \overline{A} \in \mathbb{R}^3, \qquad \overline{\mathfrak{k}} = \nabla \phi, \\
&\|\operatorname{curl} \phi\|^2_{\mathbb{R}^3} = 4\| \operatorname{axl} \operatorname{skew} \nabla \phi\|^2_{\mathbb{R}^3} = 2\|\operatorname{skew} \nabla \phi\|^2_{\mathbb{M}^{3\times 3}}, \\
&W_{\mathrm{curv}}(\nabla \phi) = \frac{\gamma+\beta}{2}\|\operatorname{dev}\operatorname{sym} \nabla \phi\|^2 + \frac{\gamma-\beta}{2}\|\operatorname{skew} \nabla \phi\|^2 \\
&\qquad + \frac{3\alpha + (\beta+\gamma)}{6}\operatorname{tr}[\nabla \phi]^2.
\end{aligned} \tag{6.2}$$

Here, f are given volume forces while u_{d} are Dirichlet boundary conditions for the displacement at $\Gamma \subset \partial\Omega$ where $\Omega \subset \mathbb{R}^3$ denotes a bounded Lipschitz domain. Surface tractions, volume couples and surface couples can be included in the standard way. The strain energy W_{mp} and the curvature energy W_{curv} are the most general isotropic quadratic forms in the *infinitesimal non-symmetric first Cosserat strain tensor* $\overline{\varepsilon} = \nabla u - \overline{A}$ and the *micropolar curvature tensor* $\overline{\mathfrak{k}} = \nabla \operatorname{axl} \overline{A} = \nabla \phi$ (curvature-twist tensor). The parameters μ, λ [MPa] are the classical Lamé moduli and α, β, γ are further micropolar curvature moduli with dimension $[\mathrm{Pa}\cdot\mathrm{m}^2] = [\mathrm{N}]$ of a force. The additional parameter $\mu_c \geq 0$ [MPa] in the strain energy is the *Cosserat couple modulus*. For $\mu_c = 0$ the two fields of displacement u and microrotations $\overline{A} \in \mathfrak{so}(3)$ decouple, and one is left formally with classical linear elasticity for the displacement u. The strong form Cosserat balance equations are given by

$$\begin{aligned}
&\operatorname{Div} \sigma = f, \quad \textit{balance of linear momentum} \\
&-\operatorname{Div} m = 4\mu_c \cdot \operatorname{axl}\operatorname{skew} \overline{\varepsilon}, \quad \textit{balance of angular momentum} \\
&\sigma = 2\mu \cdot \operatorname{sym} \overline{\varepsilon} + 2\mu_c \cdot \operatorname{skew} \overline{\varepsilon} + \lambda \cdot \operatorname{tr}[\overline{\varepsilon}] \cdot \mathbb{1} \\
&\quad = (\mu + \mu_c) \cdot \overline{\varepsilon} + (\mu - \mu_c) \cdot \overline{\varepsilon}^T + \lambda \cdot \operatorname{tr}[\overline{\varepsilon}] \cdot \mathbb{1} \\
&\quad = 2\mu \cdot \operatorname{dev}\operatorname{sym} \overline{\varepsilon} + 2\mu_c \cdot \operatorname{skew} \overline{\varepsilon} + K \cdot \operatorname{tr}[\overline{\varepsilon}] \cdot \mathbb{1}, \\
&m = (\gamma+\beta)\operatorname{dev}\operatorname{sym} \nabla\phi + (\gamma-\beta)\operatorname{skew} \nabla\phi + \frac{3\alpha + (\gamma+\beta)}{3}\operatorname{tr}[\nabla\phi]\mathbb{1}, \\
&\phi = \operatorname{axl} \overline{A}, \qquad u_{|_\Gamma} = u_{\mathrm{d}}.
\end{aligned} \tag{6.3}$$

For simplicity we do not make an explicit statement for the boundary conditions which are satisfied by the microrotations. Note only that if the microrotations $\overline{A} \in \mathfrak{so}(3)$ remain free at the boundary, we would have $m.\mathbf{n}_{|_{\partial\Omega}} = 0$. This Cosserat model can be considered with basically three different sets of moduli for the curvature energy which in each step relaxes the curvature energy. The situations are characterized by possible estimates for the curvature energy:

1. $W_{\mathrm{curv}}(\overline{\mathfrak{k}}) \geq c^+\|\overline{\mathfrak{k}}\|^2$,
2. $W_{\mathrm{curv}}(\overline{\mathfrak{k}}) \geq c^+\|\operatorname{sym} \overline{\mathfrak{k}}\|^2$,
3. $W_{\mathrm{curv}}(\overline{\mathfrak{k}}) \geq c^+\|\operatorname{dev}\operatorname{sym} \overline{\mathfrak{k}}\|^2$.

The different estimates give rise to the introduction of representative cases:

1. (*Pointwise positive case*) $\frac{1}{2}\mu L_c^2\|\nabla\phi\|^2$. This corresponds to $\alpha = \beta = 0$, $\gamma = \mu L_c^2$.
2. (*Symmetric case*) $\frac{1}{2}\mu L_c^2\|\operatorname{sym}\nabla\phi\|^2$. This corresponds to $\alpha = 0, \beta = \gamma$ and $\gamma = \frac{1}{2}\mu L_c^2$.
3. (*Conformal case*) $\frac{1}{2}\mu L_c^2\|\operatorname{dev}\operatorname{sym}\nabla\phi\|^2 = \frac{1}{2}\mu L_c^2(\|\operatorname{sym}\nabla\phi\|^2 - \frac{1}{3}\operatorname{tr}[\nabla\phi]^2)$. This corresponds to $\beta = \gamma$ and $\gamma = \frac{1}{2}\mu L_c^2$ and $\alpha = -\frac{1}{3}\mu L_c^2$. In terms of the polar ratio $\Psi = (\beta+\gamma)/(\alpha+\beta+\gamma)$, it corresponds to the limit value $\Psi = 3/2$.

All three cases are mathematically well-posed [1, 5]. The pointwise positive Case 1 is usually considered in the literature. Case 2 leads to a symmetric couple-stress tensor m, and a new motivation for Case 3 is the subject of this contribution. In a plane strain problem, all three cases coincide and only one curvature parameter matters, thus not permitting to discern any relation between the three curvature parameters.

Case 3 is called the conformal curvature case since the curvature energy is invariant under superposed infinitesimal conformal mappings, i.e., mappings $\phi_C : \mathbb{R}^3 \mapsto \mathbb{R}^3$ that satisfy $\operatorname{dev}\operatorname{sym}\nabla\phi_C = 0$. Such mappings infinitesimally preserve shapes and angles [7]. In that case, the couple stress tensor m is symmetric and trace-free. In Case 2 and Case 3, the constitutive couple stress/curvature tensor relation cannot be inverted, but the system of equations is nevertheless well-posed.

6.3 The Idea of Bounded Stiffness

Let us turn our attention to the practical aspects of the problem of determining material parameters. We investigate the question for which parameter values (μ_c, α, β, γ) the linear elastic Cosserat model can be considered to be a consistent description for a continuous solid showing size-effects. We assume the continuous solid to be available in any small size we can think of, the possibility of which is certainly included in the very definition of a continuous solid. Note that this assumption *excludes*, e.g., man made grid-structures but includes, e.g., polycrystalline material. We are investigating the situation when one or several dimensions of the specimen get small. Denoting by a such a dimension, the limit $a \to 0$ is purely formal in the sense that we are only interested in the leading order behavior for small, but not arbitrarily small a. Understanding this limiting process $a \to 0$ opens us, indirectly, the possibility to bound the stiffness of the material at smallest reasonable specimen size away from unrealistic orders of magnitude. Our conclusions are based on simple three-dimensional boundary value problems for which analytical solutions are available.

By examining the bending and torsion analytical formula and calculating the stiffness increase at small wire radius a, we are forced to conclude that the conformal curvature case $\frac{1}{2}\mu L_c^2\|\operatorname{dev}\operatorname{sym}\nabla\phi\|^2$ is the only one possible [8]. Any other combination of parameters will lead to unphysical stiffening effects. The requirement of bounded stiffness is also very natural if we compare with atomistic simulations; in such a case, all response curves will show bounded energy *and* bounded stiffness.

One should keep in mind, however, that the internal length scale L_c remains a phenomenological parameter in the Cosserat model, the value of which is not necessarily determined by a given microstructural length scale, although this is often tacitly understood. On the contrary, L_c could also be chosen large in which case the ratio L_c/a may be very large, although wires with given radius a can still be experimentally investigated.

The remainder of this paragraph is a free adaption of a statement given by Metrikine in [4, p. 740] to our situation: No material at no scale is perfectly rigid. Therefore, their stiffness is bounded at all scales. This is one of the most fundamental principles in modern physics, and any general model, which is supposed to be applicable at the complete length scale, must satisfy it. All researchers agree with that statement. Many models, however, are designed to work only at a specific length scale. Homogeneous continuum elasticity models, for example, are all applicable only at a relatively large length scale (i.e., the length scale of the corresponding mechanical processes is much larger than the characteristic length of the material microstructure). Should such models comply with bounded stiffness? There is no consensus among researchers as to how to answer this question in the case that unbounded stiffness is associated with those length scales at which the model in question is not applicable according to its initial assumptions.

The authors of this paper advocate the following answer to the above question. Imagine two models of a material which, with the same accuracy, describe static behavior of the material at a desired length scale. Imagine further that one model has bounded stiffness whereas the other has not, but its singular stiffness is associated with the length scales outside the considered length scale range. In this case, the bounded stiffness model should be preferred.

It should not only be done because it complies with bounded stiffness. More importantly, models with bounded stiffness profit from an insensitive parameter identification. This is important for experimental identification of material properties.

Whether or not the Cosserat model with conformal curvature shows bounded stiffness in all possible boundary value problems is not yet clear. We surmise that the micromorphic model [6] is a good candidate for that purpose.

6.4 Infinitesimal Conformal Mappings

The maps $u_C : \mathbb{R}^3 \mapsto \mathbb{R}^3$ that satisfy $\operatorname{dev}\operatorname{sym}\nabla u_C = 0$ are called infinitesimal conformal mappings. They form a ten-dimensional vector space and are given in closed form by

$$u_C(x) = \frac{1}{2}\big(2\langle \operatorname{axl}(\widehat{W}), x\rangle x - \operatorname{axl}(\widehat{W})\|x\|^2\big) + [\widehat{p}\mathbb{1} + \widehat{A}].x + \widehat{b}, \tag{6.4}$$

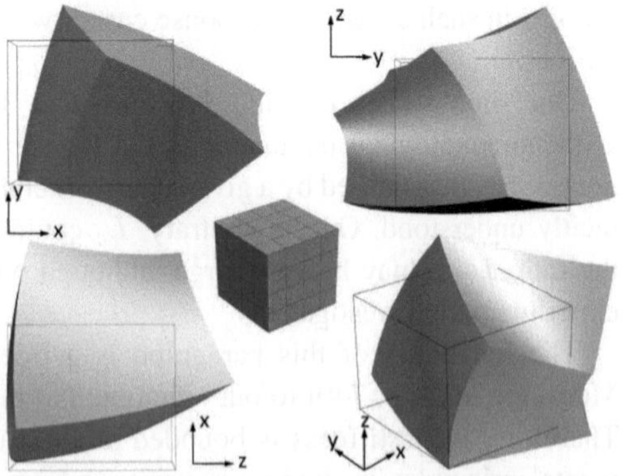

Fig. 6.1 Deformed and undeformed shape of the unit cube under infinitesimal conformal transformation u_C as boundary condition using the linear elastic Cosserat model and Case 3 assumption with $\widehat{W}_{12} = \widehat{W}_{13} = \widehat{W}_{23} = 3$, $\widehat{p} = -4$, $\widehat{A}_{12} = \widehat{A}_{13} = \widehat{A}_{23} = 4$ and $\widehat{b} = 0$, DOFs $= 4300$, quadratic elements

where $\widehat{W}, \widehat{A} \in \mathfrak{so}(3)$, $\widehat{b} \in \mathbb{R}^3$ and $\widehat{p} \in \mathbb{R}$ are arbitrary constants. The conformal mappings may be used to construct a universal solution of the Cosserat model if the conformal curvature expression (Case 3) is considered. The solution is invariant of the Cosserat coupling constant μ_c and the internal length scale L_c.[1] To understand this claim, let us consider the boundary value problem of linear Cosserat elasticity in strong form with conformal curvature:

$$\begin{aligned} \operatorname{Div}\sigma(\nabla u, \overline{A}) &= f, \qquad -\operatorname{Div} m = 4\mu_c \cdot \operatorname{axl}(\operatorname{skew}\nabla u - \overline{A}), \\ \sigma &= 2\mu \cdot \operatorname{dev}\operatorname{sym}\nabla u + 2\mu_c \cdot \operatorname{skew}(\nabla u - \overline{A}) + K \cdot \operatorname{tr}[\nabla u] \cdot \mathbb{1}, \\ m &= \mu L_c^2 \cdot \operatorname{dev}\operatorname{sym}\nabla\operatorname{axl}(\overline{A}), \qquad u_{|\partial\Omega} = u_C, \end{aligned} \tag{6.5}$$

where K is the bulk modulus. Inserting for u an *infinitesimal conformal map* u_C (Fig. 6.1) which is defined by (6.4) and choosing $\overline{A}(x) = \operatorname{anti}(\frac{1}{2}\operatorname{curl} u(x))$ simplifies the equations to

$$\begin{aligned} \operatorname{Div}\sigma(\nabla u, \overline{A}) &= \widehat{f}, \qquad -\operatorname{Div} m = 0, \\ \sigma &= 2\mu \cdot \operatorname{dev}\operatorname{sym}\nabla u + K \cdot \operatorname{tr}[\nabla u] \cdot \mathbb{1}, \\ m &= \mu L_c^2 \operatorname{dev}\operatorname{sym}\nabla\left[\frac{1}{2}\operatorname{curl} u(x)\right]. \end{aligned} \tag{6.6}$$

$\operatorname{Div}\sigma(\nabla u_C) = 3K\operatorname{axl}(\widehat{W})$, for $\widehat{p}$ is constant and $\operatorname{Div}[\langle\widehat{k}, x\rangle\mathbb{1}] = \widehat{k}$. Since the boundary value problem $\operatorname{Div}\sigma(\nabla u) = 3K\operatorname{axl}(\widehat{W})$, $u_{|\Gamma}(x) = u_C(x)$ for a given constant $\widehat{W} \in \mathfrak{so}(3)$ admits a unique solution, this solution is already given by $u(x) = u_C(x)$. We have therefore obtained an inhomogeneous, three-dimensional analytical solution for the boundary value problem of linear Cosserat elasticity with constant body forces $\widehat{f} = 3K\operatorname{axl}(\widehat{W})$. To gain further understanding of the conformal Cosserat model (Case 3) we subject a regular and rectangular network of beams

[1] Here, even a strong variation in shear modulus $\mu(x)$ would be allowed (as well as a strong variation in the couple modulus $\mu_c(x)$ and internal length scale $L_c(x)$. Only the bulk modulus K must be constant.

to an infinitesimal conformal displacement (6.4). In our comparison, we use $\widehat{b} = 0$, $\widehat{A} = 0$ and some generic values for $\widehat{p}$ and $\widehat{W}$. The area of the squared structure is 1 and the beams are characterized by a quadratic cross-section of dimension 0.05. Thus, the area of the cross-section amounts to 0.0025 and the moment of inertia to 5.208×10^{-7} against bending. The Young's modulus is set to $E = 1$. We use the Bernoulli beam theory of second order. Thus, displacements and rotations are limited to a reasonable amount. First, only nodes on the boundary are conformally displaced, all other organize themselves by minimizing strain and curvature energy of the beams. In Fig. 6.2, one can see the initial rectangular beam structure, the boundary conditions and the displacement vectors bringing all nodes of the boundary into the conformally corresponding points. The nodes within the structure meet the condition, which is fulfilled for all beams balance of momentum and balance of angular momentum.

The right picture in Fig. 6.2 indicates that curvature appears nearly everywhere in all beams. The maximum value of this curvature is about 6.24. Now, all nodes of the structure are conformally displaced. Thus, only the curvature energy of the beams can be minimized. In Fig. 6.3, one can see the initial rectangular beam structure, the boundary conditions and the displacement vectors bringing all nodes into their conformally corresponding points. The beams preserve the balance of angular momentum. The right picture in Fig. 6.3 indicates that curvature appears also nearly everywhere in all beams. The maximum value of this curvature is about 7.02. While the infinitesimal conformal mapping does not give rise to a curvature contribution in the Cosserat model with conformal curvature (Case 3), we clearly see that the beam-network response is always with curvature. This allows us to already con-

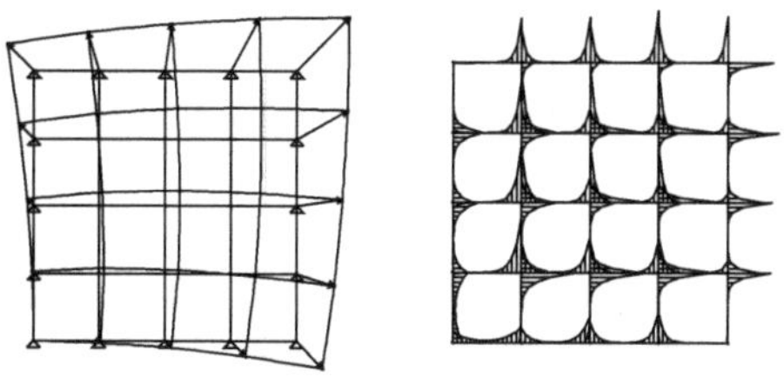

Fig. 6.2 (*Left*) Initial structure with boundary conditions, conformal displacement of boundary nodes and deformed mesh. (*Right*) Trend of curves (plotted on undeformed mesh) indicates the curvature of beams

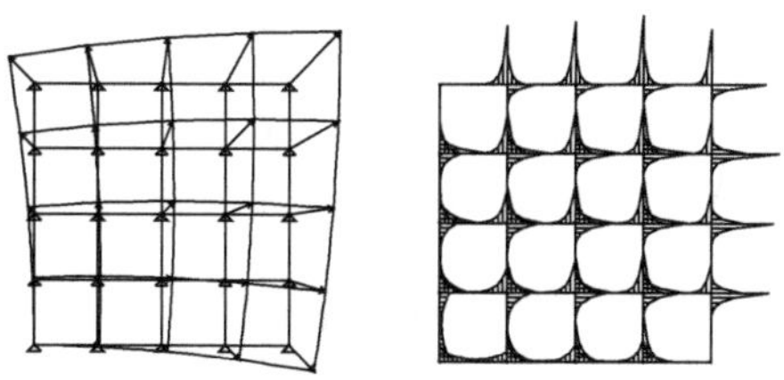

Fig. 6.3 (*Left*) Initial structure with boundary conditions, conformal displacement of all nodes and deformed mesh. (*Right*) Trend of curves (plotted on undeformed mesh) indicates the curvature of beams

clude that *the conformal Cosserat model cannot be identified with a homogenized beam model*. We rather expect a homogenized beam model to give rise to a uniform positive definite curvature expression as embodied in Case 1 of our classification.

Notation

Let $\Omega \subset \mathbb{R}^3$ be a bounded domain with Lipschitz boundary $\partial\Omega$ and let Γ be a smooth subset of $\partial\Omega$ with non-vanishing two-dimensional Hausdorff measure. For $a, b \in \mathbb{R}^3$ we let $\langle a, b\rangle_{\mathbb{R}^3}$ denote the scalar product on $\mathbb{R}^3$ with associated vector norm $\|a\|^2_{\mathbb{R}^3} = \langle a, a\rangle_{\mathbb{R}^3}$. We denote by $\mathbb{M}^{3\times3}$ the set of real 3×3 second order tensors, written with capital letters and Sym denotes symmetric second orders tensors. The standard Euclidean scalar product on $\mathbb{M}^{3\times3}$ is given by $\langle X, Y\rangle_{\mathbb{M}^{3\times3}} = \operatorname{tr}[XY^T]$, and thus the Frobenius tensor norm is $\|X\|^2 = \langle X, X\rangle_{\mathbb{M}^{3\times3}}$. In the following, we omit the indices $\mathbb{R}^3, \mathbb{M}^{3\times3}$. The identity tensor on $\mathbb{M}^{3\times3}$ will be denoted by $\mathbb{1}$, so that $\operatorname{tr}[X] = \langle X, \mathbb{1}\rangle$. We set $\operatorname{sym}(X) = \frac{1}{2}(X^T + X)$ and $\operatorname{skew}(X) = \frac{1}{2}(X - X^T)$ so that $X = \operatorname{sym}(X) + \operatorname{skew}(X)$. For $X \in \mathbb{M}^{3\times3}$ we set for the deviatoric part $\operatorname{dev} X = X\frac{1}{3}\operatorname{tr}[X]\mathbb{1} \in \mathfrak{sl}(3)$ where $\mathfrak{sl}(3)$ is the Lie-algebra of traceless matrices. The set $\operatorname{Sym}(n)$ denotes all symmetric $n \times n$-matrices. The Lie-algebra of $\operatorname{SO}(3) := \{X \in \operatorname{GL}(3) \mid X^TX = \mathbb{1}, \det X = 1\}$ is given by the set $\mathfrak{so}(3) := \{X \in \mathbb{M}^{3\times3} \mid X^T = -X\}$ of all skew symmetric tensors. The canonical identification of $\mathfrak{so}(3)$ and $\mathbb{R}^3$ is denoted by $\operatorname{axl} \overline{A} \in \mathbb{R}^3$ for $\overline{A} \in \mathfrak{so}(3)$. Note that $(\operatorname{axl} \overline{A}) \times \xi = \overline{A}.\xi$ for all $\xi \in \mathbb{R}^3$ so that

$$\operatorname{axl}\begin{pmatrix} 0 & \alpha & \beta \\ -\alpha & 0 & \gamma \\ -\beta & -\gamma & 0 \end{pmatrix} := \begin{pmatrix} -\gamma \\ \beta \\ -\alpha \end{pmatrix}, \qquad \overline{A}_{ij} = \sum_{k=1}^{3} -\varepsilon_{ijk} \cdot \operatorname{axl} \overline{A}_k, \tag{6.7}$$

$$\|\overline{A}\|^2_{\mathbb{M}^{3\times3}} = 2\|\operatorname{axl} \overline{A}\|^2_{\mathbb{R}^3}, \qquad \langle \overline{A}, \overline{B}\rangle_{\mathbb{M}^{3\times3}} = 2\langle \operatorname{axl} \overline{A}, \operatorname{axl} \overline{B}\rangle_{\mathbb{R}^3},$$

where ε_{ijk} is the totally antisymmetric permutation tensor. Here, $\overline{A}.\xi$ denotes the application of the matrix $\overline{A}$ to the vector ξ and $a \times b$ is the usual cross-product. Moreover, the inverse of axl is denoted by anti and defined by

$$\begin{pmatrix} 0 & \alpha & \beta \\ -\alpha & 0 & \gamma \\ -\beta & -\gamma & 0 \end{pmatrix} := \operatorname{anti}\begin{pmatrix} -\gamma \\ \beta \\ -\alpha \end{pmatrix}, \tag{6.8}$$
$$\operatorname{axl}(\operatorname{skew}(a \otimes b)) = -\frac{1}{2} a \times b,$$

and $2\operatorname{skew}(b \otimes a) = \operatorname{anti}(a \times b) = \operatorname{anti}(\operatorname{anti}(a).b)$. Moreover, $\operatorname{curl} u = 2\operatorname{axl}(\operatorname{skew} \nabla u)$.

References

1. Jeong, J., Neff, P.: Existence, uniqueness and stability in linear Cosserat elasticity for weakest curvature conditions. Math. Mech. Solids (2008). doi:10.1177/1081286508093581
2. Jeong, J., Ramezani, H., Münch, I., Neff, P.: A numerical study for linear isotropic Cosserat elasticity with conformally invariant curvature. Z. Angew. Math. Mech. **89**(7), 552–569 (2009)
3. Lakes, R.S.: Experimental methods for study of Cosserat elastic solids and other generalized elastic continua. In: Mühlhaus, H.B. (ed.) Continuum Models for Materials with Microstructure, pp. 1–25. Wiley, New York (1995)
4. Metrikine, A.V.: On causality of the gradient elasticity models. J. Sound Vib. **297**, 727–742 (2006)
5. Neff, P.: The Cosserat couple modulus for continuous solids is zero viz the linearized Cauchy-stress tensor is symmetric. Z. Angew. Math. Mech. **86**, 892–912 (2006)
6. Neff, P., Forest, S.: A geometrically exact micromorphic model for elastic metallic foams accounting for affine microstructure. Modelling, existence of minimizers, identification of moduli and computational results. J. Elast. **87**, 239–276 (2007)
7. Neff, P., Jeong, J.: A new paradigm: the linear isotropic Cosserat model with conformally invariant curvature energy. Z. Angew. Math. Mech. **89**(2), 107–122 (2009)
8. Neff, P., Jeong, J., Fischle, A.: Stable identification of linear isotropic Cosserat parameters: bounded stiffness in bending and torsion implies conformal invariance of curvature. Acta Mech. (2009). doi:10.1007/s00707-009-0230-z
9. Neff, P., Jeong, J., Münch, I., Ramezani, H.: Mean field modeling of isotropic random Cauchy elasticity versus microstretch elasticity. Z. Angew. Math. Phys. **3**(60), 479–497 (2009)
10. Neff, P., Jeong, J., Ramezani, H.: Subgrid interaction and micro-randomness—novel invariance requirements in infinitesimal gradient elasticity. Int. J. Solids Struct. **46**, 4261–4272 (2009)

Chapter 7
Application of Generalized Continuum Theory to the Problem of Vibration Decay in the Complex Mechanical Structures

Vladimir Palmov

Abstract A generalized continuum approach is considered for structures in which a set of oscillators is attached to each material point of a linear elastic body. Dynamical consequences are given concerning traveling waves.

7.1 Introduction

One of the significant problems of modern engineering is the problem of vibrations of such structures as airplanes, trains, industrial buildings, vessels, ships, submarines, rockets, etc. A direct approach to this problem, especially in the high frequency range, is very complex due to a large number of structural elements. The specific feature of the vibration field of these complex structures is a finite value of spatial decay. This effect cannot be explained by friction because this is too small. Instead, we intend to apply the theory of generalized continua to this problem.

The analysis of media with complex structure has attracted much attention. The simplest among the media with complex structure is the Cosserat medium [1]. The Mindlin medium with a microstructure is more sophisticated. Of extreme complexity is the multipolar mechanics developed by Green and Rivlin [2].

An essential feature of the above mentioned theories is the concept of a material point. Classical continuum mechanics deals with media of ordinary structure, that is, a material point possesses three translational degrees of freedom. In contrast to this, a point in the Cosserat theory has all of the degrees of freedom of a rigid body, i.e., six degrees of freedom. In the Mindlin theory of a medium with microstructure, each point possesses the degrees of freedom of a classical solid with a uniform strain, i.e., twelve degrees of freedom. In the multipolar mechanics, a point is governed by n

V. Palmov (✉)
St. Petersburg Polytechnical University, St. Petersburg, Russia
e-mail: palmov@mail.ru

G.A. Maugin, A.V. Metrikine (eds.), *Mechanics of Generalized Continua*, Advances in Mechanics and Mathematics 21,
DOI 10.1007/978-1-4419-5695-8_7, © Springer Science+Business Media, LLC 2010

kinematic parameters, where n is finite but can be arbitrarily large. In what follows, we consider one of the simplest media of this type.

7.2 Equations of Dynamics of Continuous Medium with Complex Structure

We postulate the existence of a carrier medium and assume that it is governed by the Lamé equation of the classical theory of elasticity, i.e.,

$$(\lambda+\mu)\nabla\nabla\cdot\mathbf{u}+\mu\Delta\mathbf{u}-\rho\ddot{\mathbf{u}}+\mathbf{K}=0, \tag{7.1}$$

where ρ is the mass density, λ and μ are the Lamé moduli of elasticity, $\mathbf{u}$ is the vector of displacement of points in the carrier medium, and $\mathbf{K}$ is the intensity of external volume forces.

A set consisting of an infinite number of isotropic non-interacting oscillators with continuously distributed eigenfrequencies is attached to each point of the carrier medium. The equation of motion of a generic oscillator with an eigenfrequency k is as follows

$$m(k)\ddot{\mathbf{v}}_k+c(k)\big[1+R_k(\partial/\partial t)\big](\mathbf{v}_k-\mathbf{u})=\mathbf{Q}_k. \tag{7.2}$$

Here $\mathbf{v}_k$ is the vector of absolute displacement of the oscillator mass and $\mathbf{Q}_k$ is an external force acting on the oscillator mass. The value of $m(k)\,\mathrm{d}k$ is equal to the mass of the oscillators having the eigenfrequencies in the interval $(k,k+\mathrm{d}k)$ per unit volume. Thus, the total mass density of the oscillators attached to the carrier structure is given by

$$m=\int_0^\infty m(k)\,\mathrm{d}k. \tag{7.3}$$

The static rigidity of an oscillator suspension is denoted in (7.2) by $c(k)$ which is given by

$$c(k)=k^2m(k). \tag{7.4}$$

The term with $R_k(\partial/\partial t)$ is introduced in (7.2) in order to take into account the energy dissipation in the oscillator suspension. It will be shown below that this account of damping is absolutely necessary for obtaining physically meaningful results. Accounting for the damping will be achieved by means of two rheological models, namely a viscoelastic model and a model of an elastoplastic material. In the first case, $R_k(\partial/\partial t)$ denotes an operator of viscoelasticity, while in the second case it is a hysteretic operator. The term with $R_k(\partial/\partial t)$ is assumed to be small for any oscillator motion in both cases.

It is now necessary to take into account the effect of the oscillators' suspension on the carrier medium. The force of this interaction per unit volume is given by

$$\mathbf{F}=\int_0^\infty c(k)\big[1+R_k(\partial/\partial t)\big](\mathbf{v}_k-\mathbf{u})\,\mathrm{d}k. \tag{7.5}$$

As this force is a volume force it must be added to (7.1). Thus, the system of equations which governs the dynamics of the medium takes the following form:

$$\begin{aligned}(\lambda+\mu)\nabla\nabla\cdot\mathbf{u}+\mu\Delta\mathbf{u}-\rho\ddot{\mathbf{u}}-\int_0^\infty m(k)\ddot{\mathbf{v}}_k\,\mathrm{d}k+\mathbf{K}+\mathbf{Q}=0,\\ m(k)\ddot{\mathbf{v}}_k+c(k)\big[1+R_k(\partial/\partial t)\big](\mathbf{v}_k-\mathbf{u})=\mathbf{Q}_k.\end{aligned} \tag{7.6}$$

The vector $\mathbf{Q}$ is equal to the external volume force acting on the oscillators, that is,

$$\mathbf{Q}=\int_0^\infty \mathbf{Q}_k(k)\,\mathrm{d}k. \tag{7.7}$$

The boundary conditions for the introduced medium coincide with those of the classical theory of elasticity because they are prescribed only for the carrier medium.

The proposed model is useful for describing the vibration propagation in such mechanical structures as industrial buildings, vessels, airplanes, spacecrafts, etc. because all of these structures possess a primary structure and secondary systems attached to the primary structure. In the framework of the proposed method, the properties of the primary structure are described by the equations of the elasticity theory. The suspended oscillators reflect the frequency properties of the secondary systems attached to the primary structure. This model allows us to determine the global properties of the vibration field of complex structures without unnecessary details of the structure and its vibrational field.

A one-dimensional version of (7.6) was presented in [3], while the full three-dimensional one is given in [4, 5].

7.3 Analysis of Traveling Waves

Using (7.6), it is possible to study arbitrary wave propagation in the medium. But we restrict ourselves only to three typical waves: compressive, shear, and one-dimensional longitudinal waves.

In order to analyze compressional waves, we accept the following expression for the wave motion

$$\mathbf{u}=(\nabla c)\exp\big[\mathrm{i}(\mathbf{n}\cdot\mathbf{r}-\omega t)\big].$$

Substituting this into (7.6) and eliminating $\mathbf{v}_k$, we get following expression for the wave number

$$n^2=\frac{\omega^2}{\lambda+2\mu}\lambda^2,$$

where

$$\lambda^2=\rho+\int_0^\infty\frac{c(k)\,\mathrm{d}k}{k^2-\omega^2/(1+\mathrm{i}\Psi)}. \tag{7.8}$$

For analyzing shear waves, we consider that

$$\mathbf{u}=(\nabla\times\mathbf{c})\exp\big[\mathrm{i}(\mathbf{n}\cdot\mathbf{r}-\omega t)\big]$$

and then get

$$n^2 = \frac{\omega^2}{\mu}\lambda^2.$$

At last, for the case of one-dimensional longitudinal waves we take

$$\mathbf{u} = \mathbf{c}\exp\big[\mathrm{i}(\mathbf{n}\cdot\mathbf{r} - \omega t)\big]$$

and get the result

$$n^2 = \frac{\omega^2}{E}\lambda^2.$$

7.4 Analysis of Space Decay in Traveling Waves

Let us next prove that the vibration decay does not depend primarily on the oscillators' damping and that it remains finite even for vanishingly small damping. To this end, it is sufficient to show that the expression for λ^2 remains complex when $\Psi_k \to 0$. For the sake of simplicity, we assume that the damping properties of all oscillators coincide, i.e., $\Psi_k = \Psi$ does not depend upon parameter k but can depend on the excitation frequency ω.

Let us introduce a complex function

$$z = \frac{\omega}{\sqrt{1 + \mathrm{i}\Psi}}. \tag{7.9}$$

Let us agree that we take that branch of the radical in (7.9) which takes on the value $+1$ for $\Psi = 0$. The imaginary part of z is then negative for any frequency ω except $\omega = 0$. Hence, as $\Psi \to 0$ the complex function z approaches the real axis from the lower half-plane of the complex variable z. By using the newly introduced variable z in (7.9), we rewrite (7.8) as follows

$$\lambda^2 = \rho + \frac{1}{2z}\int_0^{\infty} \frac{c(|k|)}{k - z}\,\mathrm{d}k. \tag{7.10}$$

In the Cauchy integral, z always lies in the lower half-plane. As $\Psi \to 0$, the complex variable z approaches the real axis, along which the integration (7.8) proceeds. In accordance with the Sokhotsky–Plemelj formula, we obtain the following limit in (7.10) as $\Psi \to 0$

$$\lambda^2 = \rho + \frac{1}{2\omega}\left[-\pi\mathrm{i}c\big(|\omega|\big) + \text{v.p.}\int_{-\infty}^{+\infty} \frac{c(|k|)}{k - z}\,\mathrm{d}k\right]. \tag{7.11}$$

Here the principal value of the integral is to be taken.

As the imaginary part in (7.11) does not vanish, the coefficient of the spatial decay has a finite value even for vanishingly small damping in oscillators, and is determined by the dependence of the oscillators' suspension on their eigenfrequencies.

This effect is typical only for the model which accounts for the complex structure of the medium, i.e., in presence of the suspended oscillators. From a physical point of view, the effect can be explained by the fact that the suspended oscillators act as dynamic absorbers.

Note that accounting for damping in the oscillators is needed to determine their vibrational amplitudes. The second formula in (7.6) indicates that the amplitude of the oscillators b_k becomes unbounded at $\Psi = 0$.

Another sophisticated analysis shows that the result (7.11) remains valid as $\Psi_k \to 0$, with Ψ_k depending not only on the frequency ω but also the eigenfrequency k.

Let us illustrate our analysis by a particular example. Let

$$m(k) = A\left(\beta^2 + k^2\right)^{-1}, \qquad \Psi_k = \Psi, \tag{7.12}$$

where A and β are positive parameters and $\Psi_k = \Psi$ does not depend upon parameter k. Substituting (7.12) into (7.3) and evaluating the integral yield

$$m = \frac{\pi A}{2\beta}.$$

This result allows us to remove the parameter A from the forthcoming equations, retaining only m and β. Inserting now (7.12) into (7.4) and (7.8) gives

$$\lambda^2 = \rho + \int_0^\infty \frac{Ak^2\,\mathrm{d}k}{(\beta^2 + k^2)[k^2 - \omega^2/(1 + \mathrm{i}\Psi)]}.$$

Evaluating the integral by means of the calculus of residues, we arrive at the following final result

$$\lambda^2 = \rho + m\left[1 + \frac{\mathrm{i}\omega}{\beta\sqrt{1 + \mathrm{i}\Psi}}\right]^{-1}. \tag{7.13}$$

The structure of this formula completely confirms our analysis. Further, it shows a weak dependence of λ on Ψ for realistic, i.e., not very large, values of Ψ.

This example is remarkable for another reason. If one sets $\rho = 0$ and $\Psi = 0$ in (7.13), then

$$n^2 = \frac{\omega^2}{E}\frac{m}{1 + \mathrm{i}\omega/\beta}.$$

Exactly the same expression for the wave number can be obtained for the problem of linear longitudinal vibrations in a rod of the Kelvin–Voigt material, i.e., a rod with a finite value of damping. Therefore, the medium under consideration possesses the following interesting peculiarity. Though the medium is "built" from high-quality elements, such as an ideally elastic carrier medium and slightly damped oscillators, it behaves like a medium of ordinary structure with finite damping.

At first glance, this conclusion seems to be paradoxical. However, careful analysis shows that the energy dissipation has a finite value even for small damping in oscillators, which is explained by the considerable amplitudes of vibrations for the resonating oscillators.

7.5 Local Property of Vibration Field

Thus we had investigated global properties of a vibration field in a complex engineering structure. These properties are very interesting for engineers. But engineers want to know more details. For instance, they want to have full information about the vibration of any element of equipment of the engineering structure. We cannot give a direct answer on this question because, in our description, this element of equipment is totally absent from our equations. So we should improve our motivation in order to include this specific element of equipment in our equations. In order to do that, we select this element that is closely surrounded by a set of other elements of the engineering structure. We use a direct and detailed modeling of this part of the structure. For the modeling of the remaining part of the engineering structure, we use the equation of the continuous medium with a complex structure, which was presented above. We formulate appropriate boundary conditions at the boundary between the zone of continuum medium and the selected zone of direct mathematical description. Thus the solution of the formulated boundary value problem will give the full answer about the selected element in the mechanical structure.

In order to be convinced that we get thus a satisfactory result, let us consider the sequence of boundary value problems with an increasing zone of direct mathematical description of the surrounding of some element of the engineering structure. If this surrounding covers the whole engineering structure, we get the exact result for the element of interest. But we believe that we get approximately the same result even if this zone is not so large.

References

1. Cosserat, E., Cosserat, F.: Théorie des corps déformables. Hermann et Fils, Paris (1909)
2. Green, A.E., Rivlin, R.S.: Multipolar continuum mechanics. Arch. Ration. Mech. Anal. **17**, 113–147 (1964)
3. Slepyan, L.I.: Wave of deformation in a rod with flexible mounted masses. MTT, No. 5 (1967). In Russian
4. Palmov, V.A.: On a model of medium of complex structure. PMM, No. 4 (1969). In Russian
5. Palmov, V.A.: Vibration of Elastoplastic Bodies. Springer, Berlin (1998)

Chapter 8
Measuring of Cosserat Effects and Reconstruction of Moduli Using Dispersive Waves

Elena Pasternak and Arcady Dyskin

Abstract We revisit the problem of identification of effects of independent rotations and reconstruction of the parameters of Cosserat continuum. The effect of rotational degrees of freedom leads to the appearance of additional shear and rotational waves; however, only at high frequencies corresponding to the wave lengths comparable to the microstructural sizes. Such waves are difficult to detect. Rotations also affect the conventional s-waves (shear-rotational waves) making them dispersive for all frequencies. We considered propagation of planar waves in 2D granulate materials consisting of parallel cylindrical particles connected by tensile, shear and rotational springs and arranged in either square or hexagonal patterns. We deduced the s-wave dispersion relationships for both cases and determined the main terms of asymptotics of low frequency; they were found to be quadratic in the frequency. These terms, alongside with the zero order terms, can be determined by fitting of the theoretically determined frequency dependence of the phase shift between the sent and received wave to its experimentally determined counterpart. This opens a way for direct experimental identification of Cosserat effects in granulate materials.

8.1 Introduction

It has been quite a while since the first realization of importance of internal rotational degrees of freedom for the correct description of mechanics of deformable solids. In 1909, two French scientists, the Cosserat brothers [4], conjectured that internal parts of materials can experience relative rotations beyond the ones caused by the displacement field and that the effect thereof warrants investigation. They

E. Pasternak (✉) · A. Dyskin
The University of Western Australia, 35 Stirling Highway, Crawley, Perth, WA 6009, Australia
e-mail: elena@mech.uwa.edu.au

A. Dyskin
e-mail: arcady@civil.uwa.edu.au

G.A. Maugin, A.V. Metrikine (eds.), *Mechanics of Generalized Continua*, Advances in Mechanics and Mathematics 21,
DOI 10.1007/978-1-4419-5695-8_8, © Springer Science+Business Media, LLC 2010

developed a complex theory, which went unnoticed at the time, rediscovered half a century later and named the Cosserat theory (e.g., [8, 6, 7]). However, numerous attempts to detect the Cosserat effects in ordinary materials proved to be inconclusive (e.g., see the review [18]). In the 1980s and 1990s a realization came that one should look for the Cosserat effects in non-homogeneous materials. Since then the theory (with various modifications) found its acceptance in modeling of geomaterials, especially layered (e.g., [19, 9, 1]) and granular (e.g., [9, 11, 10, 3]) matter. The main obstacle to its use, however, is the inability to determine and calibrate the Cosserat elastic moduli from the experiments. The frustration of the researchers was so great that the question was raised if one should revert to the use of standard theories. As a result, the higher order theories are sometimes demoted to be used as numerical stabilizers for otherwise mesh-dependent numerical models (e.g., [16]) despite the general understanding that internal grain/block rotations do exist. As another extreme, calls have been made to abandon continuum mechanics altogether in favor of particle models that automatically reproduce complex behavior of particulate materials, requiring only few parameters [5].

The problem is that the direct manifestations of Cosserat effects are hard to see as they require detection of either microrotations or specific Cosserat wave modes which exist only at very high frequencies (e.g., [13]), both impossible to perform in highly heterogeneous materials such as, for example, geomaterials. It appears that the major impediment was the reliance on a 'head-on' approach whereby specific rotational effects (often microscopic) were targeted. (A recent example is an attempt to directly measure rotational waves in the Earth's crust, e.g., [17].) The picture becomes less gloomy if one recalls that the rotational degrees of freedom affect the propagation of much easier detectable conventional shear waves creating dispersion (frequency dependence of the wave velocity).

Experimental verification of the theoretical methods of Cosserat continuum modeling meets an additional challenge: It is difficult to produce materials with noticeable rotational effects and well controlled microstructure to allow independent determination of Cosserat parameters. An obvious choice – the particulate materials, i.e., materials consisting of particles (preferably spherical) with soft links permitting measurable rotations – is difficult to manufacture in such a way that the particle arrangement and the consistency of the links can be well controlled. We propose a way to overcome this problem by considering a 2D model of particulate material consisting of cylindrical rods in square and hexagonal packings glued together. The 2D nature of the material allows much greater control of the particle placement and linking. To this end, we develop a 2D Cosserat continuum for this packings employing on the homogenization procedure based on differential expansions [15] and then derive the dispersion relationships for the shear waves.

8.2 Cosserat Continuum Models of Particulate 2D Materials

We consider a material with the 2D particles in the shape of circular cylinders in square and hexagonal packings, see Fig. 8.1. The bonding between the particles is

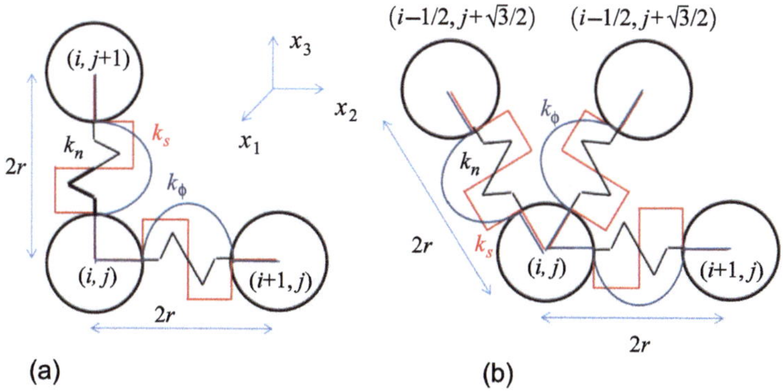

Fig. 8.1 The cells of periodicity of square (**a**) and hexagonal (**b**) packings of the cylindrical particles

presumed to be elastic and represented as a combination of in-plane tensile, shear and rotational springs with stiffnesses k_n, k_s, k_θ, respectively.

We derive the constitutive relationships by homogenizing the discrete expression for elastic energy. The latter consists of the sum of energies of deformed links, vertical and horizontal in the square packing and horizontal and two inclined in the hexagonal packing. The corresponding link energies associated with node (i, j) for the square packing read:

$$\begin{aligned}
W_{\text{square}}(i,j) &= \frac{1}{2}k_n\big[u_2(i+1,j) - u_2(i,j)\big]^2 \\
&+ \frac{1}{2}k_s\big[u_3(i+1,j) - u_3(i,j) - r\times\big(\phi_1(i+1,j)+\phi_1(i,j)\big)\big]^2 \\
&+ \frac{1}{2}k_\phi\big[\phi_1(i+1,j) - \phi_1(i,j)\big]^2 \\
&+ \frac{1}{2}k_n\big[u_3(i,j+1) - u_2(i,j)\big]^2 \\
&+ \frac{1}{2}k_s\big[u_2(i,j+1) - u_2(i,j) + r\times\big(\phi_1(i,j+1)+\phi_1(i,j)\big)\big]^2 \\
&+ \frac{1}{2}k_\phi\big[\phi_1(i,j+1) - \phi_1(i,j)\big]^2.
\end{aligned} \tag{8.1}$$

For the hexagonal packing, there are 3 expressions related to different links:

• *Horizontal link*

$$\begin{aligned}
W(i,j)_{1,n} &= \frac{1}{2}k_n\big[u_2(i+1,j) - u_2(i,j)\big]^2, \\
W(i,j)_{1,\phi} &= \frac{1}{2}k_\phi\big[\phi_1(i+1,j) - \phi_1(i,j)\big]^2, \\
W(i,j)_{1,s} &= \frac{1}{2}k_s\big[u_3(i+1,j) - u_3(i,j) - r\times\big(\phi_1(i+1,j)+\phi_1(i,j)\big)\big]^2.
\end{aligned} \tag{8.2}$$

• *Link* $(i,j) \to (i+1/2, j+\sqrt{3}/2)$

$$
\begin{aligned}
W(i,j)_n =& \frac{k_n}{2}\left[\frac{1}{2}u_2\left(i+\frac{1}{2}, j+\frac{\sqrt{3}}{2}\right) + \frac{\sqrt{3}}{2}u_3\left(i+\frac{1}{2}, j+\frac{\sqrt{3}}{2}\right)\right.\\
&\left. - \frac{1}{2}u_2(i,j) - \frac{\sqrt{3}}{2}u_3(i,j)\right]^2,\\
W(i,j)_\phi =& \frac{k_\phi}{2}\left[\phi_1\left(i+\frac{1}{2}, j+\frac{\sqrt{3}}{2}\right) - \phi_1(i,j)\right]^2,\\
W(i,j)_s =& \frac{k_s}{2}\left[-\frac{\sqrt{3}}{2}u_2\left(i+\frac{1}{2}, j+\frac{\sqrt{3}}{2}\right) + \frac{1}{2}u_3\left(i+\frac{1}{2}, j+\frac{\sqrt{3}}{2}\right)\right.\\
& + \frac{\sqrt{3}}{2}u_2(i,j) - \frac{1}{2}u_3(i,j)\\
&\left. - r \times \left(\phi_1\left(i+\frac{1}{2}, j+\frac{\sqrt{3}}{2}\right) + \phi_1(i,j)\right)\right]^2.
\end{aligned}
\tag{8.3}
$$

• *Link* $(i,j) \to (i-1/2, j+\sqrt{3}/2)$

$$
\begin{aligned}
W(i,j)_{3,n} =& \frac{k_n}{2}\left[-\frac{1}{2}u_2\left(i-\frac{1}{2}, j+\frac{\sqrt{3}}{2}\right) + \frac{\sqrt{3}}{2}u_3\left(i-\frac{1}{2}, j+\frac{\sqrt{3}}{2}\right)\right.\\
&\left. + \frac{1}{2}u_2(i,j) - \frac{\sqrt{3}}{2}u_3(i,j)\right]^2,\\
W(i,j)_{3,\phi} =& \frac{k_\phi}{2}\left[\phi_1\left(i-\frac{1}{2}, j+\frac{\sqrt{3}}{2}\right) - \phi_1(i,j)\right]^2,\\
W(i,j)_{3,s} =& \frac{k_s}{2}\left[-\frac{\sqrt{3}}{2}u_2\left(i-\frac{1}{2}, j+\frac{\sqrt{3}}{2}\right) - \frac{1}{2}u_3\left(i-\frac{1}{2}, j+\frac{\sqrt{3}}{2}\right)\right.\\
& + \frac{\sqrt{3}}{2}u_2(i,j) + \frac{1}{2}u_3(i,j)\\
&\left. - r \times \left(\phi_1\left(i-\frac{1}{2}, j+\frac{\sqrt{3}}{2}\right) + \phi_1(i,j)\right)\right]^2.
\end{aligned}
\tag{8.4}
$$

Here $2r$ is the distance between the particle centers, u_2, u_3, ϕ_1 are the corresponding components of displacement and rotation.

We apply the homogenization procedure based on differential expansions [15] to the energy and then introduce the deformation measures [14]

$$
\begin{aligned}
\gamma_{22} &= \frac{\partial u_2}{\partial x_2}, \qquad \gamma_{23} = \frac{\partial u_3}{\partial x_2} - \phi_1, \qquad \gamma_{32} = \frac{\partial u_2}{\partial x_3} + \phi_1,\\
\gamma_{33} &= \frac{\partial u_3}{\partial x_3}, \qquad \kappa_{21} = \frac{\partial \phi_1}{\partial x_2}, \qquad \kappa_{31} = \frac{\partial \phi_1}{\partial x_3},
\end{aligned}
\tag{8.5}
$$

where γ_{ij}, κ_{ij} are strains and curvatures. We now determine the constitutive relations by differentiating the homogenized energy over the deformation measures [14] and obtain

$$
\begin{aligned}
\sigma_{22} &= \frac{1}{L} k_n \gamma_{22}, \qquad & \sigma_{33} &= \frac{1}{L} k_n \gamma_{33}, \qquad & \sigma_{23} &= \frac{1}{L} k_s \gamma_{23}, \\
\sigma_{32} &= \frac{1}{L} k_s \gamma_{32}, \qquad & \mu_{21} &= \frac{1}{L} k_\phi \kappa_{21}, \qquad & \mu_{31} &= \frac{1}{L} k_\phi \kappa_{31}
\end{aligned}
\tag{8.6}
$$

for the square packing, and

$$
\begin{aligned}
\sigma_{22} &= \frac{\sqrt{3}}{4L} \left[\gamma_{22}(3k_n + k_s) + \gamma_{33}(k_n - k_s) \right], \\
\sigma_{33} &= \frac{\sqrt{3}}{4L} \left[\gamma_{22}(k_n - k_s) + \gamma_{33}(3k_n + k_s) \right], \\
\sigma_{23} &= \frac{\sqrt{3}}{4L} \left[\gamma_{23}(k_n + 3k_s) + \gamma_{32}(k_n - k_s) \right], \\
\sigma_{32} &= \frac{\sqrt{3}}{4L} \left[\gamma_{23}(k_n - k_s) + \gamma_{32}(k_n + 3k_s) \right], \\
\mu_{21} &= \frac{\sqrt{3}}{L} k_\phi \kappa_{21}, \qquad \mu_{31} = \frac{\sqrt{3}}{L} k_\phi \kappa_{31}
\end{aligned}
\tag{8.7}
$$

for the hexagonal packing.

Here L is the size of the unit cell, and σ_{ij}, μ_{ij} are stress and moment stress tensors. After either [2] or [12] are substituted into the Cosserat equations of motion, one obtains

$$
\begin{aligned}
& k_n \frac{\partial^2 u_2}{\partial x_2^2} + k_s \left(\frac{\partial^2 u_2}{\partial x_3^2} + \frac{\partial \phi_1}{\partial x_3} \right) = \tilde{\rho} \frac{\partial^2 u_2}{\partial t^2}, \\
& k_n \frac{\partial^2 u_3}{\partial x_3^2} + k_s \left(\frac{\partial^2 u_3}{\partial x_2^2} - \frac{\partial \phi_1}{\partial x_2} \right) = \tilde{\rho} \frac{\partial^2 u_3}{\partial t^2}, \\
& k_\phi \left(\frac{\partial^2 \phi_1}{\partial x_2^2} + \frac{\partial^2 \phi_1}{\partial x_3^2} \right) + k_s \left(\frac{\partial u_3}{\partial x_2} - \frac{\partial u_2}{\partial x_3} - 2\phi_1 \right) = \tilde{J} \frac{\partial^2 \phi_1}{\partial t^2}
\end{aligned}
\tag{8.8}
$$

for the square packing, and

$$
\begin{aligned}
& k_n \left(3 \frac{\partial^2 u_2}{\partial x_2^2} + 2 \frac{\partial^2 u_3}{\partial x_2 \partial x_3} + \frac{\partial^2 u_2}{\partial x_3^2} \right) \\
& \quad + k_s \left(\frac{\partial^2 u_2}{\partial x_2^2} - 2 \frac{\partial^2 u_3}{\partial x_2 \partial x_3} + 3 \frac{\partial^2 u_2}{\partial x_3^2} + 4 \frac{\partial \varphi_1}{\partial x_3} \right) = \frac{4\tilde{\rho}}{\sqrt{3}} \frac{\partial^2 u_2}{\partial t^2}, \\
& k_n \left(\frac{\partial^2 u_3}{\partial x_2^2} + 2 \frac{\partial^2 u_2}{\partial x_2 \partial x_3} + 3 \frac{\partial^2 u_3}{\partial x_3^2} \right) \\
& \quad + k_s \left(3 \frac{\partial^2 u_3}{\partial x_2^2} - 2 \frac{\partial^2 u_2}{\partial x_2 \partial x_3} + \frac{\partial^2 u_3}{\partial x_3^2} - 4 \frac{\partial \varphi_1}{\partial x_3} \right) = \frac{4\tilde{\rho}}{\sqrt{3}} \frac{\partial^2 u_3}{\partial t^2}, \\
& k_\phi \left(\frac{\partial^2 \phi_1}{\partial x_2^2} + \frac{\partial^2 \phi_1}{\partial x_3^2} \right) + k_s \left(\frac{\partial u_3}{\partial x_2} - \frac{\partial u_2}{\partial x_3} - 2\phi_1 \right) = \frac{\tilde{J}}{\sqrt{3}} \frac{\partial^2 \phi}{\partial t^2}
\end{aligned}
\tag{8.9}
$$

for the hexagonal packing.

Here $\tilde{\rho}$, $\tilde{J}$ are the density and rotational inertia per unit volume.

8.3 Dispersive Waves in 2D Cosserat Continuum

We are seeking the plane wave solution $\partial/\partial x_1 = \partial/\partial x_2 = 0$ of (8.8), (8.9) in the form $(u_2, u_3, \phi_1) = (C_1, C_2, C_3)\exp\{i\xi(x_2 - vt)\}$. This leads to the following characteristic equations:

$$\begin{aligned} &v^2\tilde{\rho} - k_n = 0, \\ &\tilde{\rho}\omega^2(\tilde{J}\omega^2 - 2k_s)v^4 + \omega^2(k_s^2 - \tilde{J}\omega^2 k_s - \tilde{\rho}\omega^2 k_\phi)v^2 + \omega^4 k_s k_\phi = 0 \end{aligned} \tag{8.10}$$

for the square packing, and

$$\begin{aligned} &v^2\tilde{\rho} - \frac{\sqrt{3}}{4}(3k_n + k_s) = 0, \\ &\tilde{\rho}\omega^2\left[\tilde{J}\omega^2 - 2\sqrt{3}k_s\right]v^4 \\ &\quad + \frac{1}{4}\omega^2\left[6k_n k_s + 6k_s^2 - \sqrt{3}\left(k_n\tilde{J}\omega^2 + 3\tilde{J}\omega^2 k_s + 4\tilde{\rho}\omega^2 k_\phi\right)\right]v^2 \\ &\quad + \frac{3}{4}\omega^4 k_\phi(k_n + 3k_s) = 0 \end{aligned} \tag{8.11}$$

for the hexagonal packing.

In both cases, the first equations correspond to the conventional non dispersive p-waves. From the second equations, we extract the long wave asymptotics. The only wave that exists in this asymptotics is a shear-rotational wave with velocity:

$$v_3^2 = \frac{k_s}{2\tilde{\rho}} - \left(\frac{\tilde{J}}{4\tilde{\rho}} - \frac{k_\phi}{2k_s}\right)\omega^2 + O(\omega^3) \tag{8.12}$$

for the square packing, and

$$v_3^2 = \frac{\sqrt{3}}{4}\frac{k_n + k_s}{\tilde{\rho}} - \left(\frac{\tilde{J}}{4\tilde{\rho}} - \frac{k_\phi}{k_n + k_s}\right)\omega^2 + O(\omega^3) \tag{8.13}$$

for the hexagonal packing.

These equations represent the dispersion relations in the low frequency (long wave) asymptotics, which can potentially be detectable by measurements without the heterogeneities attenuating the waves below the noise level.

8.4 Extension of the Phase Method for Long Wave Dispersion Measurements

Dispersion relationships (8.12), (8.13) could be used for detection of the effect of micro-rotations and the determination of the link stiffnesses constants in a particulate material if the wave velocities are measured for different frequencies. This is, however, difficult to accomplish with monochrome continuous waves, so one has to resort to pulses which contain a combination of different frequencies. We propose using the phase method, whereby the frequency dependence of phase difference $\Delta\phi(\omega)$ between the output and input signals extracted from the Fourier transform of the recorded wave forms. The phase difference is obviously proportional to the time delay between the input and output signals which, due to dispersion, can be different for different frequencies. Expressing the time delay via the length of the propagation path and the wave velocity, we write

$$\mathrm{d}\Delta\phi/\mathrm{d}\omega = \tau_{\text{delay}}, \quad \tau_{\text{delay}} = Lv(\omega)^{-1} + \tau_{\text{system}}. \tag{8.14}$$

Here τ_{system} is the time delay imposed by the wave generation/recording system.

By noting that the asymptotics (8.12), (8.13) of the dispersion relationships can be expressed in a general form as

$$v = v_0\sqrt{1-\alpha^2\omega^2},$$
$$\begin{cases} v_0^2 = \frac{k_s}{2\tilde{\rho}}, & \alpha^2 = \frac{\tilde{J}}{2k_s} - \frac{k_\phi\tilde{\rho}}{k_s^2}, \quad \text{square} \\ v_0^2 = \frac{\sqrt{3}}{4}\frac{k_n+k_s}{\tilde{\rho}}, & \alpha^2 = \frac{1}{\sqrt{3}(k_n+k_s)}[\tilde{J} - \frac{4k_\phi\tilde{\rho}}{k_n+k_s}], \quad \text{hexagonal} \end{cases} \tag{8.15}$$

one obtains the asymptotic solution of the differential equation (8.14) in the form

$$\Delta\phi \cong \left(\frac{L}{v_0} + \tau_{\text{system}}\right)\omega + \left(\frac{L\alpha^2}{6v_0}\right)\omega^3 + \text{const.} \tag{8.16}$$

The three parameters of (8.16) are determined by fitting (8.16) to the experimental curve $\Delta\phi(\omega)$ obtained from the measurements, and then the stiffnesses are determined from (8.15).

8.5 Conclusions

We considered the propagation of planar waves in 2D granulate materials consisting of parallel cylindrical particles connected by tensile, shear and rotational springs and arranged in either square or hexagonal patterns. For such materials, the derived characteristic equations show the existence of dispersion relationships in the conventional shear waves, which extends to low frequencies (long waves). In the asymptotics of low frequencies, the Cosserat effects control the terms proportional to the square of frequency—the first terms reflecting the dispersion. These terms, alongside with the terms of zero order which are the constant parts of wave velocities, can be determined by fitting of the theoretically determined frequency dependence of the phase shift between the sent and received wave to its experimentally determined counterpart. This opens a way for direct experimental identification of Cosserat effects in granulate materials.

Acknowledgements We acknowledge financial support from the Australian Research Council via Discovery Grant DP0988449. We also wish to thank Greg Sevel for his help with the paper formatting.

References

1. Adhikary, D., Dyskin, A.: A Cosserat continuum model for layered materials. Comput. Geotech. **20**(1), 15–45 (1997)
2. Atkinson, C., Leppington, F.G.: Effect of couple stresses on tip of a crack. Int. J. Solids Struct. **13**(11), 1103–1122 (1977)
3. Chang, C., Ma, L.: Elastic material constants for isotropic granular solids with particle rotation. Int. J. Solids Struct. **29**, 1001–1018 (1992)
4. Cosserat, E., Cosserat, F.: Théorie des corps déformables. Hermann et Fils, Paris (1909)
5. Cundall, P.: A discontinuous future for numerical modeling in soil and rock. In: Discrete Element Methods: Numerical Modeling of Discontinua, pp. 3–4. ASCE, Reston (2002)
6. Eringen, A.: Linear theory of micropolar elasticity. J. Math. Mech. **15**, 909–923 (1966)
7. Eringen, A., Kafadar, C.: Polar field theories. In: Continuum Physics, vol. 4, pp. 4–73. Academic Press, New York (1976)
8. Mindlin, R., Tiersten, H.: Effects of couple-stresses in linear elasticity. Arch. Ration. Mech. Anal. **11**, 415–448 (1962)
9. Mühlhaus, H.B.: Continuum models for layered and blocky rock. In: Comprehensive Rock Eng, vol. 2, pp. 209–230. Pergamon, Elmsford (1976). Invited Chapter
10. Mühlhaus, H.B., de Borst, R., Aifantis, E.: Constitutive models and numerical analyses for inelastic materials with microstructure. In: Computing Methods and Advances in Geomechanics, pp. 377–385. Balkema, Rotterdam (1991)
11. Mühlhaus, H.B., Vardoulakis, I.: The thickness of shear bands in granular materials. Géotechnique **37**, 271–283 (1987)
12. Nakamura, S., Lakes, R.: Finite element analysis of stress concentration around a blunt crack in a Cosserat elastic solid. Comput. Methods Appl. Mech. Eng. **66**, 257–266 (1988)
13. Nowacki, W.: Dynamic problems of asymmetrical elasticity. Int. J. Appl. Mech. **6**(4), 361–375 (1970)
14. Oda, M., Iwashita, K. (eds.): Mechanics of Granular Materials: An Introduction. Balkema, Rotterdam (1999)
15. Pasternak, E., Mühlhaus, H.B.: Generalised homogenisation procedures for granular materials. Eng. Math. **52**, 199–229 (2005)
16. Sluys, L., de Borst, R., Mühlhaus, H.B.: Wave propagation, localisation and dispersion in a gradient dependent medium. Int. J. Solids Struct. **30**, 1153–1171 (1993)
17. Teisseyre, R., Suchcicki, J., Teisseyre, K., Wiszniowski, J., Palangio, P.: Seismic rotation waves: basic elements of theory and recording. Ann. Geophys. **46**(4), 671–685 (2003)
18. Yoon, H., Katz, J.: Is bone a Cosserat solid? J. Mater. Sci. **18**, 1297–1305 (1983)
19. Zvolinskii, N., Shkhinek, K.: Continual model of laminar elastic medium. Mech. Solids **19**(1), 1–9 (1983)

Chapter 9
Natural Lagrangian Strain Measures of the Non-Linear Cosserat Continuum

Wojciech Pietraszkiewicz and Victor A. Eremeyev

Abstract Definitions of the Lagrangian stretch and wryness tensors in the non-linear Cosserat continuum are discussed applying three different methods. The resulting unique strain measures have several distinguishing features and are called the natural ones. They are expressed through the translation vector and either the rotation tensor or various finite rotation vector fields. The relation of the natural strain measures to those proposed in the representative literature is reviewed.

9.1 Introduction

The stretch and wryness tensors of the non-linear Cosserat continuum were originally defined by Cosserats [2] through components of some fields in the common Cartesian frame. Today their approach is hardly readable. During the last 50 years, the strain measures have been redefined by different authors in various forms using, for example, (a) components in two different curvilinear coordinate systems associated with the undeformed (reference) of deformed (actual) placements of the body, (b) components in the convective coordinate system, (c) Lagrangian or Eulerian descriptions, (d) different representations of the rotation group SO(3) in terms of various finite rotation vectors, Euler angles, quaternions etc., (e) formally different tensor operations and sign conventions, as well as (f) requiring or not the strain measures to vanish in the undeformed placement of the body. Even the gradient and divergence operators as well as the Cauchy theorem influencing definitions of work-

W. Pietraszkiewicz (✉)
Institute of Fluid-Flow Machinery of the Polish Academy of Sciences, ul. Gen. J. Fiszera 14, 80-952 Gdańsk, Poland
e-mail: pietrasz@imp.gda.pl

V.A. Eremeyev
South Scientific Center of RASci & South Federal University, Milchakova str. 8a, 344090 Rostov on Don, Russia
e-mail: eremeyev.victor@gmail.com

G.A. Maugin, A.V. Metrikine (eds.), *Mechanics of Generalized Continua*, Advances in Mechanics and Mathematics 21,

DOI 10.1007/978-1-4419-5695-8_9, © Springer Science+Business Media, LLC 2010

conjugate pairs of the stress and strain measures are not defined in the same way in the literature. The strain measures introduced by Kafadar and Eringen [4] are among the most referred to in the literature, but even their derivation is not fully complete.

In this paper, we discuss three different methods of introducing the strain measures into the non-linear Cosserat continuum: (a) by a direct geometric approach, (b) defining the strain measures as the fields work-conjugate to the respective internal stress and couple-stress tensor fields, and (c) applying the principle of invariance under superposed rigid-body deformations to the strain energy density of the polar-elastic body. Each of the three ways allows one to associate different geometric and/or physical interpretations to the corresponding strain measures. In the discussion, we use the coordinate-free vector and tensor notation. Orientations of material particles in the reference and deformed placements, respectively, as well as their changes during deformation are described in the most general way by the proper orthogonal tensors. Our primary strain measures, called the natural ones, are of the relative type, for they are required to vanish in the reference placement.

In the reference (undeformed) placement $\kappa(\mathscr{B}) = B_\kappa \subset \mathscr{E}$, the material particle $X \in \mathscr{B}$ is given by its position vector $\mathbf{x} \in E$ relative to a point $o \in \mathscr{E}$ of the 3D physical space $\mathscr{E}$ and by three orthonormal directors $\mathbf{h}_a \in E$, $a, b = 1, 2, 3$, fixing orientation of X in the 3D vector space E.

In the actual (deformed) placement $\gamma(\mathscr{B}) = B_\gamma = \chi(B_\kappa) \subset \mathscr{E}$, the position of X becomes defined by the vector $\mathbf{y} \in E$, taken here for simplicity relative to the same point $o \in \mathscr{E}$, and by three orthonormal directors $\mathbf{d}_a \in E$. As a result, the finite displacement of the Cosserat continuum can be described by two following smooth mappings:

$$\mathbf{y} = \chi(\mathbf{x}) = \mathbf{x} + \mathbf{u}(\mathbf{x}), \qquad \mathbf{d}_a = \mathbf{Q}(\mathbf{x})\mathbf{h}_a, \tag{9.1}$$

where $\mathbf{u} \in E$ is the translation vector and $\mathbf{Q} = \mathbf{d}_a \otimes \mathbf{h}_a \in SO(3)$ is the proper orthogonal microrotation tensor. Two independent fields $\mathbf{u} = \mathbf{u}(\mathbf{x})$ and $\mathbf{Q} = \mathbf{Q}(\mathbf{x})$ describe translational and rotational degrees of freedom of the Cosserat continuum, respectively.

9.2 Strain Measures by Geometric Approach

Differentials of the independent kinematic fields are

$$\begin{aligned}
&\mathrm{d}\mathbf{x} = (\mathrm{Grad}\,\mathbf{x})\,\mathrm{d}\mathbf{x}, && \mathrm{d}\mathbf{y} = (\mathrm{grad}\,\mathbf{y})\,\mathrm{d}\mathbf{y} = (\mathrm{Grad}\,\mathbf{y})\,\mathrm{d}\mathbf{x} = \mathbf{F}\,\mathrm{d}\mathbf{x},\\
&\mathrm{d}\mathbf{h}_a = \mathbf{B}\,\mathrm{d}\mathbf{x} \times \mathbf{h}_a, && \mathrm{d}\mathbf{d}_a = \mathbf{C}\,\mathrm{d}\mathbf{y} \times \mathbf{d}_a,\\
&\mathbf{B} = \frac{1}{2}\mathbf{h}_a \times \mathrm{Grad}\,\mathbf{h}_a, && \mathbf{C} = \frac{1}{2}\mathbf{d}_a \times \mathrm{grad}\,\mathbf{d}_a,
\end{aligned} \tag{9.2}$$

where Grad and grad are the gradient operators in B_κ and B_γ, and $\mathbf{B}$ and $\mathbf{C}$ are the microstructure curvature tensors in the undeformed (reference) and deformed (actual) placements of the Cosserat continuum, respectively. In particular, for any

nth order tensor field $\mathbf{Z}(\mathbf{x})$ we define its gradient in B_κ to satisfy $[\text{Grad}\,\mathbf{Z}(\mathbf{x})]\mathbf{a} = (\text{d}/\text{d}t)\mathbf{Z}(\mathbf{x}+t\mathbf{a})|_{t=0}$ for any $t \in \mathbb{R}, \mathbf{a} \in E$.

The relative changes of lengths and orientations of the Cosserat continuum during deformation are governed by differences of differentials (9.2) brought by the tensor $\mathbf{Q}$ to two comparable orientations

$$\begin{aligned} \text{d}\mathbf{y} - \mathbf{Q}\,\text{d}\mathbf{x} &= \mathbf{X}\,\text{d}\mathbf{x} = \mathbf{G}\,\text{d}\mathbf{y}, \qquad \mathbf{C}\,\text{d}\mathbf{y} - \mathbf{Q}\mathbf{B}\,\text{d}\mathbf{x} = \boldsymbol{\Phi}\,\text{d}\mathbf{x} = \boldsymbol{\Delta}\,\text{d}\mathbf{y}, \\ \mathbf{Q}^T\text{d}\mathbf{y} - \text{d}\mathbf{x} &= \mathbf{E}\,\text{d}\mathbf{x} = \mathbf{Y}\,\text{d}\mathbf{y}, \qquad \mathbf{Q}^T\mathbf{C}\,\text{d}\mathbf{y} - \mathbf{B}\,\text{d}\mathbf{x} = \boldsymbol{\Gamma}\,\text{d}\mathbf{x} = \boldsymbol{\Psi}\,\text{d}\mathbf{y}, \\ \mathbf{E} = \mathbf{Q}^T\mathbf{F} - \mathbf{I}, &\qquad \boldsymbol{\Gamma} = \mathbf{Q}^T\mathbf{C}\mathbf{F} - \mathbf{B} = -\frac{1}{2}\boldsymbol{\varepsilon} : \left(\mathbf{Q}^T\,\text{Grad}\,\mathbf{Q}\right), \end{aligned} \tag{9.3}$$

where $\mathbf{E}$, $\boldsymbol{\Gamma}$ are the relative Lagrangian stretch and wryness tensors, $\mathbf{G}$, $\boldsymbol{\Delta}$ are the relative Eulerian strain measures, while $\mathbf{X}$, $\boldsymbol{\Phi}$, $\mathbf{Y}$, $\boldsymbol{\Psi}$ are the relative two-point deformation measures. In (9.3), $\mathbf{I}$ is the identity (metric) tensor of $E \otimes E$, the 3rd-order skew tensor $\boldsymbol{\varepsilon} = -\mathbf{I} \times \mathbf{I}$ is the Ricci tensor of $E \otimes E \otimes E$, and the double dot-product : of two 3rd-order tensors $\mathbf{A}$, $\mathbf{P}$ represented in the base $\mathbf{h}_a$ is defined as $\mathbf{A} : \mathbf{P} = A_{\text{amn}}P_{\text{mnb}}\mathbf{h}_a \otimes \mathbf{h}_b$.

Let us note some interesting features of the relative Lagrangian strain measures:

1. They are given in the common coordinate-free notation; their various component representations can easily be generated, if necessary.
2. Definitions of the measures are valid for finite translations and rotations as well as for unrestricted stretches and changes of microstructure orientation of the Cosserat body.
3. The measures are expressed in terms of the rotation tensor $\mathbf{Q}$; for any specific parametrization of the rotation group $\text{SO}(3)$ by various finite rotation vectors, Euler angles, quaternions, etc. appropriate expressions for the measures can easily be found, if necessary.
4. The measures vanish in the rigid-body deformation $\mathbf{y} = \mathbf{O}\mathbf{x} + \mathbf{a}$, $\mathbf{d}_a = \mathbf{O}\mathbf{h}_a$ with a constant vector $\mathbf{a}$ and a constant proper orthogonal tensor $\mathbf{O}$ defined for the whole body.
5. In the absence of deformation from the reference placement, that is, when $\mathbf{F} = \mathbf{Q} = \mathbf{I}$, the measures vanish identically.
6. The measures are not symmetric, in general: $\mathbf{E}^T \neq \mathbf{E}$, $\boldsymbol{\Gamma}^T \neq \boldsymbol{\Gamma}$.

In our purely geometric approach, there is no need for discussing whether these measures might be defined as transposed ones or with opposite signs. The derivation process itself is concise, elegant, direct, and seems to be the most complete one in the literature.

9.3 Principle of Virtual Work and Work-Conjugate Strain Measures

Already Reissner [7] noted that the internal structure of two local equilibrium equations of the Cosserat elastic body requires specific two strain measures expressed in terms of independent translation and rotation vectors as the only field variables. We develop this idea here in the general case of the non-linear Cosserat continuum using the coordinate-free approach.

If the local coordinate-free form of the equilibrium conditions of the Cosserat continuum derived in Appendix of [5] are multiplied by two arbitrary smooth vector fields $\mathbf{v}, \boldsymbol{\omega} \in E$, then we generate the integral identity

$$\int_{B_\kappa} \{(\mathrm{Div}\,\mathbf{T} + \mathbf{f}) \cdot \mathbf{v} + [\mathrm{Div}\,\mathbf{M} - \mathrm{ax}(\mathbf{F}\mathbf{T} - \mathbf{T}^T\mathbf{F}^T) + \mathbf{m}] \cdot \boldsymbol{\omega}\}\, \mathrm{d}v - \int_{\partial B_{\kappa f}} \{(\mathbf{n}\mathbf{T} - \mathbf{t}^*) \cdot \mathbf{v} + (\mathbf{n}\mathbf{M} - \mathbf{m}^*) \cdot \boldsymbol{\omega}\}\, \mathrm{d}a = 0. \tag{9.4}$$

Here $\mathbf{f}$ and $\mathbf{m}$ are the volume force and couple vectors applied at any point $y = \chi(x)$ of the deformed body, but measured per unit volume of B_κ, $\mathbf{t}^*(\mathbf{x})$ and $\mathbf{m}^*(\mathbf{x})$ the external force and couple vector fields prescribed on the part $\partial B_{\gamma f}$, but measured per unit area of $\partial B_{\kappa f}$, $\mathbf{t}_{(n)} = \mathbf{n}\mathbf{T}_{(n)}$ and $\mathbf{m}_{(n)} = \mathbf{n}\mathbf{M}$ the surface traction and couple vector fields applied at any point of $\partial P_\gamma \in \partial B_\gamma$, but measured per unit area of $\partial P_\kappa \in \partial B_\kappa$, expressible as linear functions of the nominal type stress $\mathbf{T}$ and couple-stress $\mathbf{M}$ tensors, respectively, $\mathbf{n}$ the unit vector externally normal to ∂P_κ, $\mathrm{ax}\,\mathbf{A}$ the axial vector of the skew tensor $\mathbf{A}$, and the divergence operator Div is defined to satisfy $[\mathrm{Div}\,\mathbf{Z}(\mathbf{x})]\mathbf{a} = \mathrm{Div}[\mathbf{Z}(\mathbf{x})\mathbf{a}]$ for any $\mathbf{a} \in E$.

The vector field $\mathbf{v}$ can be interpreted as the kinematically admissible virtual translation $\mathbf{v} \equiv \delta\mathbf{y}$ and the vector field $\boldsymbol{\omega}$ as the kinematically admissible virtual rotation $\boldsymbol{\omega} \equiv \mathrm{ax}(\delta\mathbf{Q}\mathbf{Q}^T)$ in B_γ such that $\mathbf{v} = \boldsymbol{\omega} = \mathbf{0}$ on $\partial B_{\kappa d} = \partial B_\kappa \backslash \partial B_{\kappa f}$, where δ is the symbol of virtual change (variation). Then using the divergence theorem Identity (9.4) can be transformed into the principle of virtual work of the non-linear Cosserat continuum

$$\int_{B_\kappa} [\mathbf{T}^T : (\mathrm{Grad}\,\mathbf{v} - \boldsymbol{\Omega}\mathbf{F}) + \mathbf{M}^T : \mathrm{Grad}\,\boldsymbol{\omega}]\, \mathrm{d}v = \int_{B_\kappa} (\mathbf{f} \cdot \mathbf{v} + \mathbf{m} \cdot \boldsymbol{\omega})\, \mathrm{d}v + \int_{\partial B_{\kappa f}} (\mathbf{t}^* \cdot \mathbf{v} + \mathbf{m}^* \cdot \boldsymbol{\omega})\, \mathrm{d}a. \tag{9.5}$$

But we can show that $\delta\mathbf{E} = \mathbf{Q}^T(\mathrm{Grad}\,\mathbf{v} - \boldsymbol{\Omega}\mathbf{F})$ and $\delta\boldsymbol{\Gamma} = \mathbf{Q}^T\,\mathrm{Grad}\,\boldsymbol{\omega}$, where $\boldsymbol{\Omega} = \boldsymbol{\omega} \times \mathbf{I}$, and the internal virtual work density under the first volume integral of (9.5) can now be given by the expressions

$$\sigma = \mathbf{T}^T : (\mathbf{Q}\delta\mathbf{E}) + \mathbf{M}^T : (\mathbf{Q}\delta\boldsymbol{\Gamma}) = \mathbf{S} : \delta\mathbf{E} + \mathbf{K} : \delta\boldsymbol{\Gamma}, \tag{9.6}$$

where $\mathbf{S} = \mathbf{Q}^T\mathbf{T}^T$, $\mathbf{K} = \mathbf{Q}^T\mathbf{M}^T$ are the stress and couple-stress tensors whose natural components are referred entirely to the reference (undeformed) placement. The stress measures $\mathbf{S}$, $\mathbf{K}$ require the relative Lagrangian strain measures $\mathbf{E}$, $\boldsymbol{\Gamma}$ as their work-conjugate counterparts.

9.4 Invariance of the Polar-Elastic Strain Energy Density

In the polar-elastic body, the constitutive relations are defined through the strain energy density W_κ per unit volume of the undeformed placement κ. In general, the density W_κ can be assumed in the following form:

$$W_\kappa = W_\kappa(\mathbf{y}, \mathbf{F}, \mathbf{Q}, \operatorname{Grad}\mathbf{Q}; \mathbf{x}, \mathbf{B}). \tag{9.7}$$

But W_κ in (9.7) should satisfy the principle of invariance under the superposed rigid-body deformations. After appropriate transformations we can show that this requires the form of W_κ to be reduced to

$$W_\kappa = W_\kappa(\mathbf{E} + \mathbf{I}, \mathbf{I} \times \boldsymbol{\Gamma}; \mathbf{x}, \mathbf{B}) = \overline{W}_\kappa(\mathbf{E}, \boldsymbol{\Gamma}; \mathbf{x}, \mathbf{B}). \tag{9.8}$$

This again confirms that the relative Lagrangian strain measures $\mathbf{E}$, $\boldsymbol{\Gamma}$ are required to be the independent fields in the polar-elastic strain energy density in order it to be invariant under the superposed rigid-body deformation.

The geometric approach, the structure of equilibrium conditions and invariance of the polar-elastic strain energy density all require the tensors $\mathbf{E}$, $\boldsymbol{\Gamma}$ as the most appropriate Lagrangian strain measures for the non-linear Cosserat continuum. We call the measures the *natural* stretch and wryness tensors, respectively.

9.5 The Vectorial Parameterization

While the three components of $\mathbf{u}$ in (9.3) are all independent, the nine components of $\mathbf{Q}$ in (9.3) are subject to six constraints following from the orthogonality conditions $\mathbf{Q}^{-1} = \mathbf{Q}^T$, $\det\mathbf{Q} = +1$, so that only three rotational parameters of $\mathbf{Q}$ are independent. In many applications, it is more convenient to use the strain measures expressed in terms of six displacement parameters all of which are independent.

In the literature, many techniques how to parameterize the rotation group $SO(3)$ were developed, which can roughly be classified as vectorial and non-vectorial ones. Various finite rotation vectors as well as the Cayley–Gibbs and exponential map parameters are examples of the vectorial parametrization, for they all have three independent scalar parameters as Cartesian components of a generalized vector in the 3D vector space E. The non-vectorial parameterizations are expressed either in terms of three scalar parameters that cannot be treated as vector components, such as Euler-type angles, for example, or through more scalar parameters subject to additional constraints, such as unit quaternions, Cayley–Klein parameters, or direction cosines. Each of these expressions may appear to be more convenient than others when solving specific problems of the non-linear Cosserat continuum.

The microrotation tensor $\mathbf{Q}$ represents the isometric and orientation-preserving transformation of the 3D vector space E into itself. By the Euler theorem, such a transformation can be expressed in terms of the angle of rotation ϕ about the axis of rotation described by the eigenvector $\mathbf{e}$ corresponding to the real eigenvalue $+1$ of $\mathbf{Q}$ such that $\mathbf{Q}\mathbf{e} = +\mathbf{e}$, $\cos\phi = \frac{1}{2}(\operatorname{tr}\mathbf{Q} - 1)$, $\sin\phi\mathbf{e} = \frac{1}{2}\operatorname{ax}(\mathbf{Q} - \mathbf{Q}^T)$, where $\operatorname{tr}\mathbf{A}$ is the trace of the second-order tensor $\mathbf{A}$. In terms of $\mathbf{e}$ and ϕ, the microrotation tensor $\mathbf{Q}$ can be expressed by the Gibbs [3] formula

$$\mathbf{Q} = \cos\phi\,\mathbf{I} + (1 - \cos\phi)\mathbf{e}\otimes\mathbf{e} + \sin\phi\,\mathbf{e}\times\mathbf{I}. \tag{9.9}$$

In the vectorial parametrization of $\mathbf{Q}$, one introduces a scalar function $p(\phi)$ generating three components of the finite rotation vector $\mathbf{p} \in E$ defined as $\mathbf{p} = p(\phi)\mathbf{e}$, see, for example, [1]. The generating function $p(\phi)$ has to be an odd function of ϕ with the limit behavior $\lim_{\phi\to 0}(p(\phi)/\phi) = \kappa$, where κ is a positive real normalization factor (usually 1 or $1/2$), and $p(0) = 0$. Then the tensor $\mathbf{Q}$ can be represented as

$$\mathbf{Q} = \cos\phi\,\mathbf{I} + \frac{1-\cos\phi}{p^2}\mathbf{p}\otimes\mathbf{p} + \frac{\sin\phi}{p}\mathbf{p}\times\mathbf{I}. \tag{9.10}$$

Taking the gradient of (9.10) and substituting it into (9.3), after appropriate transformations the natural Lagrangian stretch $\mathbf{E}$ and wryness $\mathbf{\Gamma}$ tensors can be represented in terms of the finite rotation vector $\mathbf{p}$ by the general relations

$$\mathbf{E} = \left[\cos\phi\,\mathbf{I} + \frac{1-\cos\phi}{p^2}\mathbf{p}\otimes\mathbf{p} - \frac{\sin\phi}{p}\mathbf{p}\times\mathbf{I}\right](\mathbf{I} + \operatorname{Grad}\mathbf{u}) - \mathbf{I}, \tag{9.11}$$

$$\mathbf{\Gamma} = \left[\frac{\sin\phi}{p}\mathbf{I} + \frac{1}{p^2}\left(\frac{1}{p'} - \frac{\sin\phi}{p}\right)\mathbf{p}\otimes\mathbf{p} - \frac{1-\cos\phi}{p^2}\mathbf{p}\times\mathbf{I}\right]\operatorname{Grad}\mathbf{p}. \tag{9.12}$$

Among definitions of $\mathbf{p}$ used in the literature, let us mention the finite rotation vectors defined as

$$\boldsymbol{\theta} = 2\tan\frac{\phi}{2}\mathbf{e}, \qquad \boldsymbol{\phi} = \phi\mathbf{e}, \qquad \boldsymbol{\varpi} = \sin\phi\mathbf{e}, \qquad \boldsymbol{\rho} = \tan\frac{\phi}{2}\mathbf{e}, \tag{9.13}$$

$$\boldsymbol{\sigma} = 2\sin\frac{\phi}{2}\mathbf{e}, \qquad \boldsymbol{\mu} = 4\tan\frac{\phi}{4}\mathbf{e}, \qquad \boldsymbol{\beta} = 4\sin\frac{\phi}{4}\mathbf{e}, \tag{9.14}$$

where the generating functions are $\theta = 2\tan(\phi/2)$, $\phi, \varpi = \sin\phi$, $\rho = \tan(\phi/2)$, $\sigma = 2\sin(\phi/2)$, $\mu = 4\tan(\phi/4)$, and $\beta = 4\sin(\phi/4)$, respectively.

The explicit formulae for $\mathbf{E}$ and $\mathbf{\Gamma}$ expressed in terms of the corresponding finite rotation vectors (9.13) and (9.14) are summarized in Tables 9.1 and 9.2, see [6].

When the values of $\mathbf{u}$ and ϕ as well as their spatial gradients are infinitesimal $\|\mathbf{u}\| \ll 1$, $\|\operatorname{Grad}\mathbf{u}\| \ll 1$, $|\phi| \ll 1$, $\|\operatorname{Grad}\phi\| \ll 1$, we also have $\sin\phi \approx \phi$, $\cos\phi \approx 1$, and $p(\phi) \approx \kappa\phi$. Then it follows that $\mathbf{p} \approx \kappa\boldsymbol{\vartheta}$, $\mathbf{Q} \approx \mathbf{I} + \boldsymbol{\vartheta}\times\mathbf{I}$, where $\boldsymbol{\vartheta} = \phi\mathbf{e}$ is now the infinitesimal rotation vector. Then from (9.11) and (9.12) we obtain $\mathbf{E} \approx \boldsymbol{e} \equiv \operatorname{Grad}\mathbf{u} - \boldsymbol{\vartheta}\times\mathbf{I}$, $\mathbf{\Gamma} \approx \boldsymbol{\gamma} \equiv \operatorname{Grad}\boldsymbol{\vartheta}$. The infinitesimal strain measures $\boldsymbol{e}$, $\boldsymbol{\gamma}$ or their transpose were used in many papers and books on linear Cosserat continuum.

Table 9.1 The natural Lagrangian stretch tensor for different finite rotations vectors

$\mathbf{p}$	$\phi \in$	$\mathbf{E}$
$\boldsymbol{\theta} \equiv 2\tan\frac{\phi}{2}\mathbf{e}$	$(-\pi, \pi)$	$[1+\frac{\theta^2}{4}]^{-1}[(1-\frac{\theta^2}{4})\mathbf{I} + \frac{1}{2}\boldsymbol{\theta}\otimes\boldsymbol{\theta} - \boldsymbol{\theta}\times\mathbf{I}](\mathbf{I} + \text{Grad}\,\mathbf{u}) - \mathbf{I}$
$\boldsymbol{\phi} \equiv \phi\mathbf{e}$	$(-2\pi, 2\pi)$	$[\cos\phi\,\mathbf{I} + \frac{1-\cos\phi}{\phi^2}\boldsymbol{\phi}\otimes\boldsymbol{\phi} - \frac{\sin\phi}{\phi}\boldsymbol{\phi}\times\mathbf{I}](\mathbf{I} + \text{Grad}\,\mathbf{u}) - \mathbf{I}$
$\boldsymbol{\varpi} \equiv \sin\phi\,\mathbf{e}$	$(-\pi, \pi)$	$[\cos\phi\,\mathbf{I} + \frac{1-\cos\phi}{\varpi^2}\boldsymbol{\varpi}\otimes\boldsymbol{\varpi} - \boldsymbol{\varpi}\times\mathbf{I}](\mathbf{I} + \text{Grad}\,\mathbf{u}) - \mathbf{I}$
$\boldsymbol{\rho} \equiv \tan\frac{\phi}{2}\mathbf{e}$	$(-\pi, \pi)$	$\frac{1}{1+\rho^2}[(1-\rho^2)\mathbf{I} + 2\boldsymbol{\rho}\otimes\boldsymbol{\rho} - 2\boldsymbol{\rho}\times\mathbf{I}](\mathbf{I} + \text{Grad}\,\mathbf{u}) - \mathbf{I}$
$\boldsymbol{\sigma} \equiv 2\sin\frac{\phi}{2}\mathbf{e}$	$(-\pi, \pi)$	$[(1-\frac{1}{2}\sigma^2)\mathbf{I} + \frac{1}{2}\boldsymbol{\sigma}\otimes\boldsymbol{\sigma} - \cos\frac{\phi}{2}\boldsymbol{\sigma}\times\mathbf{I}](\mathbf{I} + \text{Grad}\,\mathbf{u}) - \mathbf{I}$
$\boldsymbol{\mu} \equiv 4\tan\frac{\phi}{4}\mathbf{e}$	$(-2\pi, 2\pi)$	$(1+\frac{\mu^2}{16})^{-2}\{[1-\frac{\mu^2}{16}(\frac{3}{8}-\frac{\mu^2}{16})]\mathbf{I} + \frac{1}{2}\boldsymbol{\mu}\otimes\boldsymbol{\mu}$ $-(1-\frac{\mu^2}{16})\boldsymbol{\mu}\otimes\mathbf{I}\}(\mathbf{I} + \text{Grad}\,\mathbf{u}) - \mathbf{I}$
$\boldsymbol{\beta} \equiv 4\sin\frac{\phi}{4}\mathbf{e}$	$(-2\pi, 2\pi)$	$\{[1-\frac{\beta^2}{2}(1-\frac{\beta^2}{16})]\mathbf{I} + \frac{1}{2}(1-\frac{\beta^2}{8})\boldsymbol{\beta}\otimes\boldsymbol{\beta}$ $-\sqrt{1-\frac{\beta^2}{16}}(1-\frac{\beta^2}{8})\boldsymbol{\beta}\times\mathbf{I}\}(\mathbf{I} + \text{Grad}\,\mathbf{u}) - \mathbf{I}$

Table 9.2 The natural Lagrangian wryness tensor for different finite rotations vectors

$\mathbf{p}$	$\phi \in$	$\boldsymbol{\Gamma}$
$\boldsymbol{\theta} \equiv 2\tan\frac{\phi}{2}\mathbf{e}$	$(-\pi, \pi)$	$[1+\frac{\theta^2}{4}]^{-1}(\mathbf{I} - \frac{1}{2}\boldsymbol{\theta}\times\mathbf{I})\,\text{Grad}\,\boldsymbol{\theta}$
$\boldsymbol{\phi} \equiv \phi\mathbf{e}$	$(-2\pi, 2\pi)$	$[\frac{\sin\phi}{\phi}\mathbf{I} + \frac{\phi-\sin\phi}{\phi^3}\boldsymbol{\phi}\otimes\boldsymbol{\phi} - \frac{1-\cos\phi}{\phi^2}\boldsymbol{\phi}\times\mathbf{I}]\,\text{Grad}\,\boldsymbol{\phi}$
$\boldsymbol{\varpi} \equiv \sin\phi\,\mathbf{e}$	$(-\pi, \pi)$	$[\mathbf{I} + \frac{1}{\varpi^2}(\frac{1}{\cos\phi} - 1)\boldsymbol{\varpi}\otimes\boldsymbol{\varpi} - \frac{1-\cos\phi}{\varpi^2}\boldsymbol{\varpi}\times\mathbf{I}]\,\text{Grad}\,\boldsymbol{\varpi}$
$\boldsymbol{\rho} \equiv \tan\frac{\phi}{2}\mathbf{e}$	$(-\pi, \pi)$	$\frac{2}{1+\rho^2}(\mathbf{I} - \boldsymbol{\rho}\times\mathbf{I})\,\text{Grad}\,\boldsymbol{\rho}$
$\boldsymbol{\sigma} \equiv 2\sin\frac{\phi}{2}\mathbf{e}$	$(-\pi, \pi)$	$[\cos\frac{\phi}{2}\mathbf{I} - \frac{1}{4\cos(\phi/2)}\boldsymbol{\sigma}\otimes\boldsymbol{\sigma} - \frac{1}{2}\boldsymbol{\sigma}\times\mathbf{I}]\,\text{Grad}\,\boldsymbol{\sigma}$
$\boldsymbol{\mu} \equiv 4\tan\frac{\phi}{4}\mathbf{e}$	$(-2\pi, 2\pi)$	$[1+\frac{\mu^2}{16}]^{-2}[(1-\frac{\mu^2}{16})\mathbf{I} + \frac{1}{8}\boldsymbol{\mu}\otimes\boldsymbol{\mu} - \frac{1}{2}\boldsymbol{\mu}\times\mathbf{I}]\,\text{Grad}\,\boldsymbol{\mu}$
$\boldsymbol{\beta} \equiv 4\sin\frac{\phi}{4}\mathbf{e}$	$(-2\pi, 2\pi)$	$[\sqrt{1-\frac{\beta^2}{16}}(1-\frac{\beta^2}{8})\mathbf{I} + \frac{1-(1-\beta^2/8)(1-\beta^2/16)}{\beta^2\sqrt{1-\beta^2/16}}\boldsymbol{\beta}\otimes\boldsymbol{\beta}$ $-\frac{1}{2}(1-\frac{\beta^2}{16})\boldsymbol{\beta}\times\mathbf{I}]\,\text{Grad}\,\boldsymbol{\beta}$

9.6 Review of Some Other Lagrangian Non-Linear Strain Measures

In Table 9.3, we present a review of various definitions of the Lagrangian strain measures proposed in 14 representative papers in the field. In those works, different notation, sign conventions, notions of gradient and divergence operators, coordinate systems, description of rotations, etc. are applied. To compare them with our natural strain measures $(9.3)_3$, we bring them into the common coordinate-free form using the microrotation tensor $\mathbf{Q}$, see [5].

The results summarized in Table 9.1 show that the stretch and wryness tensors introduced in many papers do not agree with each other and with our Lagrangian

Table 9.3 Definitions of the stretch and wryness tensors

Paper	The stretch tensor	The wryness tensor
Kafadar and Eringen (1971)	$\mathbf{F}^T\mathbf{Q}$	$-\frac{1}{2}\boldsymbol{\varepsilon} : (\mathbf{Q}^T \operatorname{Grad} \mathbf{Q})$
Stojanovic (1972)	$\mathbf{F}^T\mathbf{F}$	$\mathbf{F}^T \frac{1}{2}\boldsymbol{\varepsilon} : (\mathbf{Q} \operatorname{Grad} \mathbf{Q}^T)$
Besdo (1974)	$\mathbf{Q} - \mathbf{I}$	$\mathbf{F}[\frac{1}{2}\boldsymbol{\varepsilon} : (\mathbf{F}^{-1} \operatorname{Grad} \mathbf{F}) + \mathbf{B}] - \mathbf{Q}(\boldsymbol{\Gamma} + \mathbf{B})$
Shkutin (1980)	$\mathbf{F}^T\mathbf{Q} - \mathbf{I}$	$-\frac{1}{2}[\boldsymbol{\varepsilon} : (\mathbf{Q}^T \operatorname{Grad} \mathbf{Q})]^T + \mathbf{B}^T$
Badur and Pietraszkiewicz (1986)	$\mathbf{Q}^T\mathbf{F}$	$\frac{1}{2}\boldsymbol{\varepsilon} : (\mathbf{Q}^T \operatorname{Grad} \mathbf{Q})$
Reissner (1987)	$\mathbf{F}^T\mathbf{Q}$	$-\frac{1}{2}[\boldsymbol{\varepsilon} : (\mathbf{Q}^T \operatorname{Grad} \mathbf{Q})]^T$
Zubov (1990)	$\mathbf{F}^T\mathbf{Q}$	$-\frac{1}{2}[\boldsymbol{\varepsilon} : (\mathbf{Q}^T \operatorname{Grad} \mathbf{Q})]^T$
Dłużewski (1993)	$\mathbf{Q}^T\mathbf{F}$	$\mathbf{Q}^T \operatorname{Grad} \boldsymbol{\phi}$
Merlini (1997)	$\mathbf{F} - \mathbf{Q}$, $\mathbf{Q}^T\mathbf{F} - \mathbf{I}$	$-\mathbf{Q}\frac{1}{2}\boldsymbol{\varepsilon} : (\mathbf{Q}^T \operatorname{Grad} \mathbf{Q})$, $-\frac{1}{2}\boldsymbol{\varepsilon} : (\mathbf{Q}^T \operatorname{Grad} \mathbf{Q})$
Steinmann and Stein (1997)	$\mathbf{Q}^T\mathbf{F}$	$-\frac{1}{2}\boldsymbol{\varepsilon} : (\mathbf{Q}^T \operatorname{Grad} \mathbf{Q})$
Nikitin and Zubov (1998)	$\mathbf{Q}^T\mathbf{F}$	$-\frac{1}{2}\boldsymbol{\varepsilon} : (\mathbf{Q}^T \operatorname{Grad} \mathbf{Q})$
Grekova and Zhilin (2001)	$\mathbf{F}^T\mathbf{Q}$	$\frac{1}{2}\boldsymbol{\varepsilon} : (\mathbf{Q}^T \operatorname{Grad} \mathbf{Q})$
Nistor (2002)	$\mathbf{F}^T\mathbf{Q}$	$-\frac{1}{2}[\boldsymbol{\varepsilon} : (\mathbf{Q}^T \operatorname{Grad} \mathbf{Q})]^T$
Ramezani and Naghdabadi (2007)	$\mathbf{F}^T\mathbf{Q}$	$\frac{1}{2}\boldsymbol{\varepsilon} : (\mathbf{Q}^T \operatorname{Grad} \mathbf{Q})$
The present paper and [5]	$\mathbf{Q}^T\mathbf{F} - \mathbf{I}$	$-\frac{1}{2}\boldsymbol{\varepsilon} : (\mathbf{Q}^T \operatorname{Grad} \mathbf{Q})$

strain measures defined in (9.3). Most definitions differ only by transpose of the measures, or by opposite signs, or the measures do not vanish in the absence of deformation. Such differences are not essential for the theory, although one should be aware of them. But we have also discovered a few strain measures which are incompatible with our Lagrangian stretch and wryness tensors. One should avoid such incompatible strain measures when analyzing problems of physical importance using the Cosserat continuum model.

References

1. Bauchau, O.A., Trainelli, L., The vectorial parameterization of rotation. Nonlinear Dyn. **32**, 71–92 (2003)
2. Cosserat, E., Cosserat, F., Théorie des corps déformables. Hermann et Fils, Paris (1909)
3. Gibbs, J.W., Vector Analysis. Yale University Press, New Haven (1901)
4. Kafadar, C.B., Eringen, A.C., Micropolar media—I. The classical theory. Int. J. Eng. Sci. **9**, 271–305 (1971)
5. Pietraszkiewicz, W., Eremeyev, V.A., On natural strain measures of the non-linear micropolar continuum. Int. J. Solids Struct. **46**(34), 774–787 (2009)
6. Pietraszkiewicz, W., Eremeyev, V.A., On vectorially parameterized natural strain measures the non-linear Cosserat continuum. Int. J. Solids Struct. **46**(11–12), 2477–2480 (2009)
7. Reissner, E., On kinematics and statics of finite-strain force and moment stress elasticity. Stud. Appl. Math. **52**, 93–101 (1973)

Chapter 10
Practical Applications of Simple Cosserat Methods

David A. Burton and Robin W. Tucker

Abstract Motivated by the need to construct models of slender elastic media that are versatile enough to accommodated non-linear phenomena under dynamical evolution, an overview is presented of recent practical applications of simple Cosserat theory. This theory offers a methodology for modeling non-linear continua that is physically accurate and amenable to controlled numerical approximation. By contrast to linear models, where non-linearities are sacrificed to produce a tractable theory, large deformations are within the range of validity of simple Cosserat models. The geometry of slender and shell-like bodies is exploited to produce a theory that contains as few degrees of freedom as is physically reasonable. In certain regimes it is possible to include fluid-structure interactions in Cosserat rod theory in order to model, for example, drill-string dynamics, undersea riser dynamics and cable-stayed bridges in light wind-rain conditions. The formalism also lends itself to computationally efficient, effective models of microscopic carbon nanotubes and macroscopic gravitational antennae.

10.1 Overview

The pioneering efforts of Euler and Bernoulli in the seventeenth and eighteenth centuries extended Newton's laws for discrete mass points to any continuous deformable body and initiated the recognition of the independent Angular Momentum Principle. With the introduction of the concept of the stress-tensor by Cauchy in the nineteenth century the essential framework of classical (non-relativistic) elastodynamics was complete. However, the implementation of this framework for the solu-

D.A. Burton (✉)
Department of Physics, Lancaster University, Lancaster, UK
e-mail: d.burton@lancaster.ac.uk

R.W. Tucker
e-mail: r.tucker@lancaster.ac.uk

G.A. Maugin, A.V. Metrikine (eds.), *Mechanics of Generalized Continua*,
Advances in Mechanics and Mathematics 21,
DOI 10.1007/978-1-4419-5695-8_10, © Springer Science+Business Media, LLC 2010

tion of many practical engineering problems had to wait for effective mathematical tools. Naturally, the first use of the theory involved linearizations of the equations about stationary configurations commensurate with the linear response properties of particular materials. This led to the development of effective practical tools for solving linear partial differential equations on suitably shaped domains (e.g., Fourier and Laplace techniques) and their efficacy obscured for some time the need to address the fundamental non-linear nature of the basic equations or their boundary conditions. Indeed, until new mathematical tools became available, the theory was largely restricted to all pervading linear approximations. Although these had indisputable value it was finally appreciated that such approximations had limited domains of applicability and that the full theory was much richer in its scope.

If one takes the attitude that the methodology in constructing an effective model of a physical system should follow three distinct steps: formulation, analysis and interpretation then the analysis should consist solely in the application of precise mathematical processes exempt from any further ad-hoc physical simplifications [1]. The predominance of linearization methods for elastic problems involving strings, rods, beams, shells, etc. gave rise to somewhat haphazard modeling assumptions in which mathematical assumptions became unnecessarily involved with physical assumptions about the system under consideration. A typical example that pervades the majority of elementary texts even today is the derivation of the equation of motion of small transverse vibrations of an elastic string. Such derivations often assume that each material point is confined to a plane through its equilibrium position perpendicular to the line joining the ends of the string. Very few strings can execute purely transverse motion (without longitudinal deformation) and most derivations ignore the role played by the inextensibility of the string by assuming that the tension in the string is essentially constant. Such dubious assumptions are unnecessary since the wave equation follows most naturally from a systematic perturbation scheme applied to the exact (non-linear) equations of motion [1].

In recent years, an efficient approach to modeling the non-linear behavior of certain special classes of continuous mechanical structures has been developed. Based on the early work of the Cosserat brothers and others, the mathematical theory of Cosserat media is now well established. Antman [1] has formulated a complete theory of the mechanics of *slender* structures (rods) and *thin* structures (shells) in a recent book. The essential idea is to exploit the geometry of the structure to reduce the dimension of its configuration space. The genesis of the Cosserat approach lies in the application of the basic laws of Continuum Mechanics to a slender material continuum and the disengagement of these physical principles from the constitutive relations that model the material properties of particular structures. *Then the analysis of the model is based on mathematical methods in which approximations may be controlled.* Today one can often assess the validity of a particular approximation scheme in the absence of exact analytical solutions using computational techniques. The methods also permit an exploration of non-differentiable solutions (shock phenomena) that lie outside the remit of approaches based on the discretization of classical *differential* equations.

By contrast, one cannot a priori assume that approximations based on small displacements *about particular configurations* will necessarily provide accurate dynamical predictions over extended periods of time and one may cite ample evidence where linearized equations will completely miss dynamical behavior that arises from the underlying equations of motion, boundary conditions or material properties. The full (non-linear) equations of motion may admit distinct classes of solutions that are parametrically connected in the space of all solutions [14, 13, 2]. By contrast, stable solutions of the linearized equations are restricted to domains about particular configurations. Unstable linearized solutions signal that the approximation eventually breaks down and recourse to the full equations of motion becomes mandatory in order to understand how the system develops in time from prescribed initial and boundary conditions.

An example of this situation [15] arises in modeling the dynamics of on-shore drill-strings. Traditional models can be constructed based on Euler–Bernoulli linearized beam theory for planar vibrations coupled to axial vibrations of a heavy drill-string confined to the vertical bore cavity of an active drilling assembly. The drill-string is driven in a rotary fashion from the top and reacts to non-linear friction at the attached drill-bit. Thus the boundary conditions are generally non-linear functions of the drill-string configuration and their linearization limits the temporal evolution to a superposition of small amplitude stable eigenmodes. Without such a linearization, the drill-string can execute a complex motion in space eventually making contact with the bore-cavity along its length. A similar phenomenon occurs if one executes the evolution as a perturbation of the drill-string about a distinct *whirling motion* [10] in which the entire string vibrates in a plane rotating about a vertical axis in the bore cavity. The existence of these different dynamical configurations and the attendant clash with the bore-cavity renders a linearized approach ineffective and recourse to a Cosserat model is both simpler and more reliable. This not only accommodates naturally the coupling between lateral, axial and torsional motions but enables the inherently non-linear interaction of the drill-string with the bore cavity to be modeled.

In the context of vertical steel marine risers attached to floating platforms, similar boundary interactions can induce parametric excitations that feed energy among the various elastic modes of vibration of the riser (axial, lateral and torsional motion about its length) in unsuspected ways and the use of a Cosserat description eliminates the need for ad-hoc assumptions about the significance of the various energy transfer channels. For flexible risers that are not vertically constrained, the need for a non-perturbative approach becomes even more important since one often deals with forces and torques in which the riser executes larger curved motions in space as well as extended dynamical interactions along its length where contact with the sea-bed is in evidence. Problems of riser collisions, either with other risers externally or with internal elements, are further situations where linearized methods become suspect. In both cases, internal dampings of the structure (viscoelastic or possibly of memory type (hysteretic)) provide further sources of non-linearity that are accessible to a time domain evolution in the Cosserat description.

Recent applications of Cosserat methods have included the non-linear analysis of complex vibration states of drill-strings, marine risers, MEMS structures, rigid gears, gravitational antennae and carbon nanotubes. It has been shown that the inherent non-linearities in the simple Cosserat theory of rods [1] can be exploited for the design of space-based slender elastic gravitational wave antennae [17, 16]. Such structures are more sensitive to perturbation by gravitational waves and respond to broader frequency ranges than linearized elasticity models might suggest.

A Cosserat model of a drill-string is a flexible Cosserat rod connected at one end to a rotary drive and connected at the other end to a heavy drill bit subject to non-linear frictional forces. A Cosserat model of a marine riser is a flexible Cosserat rod vertically immersed in sea water and connected to the sea bed. A Cosserat model of a cable in a cable stayed bridge is a high tension Cosserat rod immersed in air. In the models above, the fluid-structure interaction in which the external forces and torques on the structure arise from its contact with a fluid play a role. One must model such forces and torques in terms of the structure configuration variables and properties of the external stimuli, and the dynamics of such stimuli are coupled to the motion of the structure by appropriate boundary conditions. In certain regimes such interactions can be rendered tractable to analysis, particularly in the presence of laminar potential flow with possible vorticity.

Long stay cables are important structural components of cable-stayed bridges. Due to their large flexibility and small structural damping, they are prone to vibration induced by motion of their supports and/or aerodynamic forces such as wind and rain loadings. Under the simultaneous occurrence of light-to-moderate wind and rain, large amplitude vibrations of stay cables have been observed in a number of cable-stayed bridges worldwide. The mechanism leading to rain–wind induced vibration in stay cables has recently become of concern to bridge engineers and scientists in various countries and, clearly, this phenomenon involves a complex interplay between the cables, rain rivulets and air. A fully detailed mathematical analysis would require the equations of multi-phase fluid dynamics, a model for accretion and fluid–solid adhesion and the continuum mechanics of an elastic structure. However, such an approach is extremely complicated, unwieldy and involves considerable computing power. Fortunately, in low ambient wind speeds, experiments with an artificial mobile rivulet on a fixed cylinder subject to aerodynamic loading do indicate an approach to steady rivulet oscillation. A fully dynamical Cosserat model of a cable section and a mobile rain rivulet was developed in which the complex fluid–structure interaction was approximated in two distinct ways. The first employed an approximation that permitted the use of data extrapolated from wind-tunnel measurements [8]. The second approach [7] modeled the aerodynamic interaction in terms of a sub-critical vortex description [5] which took account of the effects due to the boundary layers between the cable, rivulet and air. This was based on the observation that the instantaneous Reynolds number associated with the relative air-cable velocity is in the sub-critical range and hence dynamical vortex shedding should play a role. It is likely that the fluid vorticity generated by a cable will interact with other cables downstream and contribute significantly to their ensuing motion. One advantage of the second modeling approach is that it can be naturally generalized to accommodate more than one cable. A method for constructing flows containing point vortices in the vicinity of a pair of cable sections was recently proposed in [6], and a fluid model employing vortex sheets was developed in [4].

Recent years have seen a surge of interest in carbon nanotubes with the aim to engineer nano-scale devices. Calculations of the electronic and mechanical properties of carbon nanotubes often rely on *ab-initio* techniques employing density-functional theory. However, numerical implementations of such schemes are intensive and can require considerable computing power. Cosserat methods afford a natural alternative for formulating more efficient models of nanotubes. This approach was used to successfully describe the equilibria of a nanotube under an external load [11] and investigate its radial breathing modes [3].

The following is a brief summary of the Cosserat models used to investigate the above physical systems.

10.2 Simple Cosserat Methods

The motion of a Cosserat rod may be represented in terms of the motion in space of the line of centroids of its cross-sections and the material deformation about that line. The configuration of such a rod is modeled mathematically by a space-curve with structure: i.e., as a principal SO(3) bundle over a moving space-curve. This structure defines the relative orientation of neighboring cross-sections along the rod. Specifying a unit vector $\mathbf{d}_1$ (which may be identified with the normal to the cross-section) at each point along the rod centroid enables the state of flexure to be related to the angle between this vector and the tangent to the space-curve. Specifying a second unit vector $\mathbf{d}_2$ orthogonal to the first vector (thereby placing it in the plane of the cross-section) permits an encoding of the state of bending and twist along its length. Thus a field of two mutually orthogonal unit vectors along the rod provides three continuous dynamical degrees of freedom that, together with the continuous three degrees of freedom describing a space-curve relative to some arbitrary origin in space, define a simple Cosserat rod model.

The general mathematical theory of non-linear Newtonian elasticity is well established. The dynamics of the Cosserat rod theory follows as a well defined limit of a three-dimensional continuum and is conventionally formulated in a Lagrangian picture in which material elements are labeled by s. In the following, objects associated with an un-deformed *reference* configuration are superscripted with 0.

The dynamical evolution of a (transversely homogeneous) simple Cosserat rod with reference length L^0, reference mass density, $s \in [0, L^0] \mapsto \rho^0(s)$, and cross-sectional area, $s \in [0, L^0] \mapsto A(s)$, is governed by Euler's dynamical laws:

$$\rho^0 A\ddot{\mathbf{r}} = \mathbf{n}' + \mathbf{f}, \qquad \partial_t(\rho^0 \mathscr{I}(\mathbf{w})) = \mathbf{m}' + \mathbf{r}' \times \mathbf{n} + \mathbf{l} \tag{10.1}$$

involving implicitly the triad of orthonormal vector fields $\{\mathbf{d}_1(s,t), \mathbf{d}_2(s,t), \mathbf{d}_3(s,t)\}$ (directors) over the space-curve: $s \in [0, L^0] \mapsto \mathbf{r}(s,t)$ at time t where $\mathbf{r}' = \partial_s \mathbf{r}$, $\dot{\mathbf{r}} = \partial_t \mathbf{r}$, etc. The external force and torque densities (including couplings to external gravitational, friction, electromagnetic and fluid forces) are denoted $\mathbf{f}$ and $\mathbf{l}$, and $s \in [0, L^0] \mapsto \rho^0 \mathscr{I}$ is a moment of inertia tensor. In these field equations, the *contact force fields* $\mathbf{n}$ and *contact torque fields* $\mathbf{m}$ are related to the "*strain*" *fields* $\mathbf{u}$, $\mathbf{v}$, $\mathbf{w}$ by constitutive relations. The strains are themselves defined in terms of the configuration variables $\mathbf{r}$ and $\mathbf{d}_k$ for $k = 1, 2, 3$ by the relations: $\mathbf{r}' = \mathbf{v}$,

$\mathbf{d}'_k = \mathbf{u} \times \mathbf{d}_k$, $\dot{\mathbf{d}}_k = \mathbf{w} \times \mathbf{d}_k$. The latter ensures that the triad remains orthonormal under evolution. The last equation identifies $\mathbf{w} = \frac{1}{2}\sum_{k=1}^{3} \mathbf{d}_k \times \dot{\mathbf{d}}_k$ with the local angular velocity vector field of the director triad. To close the above equations of motion constitutive relations appropriate to the rod must be specified. These relate the contact forces and torques to the strains $\mathbf{u}$, $\mathbf{v}$, their time rates of change for viscoelastic materials and their possible dependence on the history of the evolution of the structure. In general, the model accommodates continua whose characteristics (density, cross-sectional area, rotary inertia) vary with s and the thermodynamic state of the material [9]. For a system of several rods coupled to localized regions with mass or rotary inertia, one matches rod degrees of freedom according to junction conditions describing the coupling [12]. If the rod characteristics change discontinuously conditions on the contact forces and torques on either side of the discontinuity must be satisfied.

Mechanically deforming a carbon nanotube will lead to changes in the relative positions of the nanotube's carbon atoms. Bending and twisting the nanotube may lead to significant cross-section deformation outside the remit of the simple Cosserat rod theory described above. Motivated by the efficiency and wide domain of applicability of simple Cosserat rod theory, an extended shell-like model (a *Cosserat tube*) was developed [11, 3] to describe such cross-section deformation.

It is widely accepted that a carbon nanotube may be effectively modeled as an *elastic* continuum over spatial scales that greatly exceed interatomic distances. The mechanical properties of a simple Cosserat tube may be motivated using the *Kirchoff* [1] constitutive relations for a simple Cosserat rod. The balance laws (10.1) for such a rod may be recovered from an action principle [3], and a concise description of a Cosserat tube may be developed by demanding that the tube's action reduces to the rod's action when the tube's cross-section is constrained to remain in its undeformed state. The position $\mathbf{R}$ of a material point (s, σ, ζ) in the wall of a Cosserat

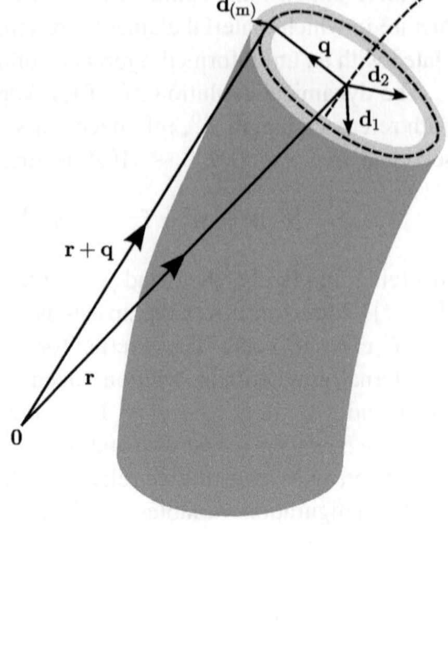

Fig. 10.1 The directors $\mathbf{d}_1$, $\mathbf{d}_2$, $\mathbf{d}_{(\mathrm{m})}$ and the position vectors $\mathbf{r}$, $\mathbf{r}+\mathbf{q}$ of the (*dashed*) space-curves representing the tube's "bulk" and the cross-section are shown. The vector fields $\mathbf{d}_{(\mathrm{m})}$ and $\mathbf{q}$ lie in the plane spanned by $\mathbf{d}_1$, $\mathbf{d}_2$

tube can be written

$$\mathbf{R}(s,\sigma,\zeta,t) = \mathbf{r}(s,t) + \mathbf{q}(s,\sigma,t) + \zeta\,\mathbf{d}_{(\mathrm{m})}(s,\sigma,t) \tag{10.2}$$

with

$$\mathbf{q}(s,\sigma,t) = q_1(s,\sigma,t)\,\mathbf{d}_1(s,t) + q_2(s,\sigma,t)\,\mathbf{d}_2(s,t)$$

where the unit vector field $\mathbf{d}_{(\mathrm{m})}$ is an additional "microstructure" director and the functions q_1, q_2, $\mathbf{d}_{(\mathrm{m})}$ encode the shape of the cross-section (see Fig. 10.1). The elastic potential $\mathscr{V}$ of the tube,

$$\begin{aligned}\mathscr{V} = \int_0^{L^0} \mathrm{d}s \int_{\mathscr{C}^0(s)} \mathrm{d}A^0 \\ \times \Bigg[\frac{1}{2}G\big(\partial_s\mathbf{R}\cdot\mathbf{d}_1 - \partial_s\mathbf{R}^0\cdot\mathbf{d}_1^0\big)^2 \\ + \frac{1}{2}G\big(\partial_s\mathbf{R}\cdot\mathbf{d}_2 - \partial_s\mathbf{R}^0\cdot\mathbf{d}_2^0\big)^2 \\ + \frac{1}{2}E\big(\partial_s\mathbf{R}\cdot\mathbf{d}_3 - \partial_s\mathbf{R}^0\cdot\mathbf{d}_3^0\big)^2 \\ + \frac{1}{2}G\big(\partial_\sigma\mathbf{R}\cdot\mathbf{d}_{(\mathrm{m})} - \partial_\sigma\mathbf{R}^0\cdot\mathbf{d}_{(\mathrm{m})}^0\big)^2 \\ + \frac{1}{2}E\big(\partial_\sigma\mathbf{R}\cdot\mathbf{d}_3\times\mathbf{d}_{(\mathrm{m})} - \partial_\sigma\mathbf{R}^0\cdot\mathbf{d}_3^0\times\mathbf{d}_{(\mathrm{m})}^0\big)^2\Bigg]\end{aligned} \tag{10.3}$$

vanishes in the reference (un-deformed) configuration, where G is the shear modulus, E is Young's modulus, ρ^0 is the mass density of the reference configuration and $\mathrm{d}A^0$ is the area element of the cross-section $\mathscr{C}^0(s)$ in the reference configuration. The variable s is the arc parameter of a space-curve running along the tube ($s \in [0, L^0]$) in the reference configuration and σ is the arc parameter of a closed space-curve within a cross-section in the reference configuration (see the dashed space-curves in Fig. 10.1). The action S_{tube} for the Cosserat tube taken to be

$$S_{\mathrm{tube}}[\mathbf{r},\mathbf{d}_1,\mathbf{d}_2,q_1,q_2,\mathbf{d}_{(\mathrm{m})}] = \int \mathrm{d}t \Bigg[\int_0^{L^0} \mathrm{d}s \int_{\mathscr{C}^0(s)} \mathrm{d}A^0 \frac{1}{2}\rho^0\partial_t\mathbf{R}\cdot\partial_t\mathbf{R} - \mathscr{V}\Bigg] \tag{10.4}$$

and equations of motion for $\mathbf{r}$, $\mathbf{d}_k$, q_1, q_2, $\mathbf{d}_{(\mathrm{m})}$ are obtained by varying S_{tube} subject to appropriate boundary conditions.

Under a constrained variation, S_{tube} reduced to an action S_{rod} for a simple Cosserat rod with Kirchoff constitutive properties. This follows by writing

$$S_{\mathrm{rod}}[\mathbf{r},\mathbf{d}_1,\mathbf{d}_2] = S_{\mathrm{tube}}[\mathbf{r},\mathbf{d}_1,\mathbf{d}_2,\check{q}_1,\check{q}_2,\check{\mathbf{d}}_{(\mathrm{m})}] \tag{10.5}$$

and enforcing the constraints

$$\mathbf{d}_{(\mathrm{m})} = \check{\mathbf{d}}_{(\mathrm{m})} \equiv \cos(\theta)\,\mathbf{d}_1 + \sin(\theta)\,\mathbf{d}_2,$$
$$q_1 = \check{q}_1 \equiv R^0 \cos(\theta), \tag{10.6}$$
$$q_2 = \check{q}_2 \equiv R^0 \sin(\theta)$$

prior to variation with respect to $\mathbf{r}$, $\mathbf{d}_1$, $\mathbf{d}_2$. The variable $\theta = \sigma / R^0$ is the angle around the rod, where the constant R^0 is the radius of the circular space-curve representing the rod's cross-section (the circular dashed curve in Fig. 10.1). The constraints (10.6) remove the final two terms from the integrand in the potential (10.3). Euler's dynamical laws (10.1) for a simple Cosserat rod with Kirchoff constitutive relations are then recovered as the Euler–Lagrange equations derived from the action S_{rod}.

10.3 Conclusion

Cosserat methods afford dynamical models of non-linear continua that are physically accurate and yet computationally amenable. By contrast to linear models, where non-linearities are sacrificed to produce a tractable theory, large deformations are within the range of validity of simple Cosserat models. The geometry of slender and shell-like bodies can be exploited to develop efficient tools for analyzing the dynamics of a broad range of important physical systems.

References

1. Antman, S.S.: Nonlinear Problems of Elasticity, 2nd edn. Applied Mathematical Sciences, vol. 107, Springer, Berlin (2005)
2. Balanov, A., Janson, N., McClintock, P.V.E., Tucker, R.W., Wang, C.: Bifurcation analysis of a neutral delay differential equation modelling the torsional motion of a driven drill string. Chaos Solitons Fractals **15**(2), 381–394 (2002)
3. Burton, D.A., Gould, T.: Dynamical model of Cosserat nanotubes. J. Phys., Conf. Ser. **62**(1), 23–33 (2007)
4. Burton, D.A., Tucker, R.W.: Geometry and dynamics of vortex sheets in 3 dimensions. Theor. Appl. Mech. (Belgrade) **28–29**, 55 (2002)
5. Burton, D.A., Hartley, D., Tucker, R.W.: Vortex-Induced Fluid Forces on Accelerating Rigid Boundaries in 2 Dimensions. Proc. 5th International Seminar on Geometry, Continua and Microstructure. Editura Academiei Romane, Sinaia (2001)
6. Burton, D.A., Gratus, J., Tucker, R.W.: Hydrodynamic forces on two moving discs. Theor. Appl. Mech. (Belgrade) **31**(2), 153–188 (2004)
7. Burton, D.A., Cao, D.Q., Tucker, R.W., Wang, C.: On the stability of stay cables under light wind and rain conditions. J. Sound Vib. **279**(1–2), 89–117 (2005)
8. Cao, D.Q., Tucker, R.W., Wang, C.: A stochastic approach to cable dynamics with moving rivulets. J. Sound Vib. **268**(2), 291–304 (2003)
9. Cao, D., Liu, D., Preston, S., Tucker, R.W.: Thermo-mechanics of Cosserat rods. In: Proc. Symposium on the Mechanics of Slender Structure (MoSS 2006) (CDROM)
10. Coomer, J., Lazarus, M., Tucker, R.W., Kershaw, D., Tegman, A.: A non-linear eigenvalue problem associated with inextensible whirling strings. J. Sound Vib. **239**(5), 969–982 (2001)

11. Gould, T., Burton, D.A.: A Cosserat rod model with microstructure. New J. Phys. **8**(137) (2006)
12. Gratus, J., Tucker, R.W.: The dynamics of Cosserat nets. J. Appl. Math. **4**, 187–226 (2003)
13. Hartley, D.H., Tucker, R.W., Tung, R., Wang, C.: On parametrically excited flexural motion of an extensible and shearable rod with a heavy attachment. Tech. Mech. **20**(2), 147 (2000)
14. Tucker, R.W., Tung, R., Wang, C.: Non-linear flexural excitations and drill-string dynamics. Extr. Math. **14**(2), 217 (1999)
15. Tucker, R.W., Wang, C.: An integrated model for drill-string dynamics. J. Sound Vib. **224**, 123–165 (1999)
16. Tucker, R.W., Wang, C.: A Cosserat detector for dynamic geometry. Rend. Semin. Mat., Univ. Politec. Torino **58**(2), 245–256 (2000)
17. Tucker, R.W., Wang, C.: Gravitational wave induced vibrations of slender structures in space. Gen. Relativ. Gravit. **35**(12), 2137–2158 (2003)

Part III
Micromorphic Media (Deformable Microstructure)

Chapter 11
Requirements on Periodic Micromorphic Media

Ralf Jänicke and Stefan Diebels

Abstract In order to investigate the properties of microstructured materials, the underlying heterogeneous material is commonly replaced by a homogeneous material involving additional degrees of freedom. Making use of an appropriate homogenization methodology, the present contribution compares deformation states predicted by the homogenization technique to the deformation state within a reference solution. The results indicate on what terms the predicted deformation modes can be clearly interpreted from the physical point of view.

11.1 Introduction

Microstructured materials are known to feature a complex mechanical behavior which is strongly dominated by the underlying microtopology. A well documented phenomenon is the stiff boundary layer effect inducing size-dependent effective material properties, e.g., [2, 17, 18]. In the literature, various approaches exist which replace the heterogeneous microcontinuum by a homogeneous macrocontinuum, enriched by additional degrees of freedom. A wide range of those approaches goes back to the fundamental considerations of the Cosserat brothers[1], and these considerations were systematically generalized by Eringen [3]. Different approaches deal with second gradient media, e.g., [8, 11, 14]. During the last decade, considerable advances in the numerical modeling of microstructured materials took place, namely by using the so-called two-level FEM or FE^2 approaches, e.g., [4, 11, 15].

In the present contribution, we make use of a second order homogenization procedure introduced in detail in [10]. For this purpose, the kinematic quantities are

R. Jänicke (✉) · S. Diebels
Chair of Applied Mechanics, Saarland University, Campus A 4.2, 66123 Saarbrücken, Germany
e-mail: r.jaenicke@mx.uni-saarland.de

S. Diebels
e-mail: s.diebels@mx.uni-saarland.de

G.A. Maugin, A.V. Metrikine (eds.), *Mechanics of Generalized Continua*,
Advances in Mechanics and Mathematics 21,
DOI 10.1007/978-1-4419-5695-8_11, © Springer Science+Business Media, LLC 2010

expressed in terms of a polynomial mean field and a periodic fluctuation on a microvolume attached to the macroscopic material point. After having reviewed the scale transition of the kinematic quantities, several numerical experiments will help analyze the significance of the deformation modes involved in the proposed approach in comparison to a reference solution. Special attention will be paid to the methodology's requirements on the attached microvolume.

11.2 Micromorphic Media

The physical picture Following the seminal works of Eringen [3], we assume the physical body $\mathscr{B}_M$ to consist of a set of deformable material points which capture a small but finite space $\mathscr{B}_m$. The microcontinuum's mapping from the material to the spatial frame is considered to be affine and reads

$$\Delta\mathbf{x}(\mathbf{X}_M, \overline{\chi}_M, t) = \overline{\chi}_M(\mathbf{X}_M, t) \cdot \Delta\mathbf{X}(\mathbf{X}_M). \tag{11.1}$$

$\overline{\chi}_M$ is defined as the microdeformation tensor, the index $(\diamond)_M$ refers to the macroscale and $(\diamond)_m$ to the microscale, respectively. By calculating the square of the deformed arc length $(\mathrm{d}\mathbf{x}_m)^2$, one has to introduce a set of three independent deformation measures. Without loss of generality, we choose the common deformation gradient $\mathbf{F}_M$, the microdeformation $\overline{\chi}_M$ and its gradient $\mathrm{Grad}\,\overline{\chi}_M$. Obviously, this set of two-field quantities does not fulfill the requirements of objectivity. Nevertheless, the given set is admissible for the following investigations because no constitutive assumptions will be met.

Scale transition of the kinematic quantities In the sequel, a consistent averaging technique will be sketched replacing a heterogeneous Cauchy medium, representing the cellular network, by a homogeneous micromorphic medium. Based on a methodology proposed by Forest and Sab [5, 7] and presented in detail in [10], one can identify the kinematic quantities of the micromorphic macrocontinuum in terms of a polynomial mean field and a periodic perturbation of the heterogeneous Cauchy microcontinuum attached to the micromorphic material point on the macroscale. To simplify matters, the attached microvolume is assumed to be a rhombic unit cell as

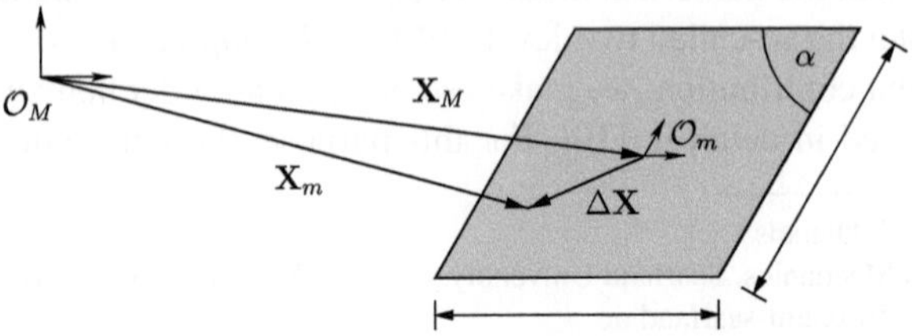

Fig. 11.1 The attached rhombic microvolume representing the cellular network. The volume centroid of $\mathscr{B}_m$ is defined by the position vector $\mathbf{X}_M$ (material frame)

depicted in Fig. 11.1. We postulate the set $(\mathbf{u}_M, \overline{\chi}_M)$ to characterize the macrostate that best fits the microscopic displacement field $\mathbf{u}_m$ in an averaged sense. Thus, we minimize the functional

$$\mathscr{F}(\mathbf{u}_M, \overline{\chi}_M) = \langle (\mathbf{u}_m - \mathbf{u}_M - (\overline{\chi}_M - \mathbf{I}) \cdot \Delta\mathbf{X})^2 \rangle \tag{11.2}$$

with the volume average $\langle \diamond \rangle = 1/V_m \int_{\mathscr{B}_M} \diamond \, \mathrm{d}V_m$ and we find

$$\langle \Delta\mathbf{u} \rangle = \mathbf{0} \quad \text{and} \quad \overline{\chi}_M = \langle \mathbf{u}_m \otimes \Delta\mathbf{X} \rangle \cdot \mathbf{G}^{-1} + \mathbf{I}, \tag{11.3}$$

where $\mathbf{G}$ is a second order geometry tensor defined by the shape of the attached microvolume. Assuming a microscopic displacement field as a polynomial of grade three, one can show that the expression

$$\begin{aligned} \Delta\mathbf{x} = {} & \mathbf{F}_M \cdot \Delta\mathbf{X} + \frac{1}{2} \operatorname{Grad} \overline{\chi}_M : (\Delta\mathbf{X} \otimes \Delta\mathbf{X}) \\ & + \frac{1}{6} \overset{*}{\chi}_M \cdot \mathbf{G}^{\underline{4}} \vdots (\Delta\mathbf{X} \otimes \Delta\mathbf{X} \otimes \Delta\mathbf{X}) + \Delta\tilde{\mathbf{x}} \end{aligned} \tag{11.4}$$

satisfies the averaging rules of (11.3), where a fourth order geometry tensor $\mathbf{G}^{\underline{4}}$ depending on the microvolume's shape has been introduced and where the difference $\overset{*}{\chi}_M = \overline{\chi}_M - \mathbf{F}_M$ has been used as an independent quantity. $\Delta\tilde{\mathbf{x}}$ represents a fluctuation field due to the microstructural periodicity. Having in mind the linear displacement field $\Delta\mathbf{x} = \mathbf{F}_M \cdot \Delta\mathbf{X} + \Delta\tilde{\mathbf{x}}$ of a so-called first-order FE^2 approach replacing a heterogeneous Cauchy microcontinuum by a homogeneous Cauchy macrocontinuum, (11.4) is pointing out the extended character of the introduced projection rule. Equation (11.4) clearly indicates the restricted character of the cubic projection link. For the 2D case, there only exist 4 independent cubic deformation modes in contrast to an unrestricted cubic polynomial allowing for 8 independent cubic deformation modes. In the sequel, the significance of the higher order deformation modes, i.e., quadratic and cubic ones, will be investigated for several perfectly periodic microstructures.

11.3 Higher Order Deformation Modes in Periodic Microstructures

The substitution of a heterogeneous microcontinuum by an extended but homogeneous macrocontinuum requires the additional macroscopic degrees of freedom to display the real deformation mechanisms of the microstructure in an adequate way. As it has been shown in the precedent section, the micromorphic kinematics enrich the microscopic displacement field by quadratic and cubic parts. In order to determine the relevance of these higher order deformation modes, a set of four perfectly periodic microstructures is subject to several numerical experiments. The analyzed

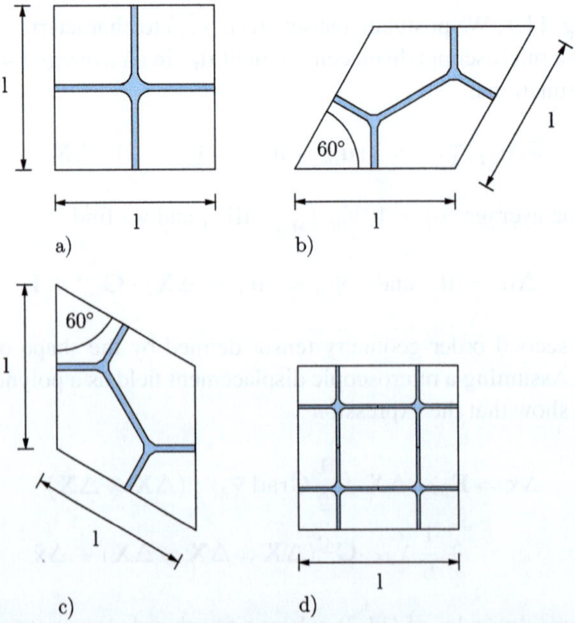

Fig. 11.2 The periodic unit cells of the investigated cellular microstructures. The structures are considered as lattices of struts which are rigidly connected in vertices. The interaction between two vertices only depends on the geometry and the material properties of the connecting struts. Microstructure (*a*) represents an 1-particle system, whereas microstructures (*b*) and (*c*) involve 2 particles, microstructure (*d*) 4 particles, respectively. The struts feature an aspect ratio $r \approx 1/20$ for (*a*) and (*d*), $r \approx 1/12$ for (*b*) and (*c*), respectively

periodic unit cells are depicted in Fig. 11.2. The given microstructures can be considered as lattices of struts which are rigidly connected in vertices. Thus, the material properties and the geometry of the struts control the interaction between two joined vertices. In analogy to systems consisting of discrete particles interacting via their boundary, one may interpret the vertices as particles and the given microstructures as n-particle systems, cf. [16], scaling n from 1 to 4. Thus, n involves the number of particles within the periodic unit cell. The experimental setup is depicted in Fig. 11.3 in a schematic way for microstructure (a) and has been accomplished in an analogous way for the microstructures (b)–(d), i.e., the particular microstructures have been microscopically resolved by finite elements. The performed deformations are assumed to be small. The cell walls follow Hooke's law.

At 4 different positions ① – ④ of the numerical experiments, the displacement field on the boundary of the labeled unit cell embedded in the microscopically resolved cellular structure is observed. In order to determine how accurate this real displacement field can be reproduced by the projection rule in (11.4), several projection polynomials have been fitted to the observed displacement field using the method of least squares. The study comprises a linear polynomial (4 independent deformation modes) in analogy to a first order FE^2 approach, a quadratic polyno-

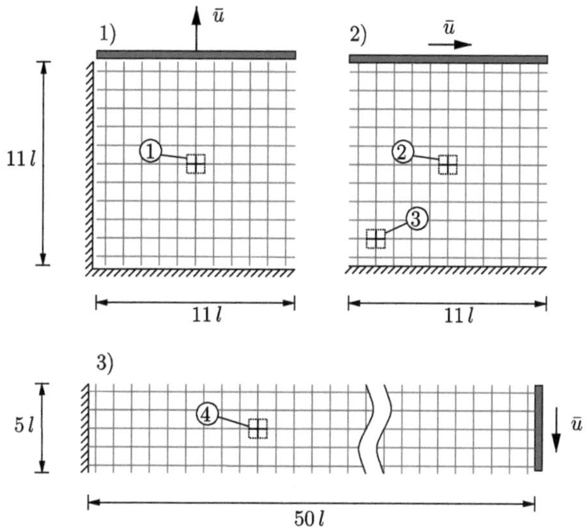

Fig. 11.3 Setup of the numerical experiments: (**1**) homogeneous tension test, (**2**) inhomogeneous shear test, (**3**) bending test, exemplary displayed for microstructure *(a)*, microstructures *(b)*–*(d)* analogously. The struts are microscopically resolved by finite elements and follow Hooke's law (Young's modulus $Y = 200$ GPa, Poisson's ratio $\nu = 0.33$). The embedded unit cells subject to the further investigations are denoted with ① – ④

mial ($4 + 6 = 10$ independent deformation modes) in analogy to a second order FE^2 approach using a second-gradient continuum on the macroscale, e.g., [11, 12], a micromorphic polynomial ($4 + 6 + 4 = 14$ independent deformation modes) as introduced earlier, and finally an unrestricted cubic one ($4+6+8 = 18$ independent deformation modes). Note that the fitting procedure is up to the well-defined polynomial orders. Thus, no periodic fluctuations are taken into account. In Figs. 11.4–11.7, the observed least squares of the fitted displacement field normalized to the linear one is plotted over the number of independent deformation modes.

11.4 Discussion

Considering Figs. 11.4–11.7, one may generally notice several points of interest:

- The increase in the fitting polynomial from a linear to a cubic one involves a decrease in the error over several orders of magnitude for the cross-like microstructure (a). In a less distinct manner, the same correlation holds for the honeycomb unit cells (b) and (c). Obviously, this is not the case for microstructure (d) showing the same symmetry as (a) but involving 4 vertices (4-particle system).
- For the microstructure (a), no higher significance of the quadratic polynomial compared to the linear one can be detected whereas, at least in the bending

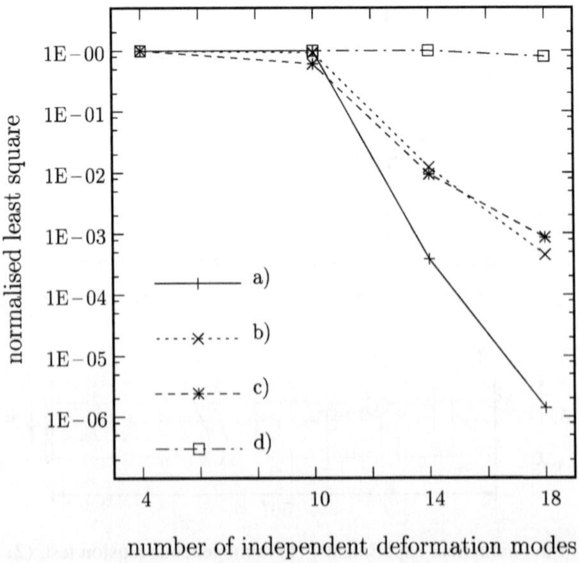

Fig. 11.4 Normalized least squares for several projection polynomials in the homogeneous tension experiment for the different microstructures at position ①. Whereas microstructure (*a*) shows a significantly decreasing error with increasing polynomial order, the decrease for (*b*) and (*c*) is less pronounced and it is nearly not observable for (*d*). Furthermore, the homogeneous deformation field does not account for differing accuracies due to the anisotropy of (*b*) and (*c*)

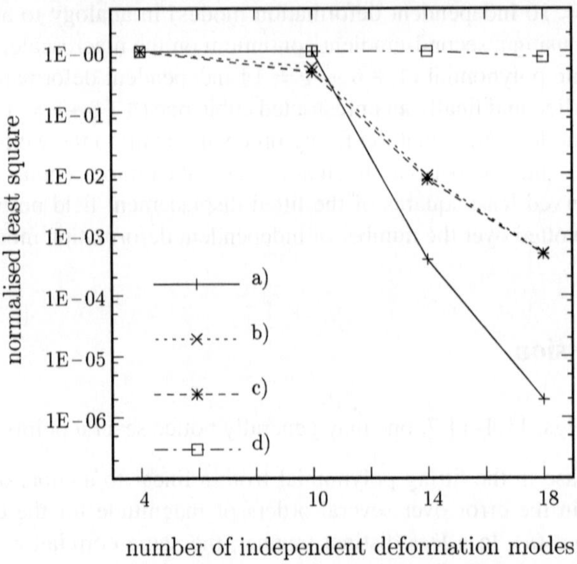

Fig. 11.5 Normalized least squares for several projection polynomials in the inhomogeneous shear experiment at the homogeneous position ②. Similar to the tension test, one can observe an dramatic decrease of the error for microstructure (*a*), a smaller but still significant decrease for (*b*) and (*c*). For (*d*), the extension to a cubic projection polynomial seems to influence the fitting accuracy only in a subordinate way

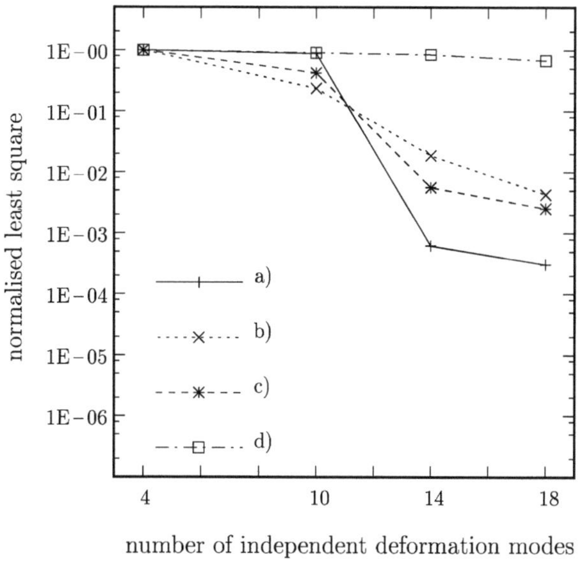

Fig. 11.6 Normalized least squares for several projection polynomials in the numerical shear experiment at the inhomogeneous position ③. Analogous to the homogeneous deformation states at ① and ②, one may observe the decrease of the error. Nevertheless, the difference between the micromorphic and the cubic polynomial is much less pronounced for microstructure (*a*). In contrast, the fitting accuracy for microstructures (*b*)–(*d*) is not seriously affected by the more inhomogeneous character of the deformation state. Merely the difference between (*b*) and (*c*) seems to be slightly more pronounced

experiment, the microstructures (b) and (c) respond to the quadratic, i.e., bending, deformation modes. This difference can be clarified taking into account the effective bending stiffnesses of the particular microstructures which depend, besides other effects, on the aspect ratio of the struts. Thus, one may expect a slightly higher effective bending stiffness for (b) and (c), $r \approx 1/12$, than for (a), $r \approx 1/20$. From the physical point of view, a low bending stiffness of the structure concentrates the bending effects close to the boundary. For (a), the bending modes decay within one layer of unit cells can not be detected at the investigated positions.

- A third point of interest is the relation between the micromorphic and the cubic polynomials at ③ and ④ for the microstructure (a). Obviously, the relevance of the complete cubic polynomial is less pronounced as it is at ① and ②.

To summarize the found observations, we want to conclude the proposed micromorphic projection rules to meet the real displacement fields of the reference computation in a significantly exacter way than the linear projection of a first order FE^2 approach does. However, the advantage of the higher order projection rules strongly depends on the topology within the investigated periodic unit cell, i.e., within the volume representative for the particular microstructure. In order to return to the analogy of vertices and particles, the present study indicates that for $(n > 1)$-particle

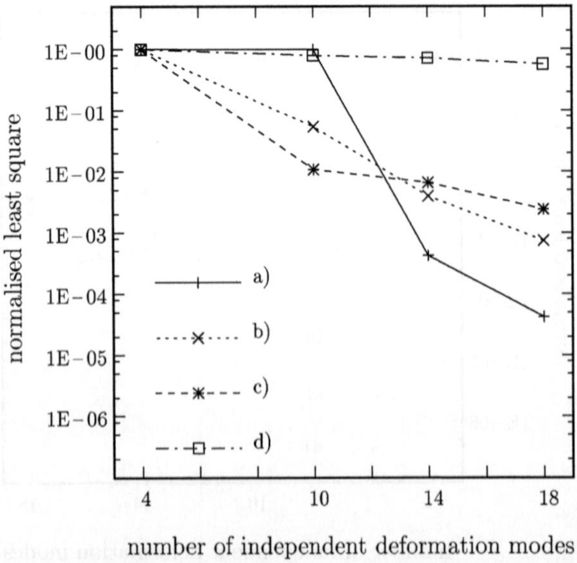

Fig. 11.7 Normalized least squares for several projection polynomials in the numerical bending experiment at position ④. Similar to position ③, the micromorphic and the cubic polynomials show a significantly higher accuracy for microstructure (*a*) than the linear and quadratic polynomials. The difference between the micromorphic and the cubic polynomials is less pronounced than at positions ① and ②. For (*b*) and (*c*), the fitting accuracy is obviously influenced by the microstructure's orientation

systems the microstructural degrees of freedom within the micromorphic continuum theory, i.e., the microrotation, the microdilatation, etc. loose their physical meaning. This result is in accordance to [16, 17] which have found similar properties identifying an effective Cosserat medium. Nevertheless, the micromorphic kinematics are able to reproduce the periodic 1-particle microstructure in a rather exact way, i.e., the deformation modes activated by the extended projection rule cover all essential deformation mechanisms of this particular microstructure. Furthermore, the second order homogenization methodology offers the fundamental advantage to reproduce size dependent boundary layer effects which has been verified in literature for several approaches, e.g., [9, 12, 13]. Thus, even assuming an arbitrary $(n > 1)$-particle system, the second order methodologies, the micromorphic as well as the second gradient approach, are a priori able to capture higher order effects. However, for the $(n > 1)$-particle case, we want to point out the present methodology to describe a phenomenological model beyond the straight interpretation based on the microtopology. Future investigations will focus on the polynomial coefficients found by the fitting technique. The question arises if one is able to restrict the full micromorphic medium to one of the subclasses proposed in [6] by an appropriate interpretation of those coefficients.

Acknowledgements The financial support by the Deutsche Forschungsgemeinschaft (DFG) under the grant DI 430/7-1 is gratefully acknowledged.

References

1. Cosserat, E., Cosserat, F.: Théorie des corps déformables. Hermann et Fils, Paris (1909)
2. Diebels, S., Steeb, H.: The size effect in foams and its theoretical and numerical investigation. Proc. R. Soc. Lond. A **458**, 2869–2883 (2002)
3. Eringen, A.C.: Microcontinuum Field Theories, vol. I: Foundations and Solids. Springer, New York (1999)
4. Feyel, F., Chaboche, J.L.: FE^2 multiscale approach for modelling the elastoviscoplastic behaviour of long fiber SiC/Ti composite materials. Comput. Methods Appl. Mech. Eng. **183**, 309–330 (2000)
5. Forest, S.: Homogenization methods and the mechanics of generalized continua—Part 2. Theor. Appl. Mech. **28**, 113–143 (2002)
6. Forest, S.: Nonlinear microstrain theories. Int. J. Solids Struct. **43**, 7224–7245 (2006)
7. Forest, S., Sab, K.: Cosserat overall modeling of heterogeneous materials. Mech. Res. Commun. **25**, 449–454 (1998)
8. Germain, P.: La méthode des puissances virtuelles en mécanique des milieux continus. Première partie: Théorie du second gradient. J. Mec. **12**, 235–274 (1973)
9. Jänicke, R., Diebels, S.: A numerical homogenisation strategy for micromorphic continua. Nuovo Cim. Soc. Ital. Fis. C **31**(1), 121–132 (2009)
10. Jänicke, R., Diebels, S., et al.: Two-scale modelling of micromorphic continua. Contin. Mech. Therm. **21**, 297–315 (2009)
11. Kouznetsova, V.G.: Computational homogenization for the multi-scale analysis of multi-phase material. PhD thesis, Technische Universiteit Eindhoven, The Netherlands (2002)
12. Kouznetsova, V.G., Geers, M.G.D., Brekelmans, W.A.M.: Size of a representative volume element in a second-order computational homogenization framework. Int. J. Multiscale Comput. Eng. **2**(4), 575–598 (2004)
13. Larsson, R., Diebels, S.: A second order homogenization procedure for multi-scale analysis based on micropolar kinematics. Int. J. Numer. Meth. Eng. **69**, 2485–2512 (2006)
14. Maugin, G.A.: Nonlocal theories or gradient-type theories: a matter of convenience? Acta Mater. **31**, 15–26 (1979)
15. Miehe, C., Koch, A.: Computational micro-to-macro transitions of discretized microstructures. Arch. Appl. Mech. **72**, 300–317 (2002)
16. Sab, K., Pradel, F.: Homogenisation of periodic Cosserat media. Int. J. Comput. Appl. Technol. **34**(1), 60–71 (2009)
17. Tekoglu, C., Onck, P.R.: Size effects in the mechanical behaviour of cellular materials. J. Math. Sci. **40**, 5911–5917 (2005)
18. Tekoglu, C., Onck, P.R.: Size effects in two-dimensional Voronoi foams: a comparison between generalized continua and discrete models. J. Mech. Phys. Solids **56**, 3541–3564 (2008)

Chapter 12
Extending Micromorphic Theory to Atomic Scale

James D. Lee, Youping Chen, and Xianqiao Wang

Abstract Micromorphic theory envisions a material body as a continuous collection of deformable particles; each possesses finite size and inner structure. It may be considered as the most successful top-down formulation of a two-level continuum model, in which the deformation is expressed as a sum of macroscopic continuous deformation and microscopic deformation of the inner structure. To enlarge the domain of applicability of the micromorphic theory, starting from many-body dynamics, we took a bottom-up approach to formulate a generalized continuum field theory in which a crystalline material is viewed as a continuous collection of lattice points while embedded within each lattice point is a group of discrete atoms. In this work, atomistic definitions and the corresponding field representations of fundamental physical quantities are introduced. The balance laws and the constitutive relations are obtained through the atomistic formulation, which naturally leads to a generalized continuum field theory. It is identical to molecular dynamics at atomic scale and can be reduced to classical continuum field theory at macroscopic scale.

12.1 Introduction to Micromorphic Theory

Microcontinuum field theories constitute extensions of the classical field theories concerned with deformations, motions, and electromagnetic interactions of material media, as continua, in microscopic space and short time scales. In terms of a

J.D. Lee (✉) · X. Wang
The George Washington University, Washington, DC, USA
e-mail: jdlee@gwu.edu

X. Wang
e-mail: xqwang@gwmail.gwu.edu

Y. Chen
University of Florida, Gainsevilla, FL, USA
e-mail: ypchen2@ufl.edu

G.A. Maugin, A.V. Metrikine (eds.), *Mechanics of Generalized Continua*, 109
Advances in Mechanics and Mathematics 21,
DOI 10.1007/978-1-4419-5695-8_12, © Springer Science+Business Media, LLC 2010

physical picture, a microcontinuum may be envisioned as a continuous collection of deformable point particles, each with finite size and inner structure. It is worthwhile to note that in the classical continuum theory a point particle is represented by a geometrical point, which is infinitesimal in size. Then the question arises: *How can one represent the intrinsic deformation of a point particle in microcontinuum*? Eringen settled this question by replacing the deformable particle with a geometric point P and some vectors attached to P, which denote the orientations and intrinsic deformations of all the material points in the deformable point particle. This is compatible with the classical picture where a material point in a continuum is endowed with physical properties such as mass density, displacement vector, electric field, stress tensor, etc. Therefore, the vectors assigned to P represent the additional degrees of freedom arising from the motions, relative to P, of all the material points in the particle. Geometrically, a particle P is identified by its position vector $\mathbf{X}$, in the reference (Lagrangian or material) state B, and vectors attached to P, representing the inner structure of P by Ξ^α $(\alpha = 1,2,3,\ldots,N)$ while N is the number of material points in the particle. The motions may be expressed as

$$\mathbf{x} = \mathbf{x}(\mathbf{X},t), \tag{12.1}$$

$$\xi^\alpha = \xi^\alpha(\mathbf{X},\Xi^\alpha,t) \quad \alpha \in [1,2,3,\ldots,N], \tag{12.2}$$

where t is the time; $\mathbf{x}$ and ξ^α, corresponding to $\mathbf{X}$ and Ξ^α, respectively, are the position vectors in the deformed (Eulerian or spatial) state b. A medium with such general motions is named microcontinuum of grade N by Eringen. In the two-level continuum model, let the position vector of a material point be decomposed as the sum of the position vector of the centroid (mass center) of the particle and the position vector of the material point relative to the centroid (cf. Fig. 1.1.2 of Eringen [6]), i.e.,

$$\mathbf{x}' = \mathbf{x} + \xi, \qquad \mathbf{X}' = \mathbf{X} + \Xi, \tag{12.3}$$

and let the motions be expressed as

$$\mathbf{x} = \mathbf{x}(\mathbf{X},t), \qquad \xi = \xi(\mathbf{X},\Xi,t). \tag{12.4}$$

If the micromotion $\xi = \xi(\mathbf{X},\Xi,t)$ is further reduced to an affine motion, i.e.,

$$\boldsymbol{\xi} = \chi_K(\mathbf{X},t)\Xi_K \quad \text{or} \quad \xi_k = \chi_{kK}(\mathbf{X},t)\Xi_K, \tag{12.5}$$

we arrive at the doorstep of the *micromorphic theory*. It is seen that the macromotion $\mathbf{x} = \mathbf{x}(\mathbf{X},t)$ accounts for the motion of the centroid of the particle; the micromotion $\xi_k = \chi_{kK}(\mathbf{X},t)\Xi_K$ accounts for the intrinsic motions of the particle; and χ_{kK} is called the microdeformation tensor. Because χ_{kK} is a second order tensor, the particle has 9 independent degrees of freedom in addition to the 3 classical translational degrees of freedom of the centroid. A unit cell or a polyatomic molecule may be viewed as a point particle in micromorphic theory [6, 7].

12.2 Balance Laws of Micromorphic Theory

The balance laws of micromorphic theory, namely conservation of mass, microinertia, energy and balance of linear momentum and momentum moments, were first derived by Eringen and Suhubi [8, 11] and Eringen [5] by means of a "microscopic space-averaging" process. Later Eringen [6] derived the balance laws in a more elegant way: balance of linear momentum and momentum moments are the consequences of the objectivity of conservation of energy:

$$\frac{\mathrm{d}\rho}{\mathrm{d}t} + \rho\nabla_{\mathbf{x}} \cdot \mathbf{v} = 0, \tag{12.6}$$

$$\rho\frac{\mathrm{d}\mathbf{i}}{\mathrm{d}t} = \rho\boldsymbol{\varphi} + \rho\boldsymbol{\varphi}^T, \tag{12.7}$$

$$\rho\frac{\mathrm{d}\mathbf{v}}{\mathrm{d}t} = \nabla_{\mathbf{x}} \cdot \mathbf{t} + \mathbf{f}, \tag{12.8}$$

$$\rho\frac{\mathrm{d}\boldsymbol{\varphi}}{\mathrm{d}t} = \nabla_{\mathbf{x}} \cdot \mathbf{m} + \mathbf{t}^T - \mathbf{s} + \boldsymbol{\upsilon} \cdot \rho\mathbf{i} \cdot \boldsymbol{\upsilon}^T + \mathbf{l}, \tag{12.9}$$

$$\rho\frac{De}{Dt} = \mathbf{m} \,\dot{:}\, \nabla_{\mathbf{x}} \otimes \boldsymbol{\upsilon} + \mathbf{t} : \nabla_{\mathbf{x}} \otimes \mathbf{v} + (\mathbf{s} - \mathbf{t})^T : \boldsymbol{\upsilon} + \nabla_{\mathbf{x}} \cdot \mathbf{q} + h, \tag{12.10}$$

where ρ is the mass density; $\rho\mathbf{i}$ is the microinertia tensor; $\rho\boldsymbol{\varphi} = \boldsymbol{\upsilon} \cdot \rho\mathbf{i}$ is the generalized spin tensor; $\mathbf{t}$ is the Cauchy stress; $\mathbf{s} = \mathbf{s}^T$ is the microstress; $\mathbf{m}$ is a third order tensor, called the moment stress; $\mathbf{v}$ is the velocity; $\boldsymbol{\upsilon}$ is a second order tensor, called the gyration tensor; $\mathbf{l}$ is a second order tensor, called the body couple density; e is the internal energy density; $\mathbf{q}$ is the heat flux; and h is the heat source.

To demonstrate that the micromorphic theory can be built on a more profound physical foundation, in this work, we follow the approach of Hardy [9] to link variables between the phase space and the physical space of a many-body system and to derive the balance laws for microcontinuum. Microscopic dynamic quantities in many-body dynamics are functions of phase-space coordinates $(\mathbf{r}, \mathbf{p})$, i.e., the positions and momenta of atoms. For a single crystal of a multi-element system

$$\mathbf{r} = \left\{\mathbf{R}^{k\alpha} = \mathbf{R}^k + \Delta\mathbf{r}^{k\alpha} \mid k = 1, 2, 3, \ldots, N_l; \alpha = 1, 2, 3, \ldots, N_a\right\}, \tag{12.11}$$

$$\mathbf{p} = \left\{m^\alpha\mathbf{V}^{k\alpha} = m^\alpha\left(\mathbf{V}^k + \Delta\mathbf{v}^{k\alpha}\right) \mid k = 1, 2, 3, \ldots, N_l; \; \alpha = 1, 2, 3, \ldots, N_a\right\}, \tag{12.12}$$

where N_l is the total number of unit cells in the system; N_a is the number of atoms in a unit cell; the superscript $k\alpha$ refers to the αth atom in the kth unit cell; m^α is the mass of the αth atom; $\mathbf{R}^{k\alpha}$ and $\mathbf{V}^{k\alpha}$ are the position and velocity of the $k\alpha$th atom, respectively; $\mathbf{R}^k$ and $\mathbf{V}^k$ are the position and velocity of the centroid of the kth unit cell, respectively; $\Delta\mathbf{r}^{k\alpha}$ and $\Delta\mathbf{v}^{k\alpha}$ are the position and velocity of the $k\alpha$th atom relative to the centroid of the kth unit cell, respectively.

A dynamic function $\mathbf{A}(\mathbf{r}, \mathbf{p})$ in phase space can be linked to its corresponding local density function $\mathbf{a}(\mathbf{x}, t)$ in physical space as

$$\mathbf{a}(\mathbf{x},t) = \sum_{k=1}^{N_l}\sum_{\alpha=1}^{N_a} \mathbf{A}(\mathbf{r},\mathbf{p})\delta(\mathbf{R}^k - \mathbf{x}), \tag{12.13}$$

where the δ-function can be any localization function with $\int_{\Omega(\mathbf{x})} \delta(\mathbf{R}^k - \mathbf{x})\, \mathrm{d}\mathbf{x} = 1$ [9]. With the properties of the δ-function [10], the time evolution of physical quantities can be obtained as

$$\begin{aligned}\left.\frac{\partial \mathbf{a}(\mathbf{x},t)}{\partial t}\right|_{\mathbf{x}} &= \sum_{k=1}^{N_l}\sum_{\alpha=1}^{N_a} \dot{\mathbf{A}}(\mathbf{r},\mathbf{p})\delta(\mathbf{R}^k - \mathbf{x}) \\ &\quad - \nabla_{\mathbf{x}} \cdot \sum_{k=1}^{N_l}\sum_{\alpha=1}^{N_a} \mathbf{V}^k \otimes \mathbf{A}(\mathbf{r},\mathbf{p})\delta(\mathbf{R}^k - \mathbf{x}). \end{aligned} \tag{12.14}$$

When $\mathbf{a}(\mathbf{x},t)$ is the local density of a conserved quantity, (12.14) is the corresponding microscopic balance law. The conserved quantities in micromorphic theory are

$$\rho = \sum_{k=1}^{N_l}\sum_{\alpha=1}^{N_a} m^\alpha \delta(\mathbf{R}^k - \mathbf{x}) = \sum_{k=1}^{N_l} m\delta(\mathbf{R}^k - \mathbf{x}), \tag{12.15}$$

$$\rho\mathbf{i} = \sum_{k=1}^{N_l}\sum_{\alpha=1}^{N_a} m^\alpha \Delta\mathbf{r}^{k\alpha} \times \Delta\mathbf{r}^{k\alpha}\delta(\mathbf{R}^k - \mathbf{x}), \tag{12.16}$$

$$\rho\mathbf{v} = \sum_{k=1}^{N_l}\sum_{\alpha=1}^{N_a} m^\alpha \mathbf{V}^{k\alpha}\delta(\mathbf{R}^k - \mathbf{x}) = \sum_{k=1}^{N_l} m\mathbf{V}^k\delta(\mathbf{R}^k - \mathbf{x}), \tag{12.17}$$

$$\rho\boldsymbol{\varphi} = \sum_{k=1}^{N_l}\sum_{\alpha=1}^{N_a} m^\alpha \Delta\mathbf{v}^{k\alpha} \times \Delta\mathbf{r}^{k\alpha}\delta(\mathbf{R}^k - \mathbf{x}), \tag{12.18}$$

$$\rho E = \sum_{k=1}^{N_l}\left\{\frac{1}{2}m\left(\mathbf{V}^k\right)^2 + \sum_{\alpha=1}^{N_a}\left[\frac{1}{2}m^\alpha\left(\boldsymbol{\upsilon}^k \cdot \Delta\mathbf{r}^{k\alpha}\right)^2 + U^{k\alpha}\right]\right\}\delta(\mathbf{R}^k - \mathbf{x}). \tag{12.19}$$

Notice that the micromotion of micromorphic material is affine, i.e., $\Delta\mathbf{v}^{k\alpha} = \boldsymbol{\upsilon}^k \cdot \Delta\mathbf{r}^{k\alpha}$, we obtained almost the same set of balance laws except

$$\rho\frac{\mathrm{d}\mathbf{i}}{\mathrm{d}t} = \rho\boldsymbol{\varphi} + \rho\boldsymbol{\varphi}^T - \nabla_{\mathbf{x}} \cdot \boldsymbol{\gamma}, \tag{12.20}$$

$$\rho\frac{\mathrm{d}\boldsymbol{\varphi}}{\mathrm{d}t} = \nabla_{\mathbf{x}} \cdot \mathbf{m} + \mathbf{t}^T - \mathbf{s} + \boldsymbol{\upsilon} \cdot \rho\mathbf{i} \cdot \boldsymbol{\upsilon}^T + \mathbf{l} - \nabla_{\mathbf{x}} \cdot (\boldsymbol{\upsilon} \cdot \boldsymbol{\gamma}), \tag{12.21}$$

where

$$\boldsymbol{\gamma} = \sum_{k=1}^{N_l}(\mathbf{V}^k - \mathbf{v}) \otimes \sum_{\alpha=1}^{N_a} m^\alpha \Delta\mathbf{r}^{k\alpha} \otimes \Delta\mathbf{r}^{k\alpha}\delta(\mathbf{R}^k - \mathbf{x}). \tag{12.22}$$

It is seen that $\mathbf{V}^k - \mathbf{v}$ is the difference between phase space velocity and physical space velocity and it is easy to understand why γ does not appear in the "microscopic space-averaging" process. The detailed expressions of $\mathbf{m}, \mathbf{t}, \mathbf{s}, \mathbf{q}, \mathbf{l}$ and h are referred to Chen and Lee [2, 3].

12.3 Extension of Micromorphic Theory to Atomic Scale

In the process of constructing a multiscale concurrent atomistic/continuum theory, to keep the knowledge and information at the atomistic level as much as possible, we now relax the assumption of affine motion and let the motions be back to the generality as indicated in (12.1), (12.2). Now the link between a dynamic function in phase space and its corresponding local density function in physical space can be established through a localization function and the Kronecker δ-function as [1]

$$\mathbf{a}(\mathbf{x}, \mathbf{y}^\alpha, t) = \sum_{k=1}^{N_l} \sum_{\xi=1}^{N_a} \mathbf{A}(\mathbf{r}, \mathbf{p}) \delta(\mathbf{R}^k - \mathbf{x}) \tilde{\delta}(\Delta \mathbf{r}^{k\xi} - \mathbf{y}^\alpha), \tag{12.23}$$

with

$$\tilde{\delta}(\Delta \mathbf{r}^{k\xi} - \mathbf{y}^\alpha) \equiv \begin{cases} 1, & \text{if } \xi = \alpha \text{ and } \Delta \mathbf{r}^{k\alpha} = \mathbf{y}^\alpha, \\ 0, & \text{otherwise.} \end{cases} \tag{12.24}$$

Similarly, the time evolution of a physical quantity can be obtained as:

$$\begin{aligned} \left.\frac{\partial \mathbf{a}(\mathbf{x}, \mathbf{y}^\alpha, t)}{\partial t}\right|_{\mathbf{x}, \mathbf{y}^\alpha} &= \sum_{k=1}^{N_l} \sum_{\xi=1}^{N_a} \dot{\mathbf{A}}(\mathbf{r}, \mathbf{p}) \delta(\mathbf{R}^k - \mathbf{x}) \tilde{\delta}(\Delta \mathbf{r}^{k\xi} - \mathbf{y}^\alpha) \\ &\quad - \nabla_{\mathbf{x}} \cdot \sum_{k=1}^{N_l} \sum_{\xi=1}^{N_a} \mathbf{V}^k \otimes \mathbf{A}(\mathbf{r}, \mathbf{p}) \delta(\mathbf{R}^k - \mathbf{x}) \tilde{\delta}(\Delta \mathbf{r}^{k\xi} - \mathbf{y}^\alpha) \\ &\quad - \nabla_{\mathbf{y}^\alpha} \cdot \sum_{k=1}^{N_l} \sum_{\xi=1}^{N_a} \Delta \mathbf{v}^{k\xi} \otimes \mathbf{A}(\mathbf{r}, \mathbf{p}) \delta(\mathbf{R}^k - \mathbf{x}) \tilde{\delta}(\Delta \mathbf{r}^{k\xi} - \mathbf{y}^\alpha). \end{aligned} \tag{12.25}$$

Following the pattern of (12.23), the mass density ρ^α, linear momentum density $\rho^\alpha(\mathbf{v} + \Delta \mathbf{v}^\alpha)$, angular momentum density $\rho^\alpha \boldsymbol{\psi}^\alpha$, internal energy density $\rho^\alpha e^\alpha$, interatomic force density $\mathbf{f}^\alpha$, external force density $\mathbf{f}^\alpha_{\text{ext}}$, the homogeneous part $\mathbf{t}^\alpha$ and inhomogeneous part $\boldsymbol{\tau}^\alpha$ of stress tensor, the homogeneous part $\mathbf{q}^\alpha$ and inhomogeneous part $\mathbf{j}^\alpha$ of heat flux, and heat source h are defined as

$$\rho^\alpha = \sum_{k=1}^{N_l} \sum_{\xi=1}^{N_a} m^\xi \delta(\mathbf{R}^k - \mathbf{x}) \tilde{\delta}(\Delta \mathbf{r}^{k\xi} - \mathbf{y}^\alpha), \tag{12.26}$$

$$\rho^\alpha(\mathbf{v}+\Delta\mathbf{v}^\alpha)=\sum_{k=1}^{N_l}\sum_{\xi=1}^{N_a}m^\xi(\mathbf{V}^k+\Delta\mathbf{v}^{k\xi})\delta(\mathbf{R}^k-\mathbf{x})\tilde{\delta}(\Delta\mathbf{r}^{k\xi}-\mathbf{y}^\alpha),\quad(12.27)$$

$$\rho^\alpha\boldsymbol{\psi}^\alpha=\sum_{k=1}^{N_l}\sum_{\xi=1}^{N_a}m^\xi\mathbf{V}^{k\xi}\times\mathbf{R}^{k\xi}\delta(\mathbf{R}^k-\mathbf{x})\tilde{\delta}(\Delta\mathbf{r}^{k\xi}-\mathbf{y}^\alpha),\quad(12.28)$$

$$\rho^\alpha e^\alpha=\sum_{k=1}^{N_l}\sum_{\xi=1}^{N_a}\left[\frac{1}{2}m^\xi(\widetilde{\mathbf{V}}^{k\xi})^2+U^{k\xi}\right]\delta(\mathbf{R}^k-\mathbf{x})\tilde{\delta}(\Delta\mathbf{r}^{k\xi}-\mathbf{y}^\alpha),\quad(12.29)$$

$$\mathbf{f}^\alpha=\sum_{k=1}^{N_l}\sum_{\xi=1}^{N_a}\mathbf{F}^{k\xi}\delta(\mathbf{R}^k-\mathbf{x})\tilde{\delta}(\Delta\mathbf{r}^{k\xi}-\mathbf{y}^\alpha),\quad(12.30)$$

$$\mathbf{f}^\alpha_{\text{ext}}=\sum_{k=1}^{N_l}\sum_{\xi=1}^{N_a}\mathbf{F}^{k\xi}_{\text{ext}}\delta(\mathbf{R}^k-\mathbf{x})\tilde{\delta}(\Delta\mathbf{r}^{k\xi}-\mathbf{y}^\alpha),\quad(12.31)$$

$$\begin{aligned}\mathbf{t}^\alpha=&-\sum_{k=1}^{N_l}\sum_{\xi=1}^{N_a}m^\xi\widetilde{\mathbf{V}}^k\otimes\widetilde{\mathbf{V}}^{k\xi}\delta(\mathbf{R}^k-\mathbf{x})\tilde{\delta}(\Delta\mathbf{r}^{k\xi}-\mathbf{y}^\alpha)\\&-\frac{1}{2}\sum_{k,l=1}^{N_l}\sum_{\xi,\eta=1}^{N_a}(\mathbf{R}^k-\mathbf{R}^l)\otimes\mathbf{F}^{k\xi}B(k,\xi,l,\eta,\mathbf{x},\mathbf{y}^\alpha),\end{aligned}\quad(12.32)$$

$$\begin{aligned}\boldsymbol{\tau}^\alpha=&-\sum_{k=1}^{N_l}\sum_{\xi=1}^{N_a}m^\xi\Delta\tilde{\mathbf{v}}^{k\xi}\otimes\widetilde{\mathbf{V}}^{k\xi}\delta(\mathbf{R}^k-\mathbf{x})\tilde{\delta}(\Delta\mathbf{r}^{k\xi}-\mathbf{y}^\alpha)\\&-\frac{1}{2}\sum_{k,l=1}^{N_l}\sum_{\xi,\eta=1}^{N_a}(\Delta\mathbf{r}^{k\xi}-\Delta\mathbf{r}^{l\eta})\otimes\mathbf{F}^{k\xi}B(k,\xi,l,\eta,\mathbf{x},\mathbf{y}^\alpha),\end{aligned}\quad(12.33)$$

$$\begin{aligned}\mathbf{q}^\alpha=&-\sum_{k=1}^{N_l}\sum_{\xi=1}^{N_a}\widetilde{\mathbf{V}}^k\left[\frac{1}{2}m^\xi(\widetilde{\mathbf{V}}^{k\xi})^2+U^{k\xi}\right]\delta(\mathbf{R}^k-\mathbf{x})\tilde{\delta}(\Delta\mathbf{r}^{k\xi}-\mathbf{y}^\alpha)\\&-\frac{1}{2}\sum_{k,l=1}^{N_l}\sum_{\xi,\eta=1}^{N_a}(\mathbf{R}^k-\mathbf{R}^l)\widetilde{\mathbf{V}}^{k\xi}\cdot\mathbf{F}^{k\xi}B(k,\xi,l,\eta,\mathbf{x},\mathbf{y}^\alpha),\end{aligned}\quad(12.34)$$

$$\begin{aligned}\mathbf{j}^\alpha=&-\sum_{k=1}^{N_l}\sum_{\xi=1}^{N_a}\Delta\tilde{\mathbf{v}}^{k\xi}\left[\frac{1}{2}m^\xi(\widetilde{\mathbf{V}}^{k\xi})^2+U^{k\xi}\right]\delta(\mathbf{R}^k-\mathbf{x})\tilde{\delta}(\Delta\mathbf{r}^{k\xi}-\mathbf{y}^\alpha)\\&-\frac{1}{2}\sum_{k,l=1}^{N_l}\sum_{\xi,\eta=1}^{N_a}(\Delta\mathbf{r}^{k\xi}-\Delta\mathbf{r}^{l\eta})\widetilde{\mathbf{V}}^{k\xi}\cdot\mathbf{F}^{k\xi}B(k,\xi,l,\eta,\mathbf{x},\mathbf{y}^\alpha),\end{aligned}\quad(12.35)$$

$$h^\alpha=\sum_{k=1}^{N_l}\sum_{\xi=1}^{N_a}\widetilde{\mathbf{V}}^{k\xi}\cdot\mathbf{F}^{k\xi}_{\text{ext}}\delta(\mathbf{R}^k-\mathbf{x})\tilde{\delta}(\Delta\mathbf{r}^{k\xi}-\mathbf{y}^\alpha),\quad(12.36)$$

where $\widetilde{\mathbf{V}}^{k\xi} = \mathbf{V}^{k\xi} - (\mathbf{v} + \Delta\mathbf{v}^{\xi})$, $\widetilde{\mathbf{V}}^{k} = \mathbf{V}^{k} - \mathbf{v}$ and $\Delta\widetilde{\mathbf{v}}^{k\xi} = \Delta\mathbf{v}^{k\xi} - \Delta\mathbf{v}^{\xi}$ are the differences between phase space velocities and physical space velocities; $\mathbf{F}^{k\xi}_{\text{ext}}$ is the body force, such as gravitational force and Lorentz force, acting on the $k\xi$th atom; $U^{k\xi}$ is the potential energy of the $k\xi$th atom; $\mathbf{F}^{k\xi}$ is the interatomic force acting on the $k\xi$th atom with the understanding that the total potential energy U is additive, i.e.,

$$U = \sum_{k=1}^{N_l} \sum_{\xi=1}^{N_a} U^{k\xi}, \qquad \mathbf{F}^{k\xi} = -\frac{\partial U}{\partial \mathbf{R}^{k\xi}} = -\frac{\partial U^{k\xi}}{\partial \mathbf{R}^{k\xi}} \tag{12.37}$$

and $B(k, \xi, l, \eta, \mathbf{x}, \mathbf{y}^{\alpha})$ is defined as

$$B \equiv \int_0^1 \delta(\mathbf{R}^k\lambda + \mathbf{R}^l(1-\lambda) - \mathbf{x})\tilde{\delta}(\Delta\mathbf{r}^{k\xi}\lambda + \Delta\mathbf{r}^{l\eta}(1-\lambda) - \mathbf{y}^{\alpha})\,\mathrm{d}\lambda. \tag{12.38}$$

Based on (12.25), a lengthy but straightforward process leads to the local balance laws of mass, linear momentum, angular momentum, and energy for each atom $\alpha \in [1, 2, 3, \ldots, N_a]$ at any point in the field $(\mathbf{x}, t)$ as [4]

$$\frac{\mathrm{d}\rho^{\alpha}}{\mathrm{d}t} + \rho^{\alpha}\nabla_{\mathbf{x}} \cdot \mathbf{v} + \rho^{\alpha}\nabla_{\mathbf{y}^{\alpha}} \cdot \Delta\mathbf{v}^{\alpha} = 0, \tag{12.39}$$

$$\rho^{\alpha}\frac{\mathrm{d}(\mathbf{v} + \Delta\mathbf{v}^{\alpha})}{\mathrm{d}t} = \nabla_{\mathbf{x}} \cdot \mathbf{t}^{\alpha} + \nabla_{\mathbf{y}^{\alpha}} \cdot \boldsymbol{\tau}^{\alpha}, + \mathbf{f}^{\alpha}_{\text{ext}} \tag{12.40}$$

$$\mathbf{t}^{\alpha} + \boldsymbol{\tau}^{\alpha} = (\mathbf{t}^{\alpha} + \boldsymbol{\tau}^{\alpha})^T, \tag{12.41}$$

$$\rho^{\alpha}\frac{\mathrm{d}e^{\alpha}}{\mathrm{d}t} = \mathbf{t}^{\alpha} : \nabla_{\mathbf{x}}(\mathbf{v} + \Delta\mathbf{v}^{\alpha}) + \boldsymbol{\tau}^{\alpha} : \nabla_{\mathbf{y}^{\alpha}}(\mathbf{v} + \Delta\mathbf{v}^{\alpha}) + \nabla_{\mathbf{x}} \cdot \mathbf{q}^{\alpha} + \nabla_{\mathbf{y}^{\alpha}} \cdot \mathbf{j}^{\alpha} + h^{\alpha}, \tag{12.42}$$

where the material time-rate of A^{α} is defined as

$$\frac{\mathrm{d}A^{\alpha}}{\mathrm{d}t} \equiv \frac{\partial A^{\alpha}}{\partial t} + \mathbf{v} \cdot \nabla_{\mathbf{x}} A^{\alpha} + \Delta\mathbf{v}^{\alpha} \cdot \nabla_{\mathbf{y}^{\alpha}} A^{\alpha}. \tag{12.43}$$

Similar to the situation in classical field theory, the balance of angular momentum leads to the symmetry of stress tensor $\mathbf{t}^{\alpha} + \boldsymbol{\tau}^{\alpha}$. It is seen that, from (12.32), (12.33), the symmetry of the stress tensor is automatically satisfied.

12.4 Discussions

If the point particle in micromorphic theory is reduced to a geometrical point, i.e., infinitesimal in size, then one can easily verify that $\mathbf{m} = \mathbf{l} = \boldsymbol{\upsilon} = \mathbf{i} = \boldsymbol{\varphi} = 0$, $\mathbf{t} = \mathbf{s} = \mathbf{t}^T$, and the balance laws of micromorphic theory become exactly the same as in classical continuum theory.

Each point particle in micromorphic theory has 12 degrees of freedom, i.e., 3 translational degrees of freedom for the centroid and 9 degrees of freedom for the micromotion, (12.5). In this sense, each unit cell in the multiscale concurrent atomistic/continuum theory has $3N_a$ degrees of freedom, where N_a is the number of atoms in a unit cell. The multiscale theory is a theory in physical space $(\mathbf{x}, t)$; (12.38), (12.39), (12.41) are the balance laws of mass, linear momentum, and energy for each and every atom in the unit cell; unlike classical continuum theories, the constitutive equations for the internal energy density $\rho^\alpha e^\alpha$, stress tensors $\mathbf{t}^\alpha$ and $\boldsymbol{\tau}^\alpha$, heat fluxes $\mathbf{q}^\alpha$ and $\mathbf{j}^\alpha$, (12.29), (12.32)–(12.35), for each and every atom are explicitly obtained; the only constitutive relation needed is the interatomic potential energy $U^{k\xi}$. In other words, although this theory that we derived is a field theory, we never lose the sight of atoms. This is the fundamental difference between this work and many other multiscale theories.

It is worthwhile to note that, from (12.29), the summation of internal energy density over all the atoms in a unit cell gives the internal energy per volume of a unit cell as

$$\sum_{\alpha=1}^{N_a} \rho^\alpha e^\alpha = \sum_{k=1}^{N_l}\sum_{\alpha=1}^{N_a}\left[\frac{1}{2}m^\alpha (\tilde{\mathbf{V}}^{k\alpha})^2 + U^{k\alpha}\right]\delta(\mathbf{R}^k - \mathbf{x})$$
$$= \frac{3}{2}\frac{k_B}{V} T(\mathbf{x}, t) + U_{\text{pot}}(\mathbf{x}, t), \tag{12.44}$$

where k_B is the Boltzmann constant; $T(\mathbf{x}, t)$ is the temperature; $U_{\text{pot}}(\mathbf{x}, t)$ is the potential energy density; and V is the volume of a unit cell. Then it is seen that the summation of (12.41) essentially emerges as the governing equation for temperature.

References

1. Chen, Y.: Reformulation of microscopic balance equations for multiscale materials modeling. J. Chem. Phys. **130**(13), 134706–134706-6 (2009)
2. Chen, Y., Lee, J.D.: Connecting molecular dynamics to micromorphic theory, Part I: Instantaneous mechanical variables. Physica A **322**, 359–376 (2003)
3. Chen, Y., Lee, J.D.: Connecting molecular dynamics to micromorphic theory, Part II: Balance laws. Physica A **322**, 377–392 (2003)
4. Chen, Y., Lee, J.D.: Atomistic formulation of a multiscale theory for nano/micro physics. Philos. Mag. **85**, 4095–4126 (2005)
5. Eringen, A.C.: Simple microfluids. Int. J. Eng. Sci. **2**(2), 205–217 (1964)
6. Eringen, A.C.: Microcontinuum Field Theories—I: Foundations and Solids. Springer, New York (1999)
7. Eringen, A.C.: Microcontinuum Field Theories—II: Fluent Media. Springer, New York (2001)
8. Eringen, A.C., Suhubi, E.S.: Nonlinear theory of simple micro-elastic solids. Int. J. Eng. Sci. **2**(2), 189–203 (1964)

9. Hardy, R.J.: Formulas for determining local properties in molecular-dynamics simulations: shock waves. J. Chem. Phys. **76**(1), 622–628 (1982)
10. McLennan, J.A.: Introduction to Nonequilibrium Statistical Mechanics. Prentice Hall, New Jersey (1989)
11. Suhubi, E.S., Eringen, A.C.: Nonlinear theory of simple micro-elastic solids. II. Int. J. Eng. Sci. **2**(4), 389–404 (1964)

Part IV
From the Discrete to the Continuum Description (Cosserat and Other Continua Often in Relation to Dynamical Properties, Homogenization)

Part IV

From the Discrete to the Continuum Description (Cosserat and Other Continua Often in Relation to Dynamical Properties, Homogenization)

Chapter 13
Nonlinear Theory of Cardinal Rearrangement of the Solid Body Structure in the Field of Intensive Pressure

Eron L. Aero and A.N. Bulygin

Abstract A nonlinear theory of microscopic and macroscopic strains is developed for the case of large inhomogeneous relative displacements of two sublattices making up a complex crystal lattice. The standard linear theory of acoustic and optical oscillations of a complex lattice is generalized, taking into account new additive principle of internal translational symmetry—relative shear of two sublattices leaving deformation energy invariant. As a result, the force interaction between the sublattices is characterized by a nonlinear periodic force of its mutual displacements. The theory describes large microdisplacements due to bifurcation transitions of atoms into neighboring cells. As a result, the theory predicts defect formations, switching interatomic bonds, phase transitions, formation of nanoclasters, etc. Some examples of resolutions of nonlinear equations of equilibrium are presented.

13.1 Introduction

Now considerable attention is called to the problem of structural and phase transitions in nano-structured materials, degradation of material properties under loading (ageing and fatigue) and to with it connected processes of generations of defects of the structure and other damages of it. A well-developed approach to the solution of the problem is based on an artificial introduction of concrete, previously designed elements of a damage of a crystalline structure with subsequent monitoring of their evolution under intensive power and temperature influence. A more universal approach, following a remarkable work of Cosserats [7], is based on an introduction of

E.L. Aero (✉) · A.N. Bulygin
Institute of Problems in Mechanical Engineering, Bolshoy av. 61, V.O., St. Petersburg 199178, Russia
e-mail: 16aero@mail.ru

A.N. Bulygin
e-mail: bulygin_an@mail.ru

G.A. Maugin, A.V. Metrikine (eds.), *Mechanics of Generalized Continua*, Advances in Mechanics and Mathematics 21,
DOI 10.1007/978-1-4419-5695-8_13, © Springer Science+Business Media, LLC 2010

internal degrees of freedom into a continuum model of a solid. Numerous attempts of its realizations revealed new effects in the framework of linear theory. Main new results concern an appearance of new optical oscillations, spatial–temporal dispersion of elastic features and border effects in statics.

However, this approach is not adequate enough to the new problems arising while studying the formation and control of the structure of new materials. Small variations in the internal structure described by the linear theories, simple precise variations in the macroscopic geometry of the lattice, giving only re-normalization of material constants. Hence a model of a complex structure is reduced to that of a simple one. The cardinal rearrangements of the structure appear beyond the linear approach. The transition to the essentially nonlinear equations yields a possibility to predict drastic structural rearrangements, lowering of the potential barriers, switching of interatomic connections, arising of singular defects and other damages, phase transitions.

There is a limitation of a phenomenological theory. Indeed, internal parameters, like micro-strains describing structural deviations, may be incorporated into the theory, they could be even estimated. However, their physical sense might be understood provided that an initial structural state of the body is defined, and material scales of length and time are introduced. The last is skipped in the continua theory.

It turns out that simultaneous description of discreteness and nonlinearity of the model allows us to formulate a new principle of translation invariance of the energy. A similar principle was introduced for the first time by E. and F. Cosserat for a medium with rotational degrees of freedom. A model of complex lattice may be used as a base of a suggested theory. The complex lattice consists of two sub-lattices which coincide or merge into one by a shift for a constant structural vector $\mathbf{u}_0$, appearing as a parameter of the complex lattice (Fig. 13.1). The model is known in the physics of solid state, however, it was developed in a linear and anharmonic limit. Two equations arise in the linear theory of crystalline lattice of Karman and Born-Huang [6], one for the acoustic ($\mathbf{U}$) and the other for the optical ($\mathbf{u}$) displacement. The physical mechanics of non-ideal crystalline lattice with defects has been developed in the works of Kosevich [8]. A linear nonlocal theory of complex crystalline lattice was developed by Kunin [9] where a long-range action was refined.

Here, in the local nonlinear theory [1, 3, 2, 4, 5, 10], the main attention is paid to the effects of a short-range action responsible for the cardinal variations in the structure, in particular, for the generation of defects, new phases and the so-called reconstructive transitions or changing of the class of lattice symmetry. The employment of essentially nonlinear equations gives rise to predicting the lowering of potential barriers and switching of interatomic connections.

The introduction of variations of a local topology in the theory by means of internal degrees of freedom ($\mathbf{u}$) turns out efficient, provided that the generation of the linear approach is done as follows. Consider arbitrarily large displacements of sub-lattices $\mathbf{u}$ and put an additional element of translational symmetry, typical for complex lattices but have not used before in physics of solid state. Certainly, the relative displacement of one sub-lattice for one period (or for its integer number) yields a merge of the sub-lattice with itself, and the structure of the complex lattice is reproduced. It means that its energy should be a periodic function of the

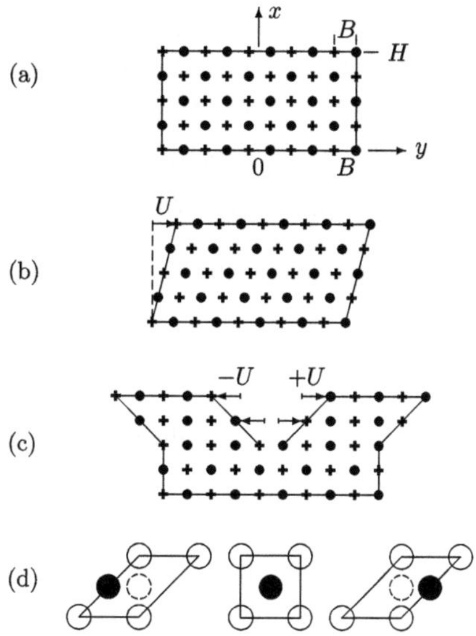

Fig. 13.1 (**a**) The complex crystalline lattice consisting of two sublattices. (**b**) Macroscopic deformation without a relative shear of sublattices. (**c**) Microdeformations at twinning. (**d**) Bifurcation of the structure of an elementary cell during microdeformation

relatively tough displacements of sub-lattices **u** which are invariant to such a translation. Certainly, a classical principle of translational symmetry is kept, that yields an invariance of the energy of the lattice to the shared translation **U** of both sub-lattices for one period of the complex lattice. The approach allows us to introduce new parameters of a crystal in micromechanics, that makes possible a description of micromechanisms of the cardinal rearrangements of the lattice such as characteristics of the short-range order, potential barriers, typical sizes of the elements of structure and interphase boundaries, bifurcation parameters.

13.2 General Equations

Let us introduce a displacement **U** of the center of inertia of a pair of atoms (elementary cell) and their relative displacement u inside the cell (due to variation of $\mathbf{u}_0$) as follows

$$\mathbf{U} = \frac{m_1\mathbf{U}_1 + m_2\mathbf{U}_2}{(m_1 + m_2)}, \qquad \mathbf{u} = \frac{\mathbf{U}_1 - \mathbf{U}_2}{a}. \tag{13.1}$$

Here $\mathbf{U}_1$ and $\mathbf{U}_2$ are the displacements of atoms with masses m_1 and m_2 of the first and the second sub-lattices, respectively, a is the period of a sub-lattice.

The development of a three-dimensional theory requires explicit introduction of an energy invariant both to the mutual displacements and rotations of sub-lattices. The basic idea is in the development of a scalar periodic function of the simplest rotation invariants of the vector field. This might be done by choosing it as

$$\Phi \to \left(1 - \cos\sqrt{u_i \alpha_{ik} u_k}\right), \quad \alpha_{ik} = a_1^{-2} k_i k_i + a_2^{-2} m_i m_i + a_3^{-2} n_i n_i, \tag{13.2}$$

where α_{ik} is a tensor of inverse periods of the lattice (a_1, a_2, a_3), while **k**, **m**, **n** are the unit vectors of crystallographic directions.

The force of interaction of neighboring atoms reads

$$P_i = \frac{\partial \Phi}{\partial u_i} - \frac{\partial \Phi}{\partial u_R}\frac{\partial u_R}{\partial u_i} - \frac{\partial u_R}{\partial u_i}\sin(u_R), \quad u_R = (u_i \alpha_{ik} u_k)^{1/2}. \tag{13.3}$$

Simpler relationships arise for the crystals of cubic symmetry ($a_1 = a_2 = a_3 - a$),

$$\begin{aligned} &\alpha_{ik} = a^{-2}(k_i k_i + m_i m_i + n_i n_i), \qquad u_R^2 = u^2; \\ &P_i = l_i \sin(u), \quad l_i = u_i/u, \\ &u = \left(u_x^2 + u_y^2 + u_z^2\right)^{1/2}. \end{aligned} \tag{13.4}$$

Here u is the absolute value of the vector of micro-displacements, l_i is its basic vector.

It is easy to see that both functions are periodic along the directions **k**, **m**, **n** with periods equal to a_1, a_2, a_3, respectively.

General equations for macro- and micro-displacements $\mathbf{U}(t, x, y, z)$, $\mathbf{u}(t, x, y, z)$ follow using the Lagrangian

$$\begin{aligned} L = \int_0^t \int_V \Big[&\frac{1}{2}\rho \dot{U}_n \dot{U}_n + \frac{1}{2}\mu \dot{u}_n \dot{u}_n - \frac{1}{2}\lambda_{ikmn} U_{(i,k)} U_{(m,n)} \\ &- \frac{1}{2} K_{ikmn} u_{i,k} u_{m,n} + s_{ik} U_{(i,k)} \Phi(u_R) - p\Phi(u_R) \Big] \mathrm{d}V \end{aligned} \tag{13.5}$$

and they read (due to the variations of $\dot{U}_i$, $\dot{u}_i$, $U_{(i,k)}$, $u_{i,k}$, u_i)

$$\rho \ddot{U}_i = \lambda_{ikmn} U_{(km),n} - s_{in}\left[\Phi(u_R)\right]_{,n}, \tag{13.6}$$

$$\begin{aligned} &\mu \ddot{u}_i = K_{ikmn} u_{k,mn} - P l_i \frac{\partial \Phi}{\partial u_R}; \\ &l_i = \frac{\partial u_R}{\partial u_i}; \qquad P = p - s_{ik} U_{(i,k)}. \end{aligned} \tag{13.7}$$

Here commas in interlinear indices denote spatial derivatives, their inclusion in parentheses denotes symmetrization, and a period on top corresponds to the temporal derivative.

The first equation is useful in rewriting the equation of mechanics of a continuum medium in the standard form:

$$\rho \ddot{U}_i = \sigma_{ik,k}, \tag{13.8}$$

where the stress tensor σ_{ik} is introduced as

$$\sigma_{ik} = \lambda_{ikmn} U_{(m,n)} - s_{ik} \Phi(u_R). \tag{13.9}$$

Equations (13.7)–(13.9) correspond to the translation invariant equations of the dynamics of the double continuum. Equation (13.9) allows us to consider the stress as a sum of an elastic part (the first term) and an inelastic part caused by a microstructure of the lattice. The latter disappears at small micro-displacements and achieves a maximum value equal to s_{ik}, when $u = \pi$, or at the shift of atoms from the potential holes to the peaks of interatomic potential barriers. Certainly, the value of the material tensor s_{ik} is the limit of inelastic stresses corresponding to the loss of stability of the lattice. The plastic strains as well as interphase transitions and other bifurcation processes are possible further. They are defined by the field of micro-displacements that is obtained from (13.7).

The value of P in (13.7) corresponds to an effective potential barrier of the rearrangement of the lattice, depending on the strain gradient $U_{i,k}$. It might be excluded expressing through σ_{ik}. This gives another representation for P

$$P = p - s_{mn}\lambda^{-1}_{mnik}\sigma_{ik}. \tag{13.10}$$

Choosing the periodic function Φ according to (13.2) and taking into account (13.3), the governing equations are rewritten as

$$\rho\ddot{U}_i = \lambda_{ikmn}U_{(mn),k} - s_{ik}l_k\sin(u_R), \quad u_R = (u_i\alpha_{ik}u_k)^{1/2}, \tag{13.11}$$

$$\mu\ddot{u}_i = K_{ikmn}u_{k,mn} - Pl_i\sin(u_R), \quad P = p - s_{ik}U_{(ik)}. \tag{13.12}$$

These equations correspond to the crystal of any class of symmetry but they are too complicated for an analysis. Therefore, a more simple case will be considered further.

13.3 Two-Component Two-Dimensional Micro-Fields in Cubic Crystals

Consider a plane problem with the strain field of the form

$$\begin{aligned} &U_x = U_x(x,y,t), \qquad U_y = U_y(x,y,t), \qquad U_z = 0,\\ &u_x = u_x(x,y,t), \qquad u_y = u_y(x,y,t), \qquad u_z = 0. \end{aligned} \tag{13.13}$$

According to (13.4) for cubic crystals $u_R = u = (u_x^2 + u_y^2)^{1/2}$, (13.11), (13.12) read as

$$\mu\ddot{u}_x = -Pl_x\sin\left(u_x^2 + u_y^2\right)^{1/2} + K_1u_{x,xx} + K_{23}u_{y,xy} + K_3u_{x,yy}, \tag{13.14}$$

$$\mu\ddot{u}_y = -Pl_y\sin\left(u_x^2 + u_y^2\right)^{1/2} + K_1u_{y,yy} + K_{23}u_{x,yx} + K_4u_{y,xx}, \tag{13.15}$$

$$\rho\ddot{U}_x = \sigma_{xx,x} + \sigma_{xy,y}, \qquad \rho\ddot{U}_y = \sigma_{yy,y} + \sigma_{yx,x}. \tag{13.16}$$

Here the components of the strain tensor are expressed according to (13.9)

$$\sigma_{xx} = \lambda_1 U_{x,x} + \lambda_2 U_{y,y} - s_{xx}\left(1 - \cos\left(u_x^2 + u_y^2\right)^{1/2}\right), \tag{13.17}$$

$$\sigma_{yy} = \lambda_2 U_{x,x} + \lambda_1 U_{y,y} - s_{yy}\left(1 - \cos\left(u_x^2 + u_y^2\right)^{1/2}\right), \tag{13.18}$$

$$\sigma_{yx} = \lambda_3 (U_{x,y} + U_{y,x}) - s_{yx}\left(1 - \cos\left(u_x^2 + u_y^2\right)^{1/2}\right). \tag{13.19}$$

Further simplifications are needed for obtaining analytical solutions.

13.4 Thin Layer Assumption. Statics

Consider a static version of (13.16) with the defining parities given by (13.17)–(13.19).

Assume that a layer is directed along the OY axis, while the solution is localized in the vicinity of this axis, then besides the conditions $\ddot{U} = \ddot{u} = 0$ a natural assumption may be suggested

$$U_x \ll U_y, \qquad u_x \ll u_y, \qquad l_x \ll l_y, \qquad \partial()/\partial y \ll \partial()/\partial x. \tag{13.20}$$

Equations (13.20) allow us to write leading order equations defining two-dimensional stationary fields of macro- and micro-strains $U_x(x,y), U_y(x,y); u_x(x,y), u_y(x,y)$. The substitution of (13.17)–(13.19) into (13.16) allows simplifying them with the help of (13.20). Neglecting the terms with derivatives $U_{x,yy}$, $U_{y,xy}$, $u_{x,y}$ in the equation for the x projection and terms $U_{y,yy}$, $u_{y,y}$ in the equation for y projection, one obtains two equations of the second order that may be integrated once with respect to x, giving

$$\lambda_1 U_{x,x} + (\lambda_2 + \lambda_3) U_{y,y} - s_{xx}\left(1 - \cos(u_y)\right) = \varepsilon_0(y), \tag{13.21}$$

$$\lambda_3 U_{y,x} - s_{yx}\left(1 - \cos(u_y)\right) = \sigma_0(y). \tag{13.22}$$

Here $\sigma_0(y), \varepsilon_0(y)$ are the constants of integration with respect to x. Then the system of two equations appears for the macroscopic fields $U_x(x,y)$, $U_y(x,y)$ connected with the micro-field $u(x,y)$. Then a consideration of the static reduction of (13.14), (13.15) is needed. Leaving the left-hand side in (13.14) and changing u with u_y in (13.15), we rewrite the system (13.14)–(13.15) as

$$0 \approx K_1 u_{x,xx} + K_{23} u_{y,xy}, \tag{13.23}$$

$$\mu \ddot{u}_y \approx -P l_y \sin(u_y) + K_1 u_{y,yy} + K_{23} u_{x,yx} + K_4 u_{y,xx}. \tag{13.24}$$

The term $u_{x,yy}$ is skipped in (13.23) due to smoothness along the layer. It is integrated once with respect to x, giving the first order equation

$$B \approx K_1 u_{x,x} + K_{23} u_{y,y}. \tag{13.25}$$

Here B is a constant of integration. Differentiating (13.25) with respect to y, one excludes from (13.23) the mixed derivative $u_{x,yx}$; this gives rise to

$$\mu\ddot{u}_y \approx -Pl_y \sin(u_y) + K_4 u_{y,xx} + K_5 u_{y,yy}. \tag{13.26}$$

Its right-hand side may be simplified using (13.23) and the assumption $l_y \approx 1$. Due to the smallness of the component u_x, the equation may be finally written as

$$\mu\ddot{u}_y \approx -P\sin(u_y) + K_5 u_{y,yy} + K_4 u_{y,xx}, \qquad K_5 = K_1 - K_{23}^2/K_1. \tag{13.27}$$

Here P is an effective potential, depending on deformation gradients according to (13.7). For cubic crystals we get

$$P = p - s_{xx}U_{x,x} + s_{yy}U_{yy} + (s_{xy} + s_{yx})(U_{x,y} + U_{y,x}). \tag{13.28}$$

Then the equations for macro- and micro-fields become coupled, which strongly complicates their analysis. However, according to (13.21) and (13.22), the deformation gradients are expressed through two arbitrary functions $\sigma_0(y)$ and $\varepsilon_0(y)$ that may be replaced with constants. Then only one of the three functions $U_{x,x}$, $U_{y,y}$, $U_{x,y}$ remains in (13.28). It simplifies a choice of useful hypothesis for obtaining solutions.

It is useful to transform to the dependence of the effective potential barrier P on the stresses according to its equivalent definition (13.10). For cubic crystals, following (13.17), (13.18) and (13.19), we have

$$P = p - \bar{\sigma}; \qquad \bar{\sigma} = \varepsilon_1\sigma_{xx} + \varepsilon_2\sigma_{yy} + \varepsilon_3\sigma_{yx} + \widehat{S}\big(1 - \cos(u)\big). \tag{13.29}$$

Here ε_1, ε_2, ε_3, $\widehat{S}$ are complicated combinations of the material constants (λ_1, λ_2, λ_3) and coefficients of striction (s_{xx}, s_{yy}, s_{yx}). The three components of stresses are expressed through two constants σ_0, ε_0 according to (13.21) and (13.22). Then the useful hypothesis may be chosen for construction of particular solutions, assuming P does not depend on the coordinates under certain boundary conditions. The stress is a constant in the one-dimensional case when only dependence of one coordinate is taken into account.

13.5 Some Particular Solutions

One of the solutions of (13.26) reads

$$\tan(u_y/4) = \frac{A\,\mathrm{cn}(xK_1/H)}{\mathrm{sn}(xK_1/H)\,\mathrm{dn}(yK_2/B)} \approx \frac{A}{\mathrm{sh}(x/L)\,\mathrm{dn}(yK_2/B)}. \tag{13.30}$$

Here $\bar{K}_1 = \bar{K}_1(\nu_1)$, $\bar{K}_2 = \bar{K}_2(\nu_2)$ are complete elliptic integrals of the first kind. The arguments ν_1, ν_2 play the role of constants of integration in the present theory, while sn(), cn(), dn() are the standard notations for the Jacobi elliptic functions.

The second equality in (13.29) is obtained in the particular case $\nu_1 \to 1$, $\nu_2 \to 0$, when the form of the elliptic functions is simplified. The function $\mathrm{sh}(x/L)$ describes a variation of the structure within a contact zone of the width L. One gets obvious limiting relationships

$$\begin{aligned} &|x| \to 0, \qquad |\,\mathrm{sh}()| \to 0, \qquad \tan(u_y/4) \to \pm\infty, \qquad u_y \to 2\pi, \\ &|x| \to \infty, \qquad |\,\mathrm{sh}()| \to \infty, \qquad \tan(u_y/4) \to 0, \qquad u_y \to 0. \end{aligned} \tag{13.31}$$

The first relationship means that structural variations in the center of contact zone $x = 0$ are maximal, and the atoms move for an integer number of the periods of the lattice. The second one means that the atoms remain in the initial state of equilibrium, $\pm u_y = 0$, far from the contact zone. These relationships are the basis for the statement of the boundary conditions.

Using the field found and substituting it into (13.24), one obtains the solution for the second component of the displacement of the form

$$K_1 u_x = K_{23} \int \bigl[\arctan(u_y/4)\bigr]\, dx - \varepsilon_0 x. \tag{13.32}$$

Using these relationships for the dependence of macro-displacements on micro-strains and stresses, one can obtain the distortion of the size and shape of the plate, calculating the micro-displacement U_i. The employment of (13.16), (13.17) and (13.18) allows us to obtain the stresses σ_{xx}, σ_{yy}, σ_{xy}. A tribute of the structure in them is described by the formula

$$1 - \cos(u) = \frac{8A^2 \,\mathrm{tn}^2(xK_1/H)\,\mathrm{dn}^2(yK_2/B)}{A^2 + \mathrm{tn}^2(xK_1/H)\,\mathrm{dn}^2(yK_2/B)}. \tag{13.33}$$

Another particular solution of (13.27) reads

$$\begin{aligned} &\mathrm{tg}(u_y/4) = A\ \mathrm{dn}(xK_1/H)\ \mathrm{dn}(yK_2/B), \\ &A^2 = 1/(1-\nu_1^2)(1-\nu_2^2). \end{aligned} \tag{13.34}$$

Some features of a super-structure are caused by the boundedness of the Jacobi function $\mathrm{dn}(*)$. Therefore, the right-hand side of the solution does not apply at infinity, though reaches large values at $\nu_1 \to 1$, $\nu_2 \to 1$. It means that $u_y < 2\pi$, and the atoms do not jump into the neighboring holes, as it happens at the appearance of singularities. The corresponding micro-structure has the form of two-dimensional periodic system of nanocrystals separated by wide boundaries, when the lattice is disordered to some extent. No singular defects arise, contrary to the previous case.

One can find more particular solutions in [1, 3, 2]. Similar solutions of (13.14) and (13.15) may be obtained in the dynamical case of plane waves propagating with constant velocities and depending on the phase variables $\chi = x - V_1 t$, $\eta = y - V_2 t$. Some dynamical problems were studied in [4, 5, 10].

13.6 Conclusions

An important feature of adequate structural rearrangements should be an essential nonlinearity of the processes that may reach the points of bifurcation of the structure and give rise to the cardinal variations in the existing properties and the arising of new ones, including possible description of the process of the defect generation. The development of a theory is possible on the base of the known model of a complicated lattice whose elementary cell contains more than one atom. Existing theories are linear as a rule, however, they could describe anharmonicities of low degree as a last resort. The presented theory allows describing some interesting mechanisms of structural rearrangements by the exact solutions of nonlinear governing equations. Some nontrivial effects are revealed for the first time, like switching of interatomic connections by lowering of potential barriers under an influence of external stresses. As a result, the theory predicts an appearance of surface defects in initially ideal structure, its fragmentation with singular boundaries, loss of stability of homogeneous strains beyond the threshold of intensive action on the body, inelastic diffusionless strains, loss of long-range and short-range orders which are the micro-mechanisms of the surface reconstruction. It is interesting to note that in the complex stressed state the lattice hardening is also possible besides an inverse effect of softening. The criteria of the transitions in the form of bifurcating dependencies of the amplitude of microscopic field on the stresses and the domain sizes at the surface are obtained in [3, 2]. Further development of the theory is required to connect with the problem of thermodynamical stability of defects both in the fields of stresses and temperatures.

Acknowledgements The work has been supported by the Russian Foundation for Basic Researches, grants Nos. 07-01-00213 and 09-01-00469a.

References

1. Aero, E.L.: Structural transitions and shear stability of polyatomic layers. Inorg. Mater. **35**(8), 860–862 (1999)
2. Aero, E.L.: Micromechanics of a double continuum in a model of a medium with variable periodic structure. J. Eng. Math. **55**, 81–95 (2002)
3. Aero, E.L.: Inhomogeneous microscopic shear strains in a complex crystal lattice subjected to larger macroscopic strains (exact solutions). Phys. Solid State **45**(8), 1557–1565 (2003)
4. Aero, E.L., Bulygin, A.N.: Strongly nonlinear theory of nanostructure formation owing to elastic and nonelastic strains in crystalline solids. Mech. Solids **42**, 807–822 (2007)
5. Aero, E.L., Bulygin, A.N.: Nonlinear theory of localized waves in complex crystalline lattices as discrete-continuum systems. Vichislit. Mech. Sploshn. Sred **1**, 14–30 (2008). In Russian
6. Born, M., Huang, K.: Dynamic Theory of Crystal Lattices. Clarendon Press, Oxford (1954)
7. Cosserat, E., Cosserat, F.: Théorie des corps déformables. Hermann, Paris (1909)
8. Kosevich, A.M.: Theory of Crystal Lattice. Vyshcha Shkola, Kharkov (1988). In Russian
9. Kunin, I.A.: Elastic Media with Microstructure. II Three-Dimensional Models. Springer, Berlin (1983)
10. Porubov, A.V., Aero, E.L., Maugin, G.A.: Two approaches to study essentially nonlinear and dispersive properties of the internal structure of materials. Phys. Rev. E. **79**, 046608 (2009)

Chapter 14
Generalized Beams and Continua. Dynamics of Reticulated Structures

Claude Boutin, Stéphane Hans, and Céline Chesnais

Abstract This paper deals with the dynamic behavior of periodic reticulated beams and materials. Through the homogenization method of periodic discrete media the macro-behavior is derived at the leading order. With a systematic use of scaling, the analysis is performed on the archetypical case of symmetric unbraced framed cells. Such cells can present a high contrast between shear and compression deformability, conversely to "massive" media. This opens the possibility of enriched local kinematics involving phenomena of global rotation, inner deformation or inner resonance, according to studied configuration and frequency range.

14.1 Introduction

This paper deals with the macroscopic dynamic behavior of periodic reticulated structures and materials widely encountered in mechanical engineering. Periodic lattices have been studied through various approaches such as transfer matrix, variational approach [8], finite difference operator, cf. [10]. Asymptotic methods of homogenization [11] initially developed for periodic media, were extended to multiple parameters and scale changes by [5] and adapted to periodic discrete structures by [3], then [9]. Those studies aim at relate the local structure and the global behavior.

The structural morphology of reticulated media makes that the basic cells can present a high contrast between shear and compression deformability (conversely to

C. Boutin (✉) · S. Hans · C. Chesnais
DGCB, FRE CNRS 3237, Ecole Nationale des Travaux Publics de l'Etat, Université de Lyon, Lyon, France
e-mail: claude.boutin@entpe.fr

S. Hans
e-mail: stephane.hans@entpe.fr

C. Chesnais
e-mail: celine.chesnais@entpe.fr

G.A. Maugin, A.V. Metrikine (eds.), *Mechanics of Generalized Continua*,
Advances in Mechanics and Mathematics 21,
DOI 10.1007/978-1-4419-5695-8_14, © Springer Science+Business Media, LLC 2010

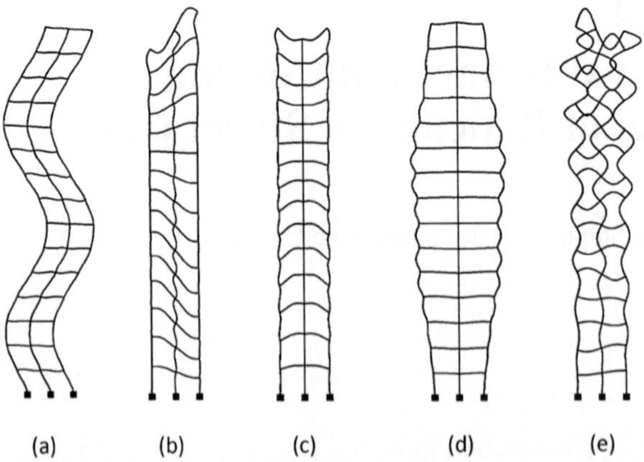

Fig. 14.1 Examples of typical and atypical dynamic behaviors of 1D framed structures

"massive" media). This opens the possibility of enriched local kinematics involving phenomena of global rotation, inner deformation or inner resonance, according to studied configuration and frequency range. A numerical illustration of these atypical situations is given in Fig. 14.1 that shows the some unusual macroscopic modes. The present study investigates and summarizes those phenomena by a systematic analysis performed on the archetypical case of symmetric unbraced framed cells [1, 7]. Assuming the cell size is small compared to the wavelength, the homogenization of periodic discrete media leads to the macro-behavior at the leading order. The paper is organized as follows. Section 14.2 gives an overview of the method and the assumptions. In Sect. 14.3, several beam modeling under transverse vibrations are established by varying the properties of the basic frame elements, and the frequency range. Section 14.4 is devoted to longitudinal vibrations. Section 14.5 focuses on the analogy between these results and the mechanics of generalized continua.

14.2 Overview of Discrete Homogenization

The analysis of periodic beam lattices is performed in two steps [12]: first, the discretization of the balance of the structure under harmonic vibrations; second, the homogenization, leading to a continuous model elaborated from the discrete description. An outline of this method is given hereafter.

Discretization of the Dynamic Balance Studied structures (Fig. 14.2) are made of plates behaving as Euler–Bernoulli beam in out-of-plane motion, and assembled with rigid connections. The motions of each extremity connected to the same node are identical and define the discrete nodal kinematic variables of the system. The discretization consists in integrating the dynamic balance (in harmonic regime) of the beams, the unknown displacements and rotations at their extremities taken as boundary conditions. Forces applied by an element on its extremities are then ex-

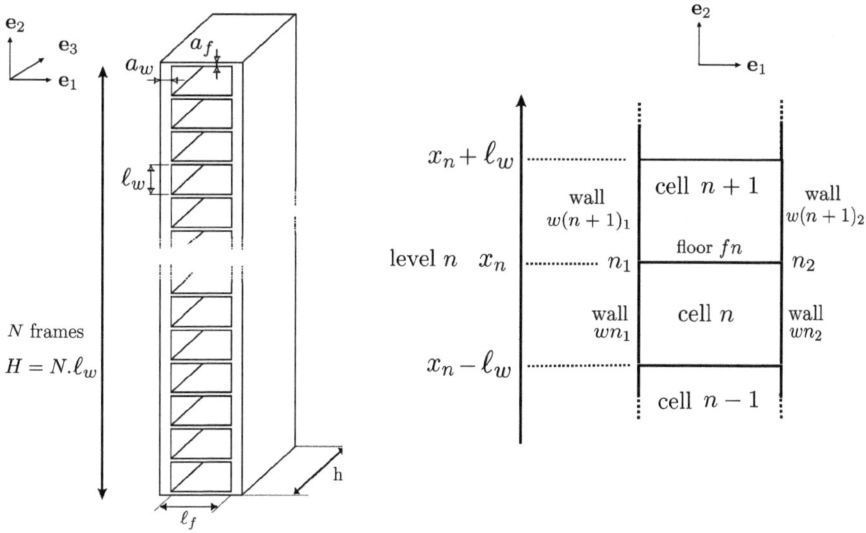

Fig. 14.2 The class of unbraced framed structures (*left*) and the basic frame and notations (*right*)

pressed as functions of the nodal variables. The balance of each element being satisfied, it remains to express the balance of forces applied to the nodes. Thus, the balance of the whole structure is rigorously reduced to the balance of the nodes.

Homogenization Method The key assumption of homogenization is that the cell size ℓ is small compared to the characteristic size L of the vibrations of the structure. Thus $\varepsilon = \ell/L \ll 1$. The existence of a macro scale is expressed by means of macroscopic space variable x. The physical variables are continuous functions of x coinciding with the discrete variables at any node, e.g., $U_\varepsilon(x = x_n) = U(\text{node } n)$. These quantities, assumed to converge when ε tends to zero, are expanded in powers of ε: $U_\varepsilon(x) = U^0(x) + \varepsilon U^1(x) + \varepsilon^2 U^2(x) + \cdots$. Similarly, all unknowns, including the modal frequency, are expanded in powers of ε. As $\ell = \varepsilon L$ is a small increment with respect to x, the variations of the variables between neighboring nodes are expressed using Taylor's series; this in turn introduces the macroscopic derivatives.

To account properly for the local physics, the geometrical and mechanical characteristics of the elements are scaled according to the powers of ε. As for the modal frequency, scaling is imposed by the balance of elastic and inertia forces at macro level. This scaling insures that each mechanical effect appears at the same order whatever the ε value is. Therefore, the same physics is kept when $\varepsilon \to 0$, i.e., for the homogenized model. Finally, the expansions in ε power are introduced in the nodal balances. Those relations, valid for any small ε, lead for each ε-order to balance equations whose resolution defines the macroscopic governing equations.

Inner Quasi-Statism and Inner Dynamics In general, the scale separation requires that, at the modal frequency of the global system, wavelengths of the compression and bending vibrations generated in each local element are much longer than the element length. In that case, the nodal forces can be developed in Taylor's series with respect to ε. This situation corresponds to a quasi static state at the local scale. Nevertheless, in higher frequency range, it may occurs that only the compression wavelength is much longer than the length elements while local resonance in bending appears. The homogenization remains possible through the expansions of the compression forces and leads to atypical descriptions with inner dynamics. Above this frequency range, the local resonance in both compression and bending makes impossible the homogenization process.

14.3 Studied Structures

We study the vibrations of structures of height $H = N.\ell_w$ constituted by a pile of a large number N of identical unbraced frames called cells (Fig. 14.2). The parameters of horizontal elements ($i = f$) and vertical elements ($i = w$) are: length ℓ_i; thickness a_i; section area A_i; inertia $I_i = a_i^3 h/12$ in direction e_3; density ρ_i; elastic modulus E_i.

The kinematic is characterized at any level n by the motions of the two nodes in the plane (e_1, e_2), i.e., the displacements in the two directions and the rotation (u_1, u_2, θ). These six variables can be replaced by (i) three variables associated to the rigid body motion of the level n: the mean transverse displacements, $U(n)$ along e_1, $V(n)$ along e_2 and the rotation $\alpha(n)$ (differential vertical nodal motion divided by ℓ_p) and (ii) three variables corresponding to its deformation: the mean and differential rotations of the nodes, $\theta(n)$ and $\Phi(n)$, and the transverse dilatation $\Delta(n)$, cf. Fig. 14.3. Because of the longitudinal symmetry the transverse and longitudinal kinematics respectively governed by (U, α, θ) and (V, Φ, Δ) are uncoupled.

A systematic study enables to identify the family dynamic behaviors by evolving gradually the properties of the frame elements and the frequency range.

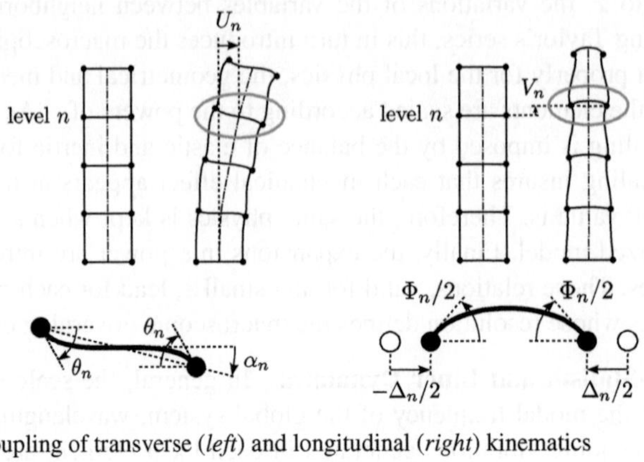

Fig. 14.3 Decoupling of transverse (*left*) and longitudinal (*right*) kinematics

14.4 Transverse Vibrations

The transverse vibrations can be classified in two categories. For the first category, the horizontal elastic force balances the horizontal translation inertia. In this situation, it can be shown that the corresponding range of frequency is such that the cell behaves in the quasi static range. Lower frequencies would lead to a static description of the structure. Conversely, at higher frequencies the horizontal elastic forces cannot balance the translation inertia so that the translation should vanish. Then the vibration results from the balance between the elastic momentum and the rotation inertia. This leads a second unusual category.

14.4.1 Low Frequency Range—Generic Beam Model

The possible beam-like behaviors are established by varying the properties of the basic frame elements. The synthesis of the different macroscopic behaviors derived according to the properties of the frame elements shows that only three mechanisms—shear, global bending, inner bending—governs the physics at the macroscale. Each of them is associated to an elastic cell parameter of stiffness: in shear K, in global bending $E_w I$, and in inner bending $2E_w I_w$. Owing to the quasi static local state, these parameters are deduced from the elastic properties of elements in statics.

The method enables to build a generic beam model involving the three kinematic variables U, α, θ and including the three mechanisms. It is governed by the following sixth order equation:

$$2E_w I_w \frac{E_w I}{K} U^{(6)}(x) - (2E_w I_w + E_w I) U^{(4)}(x) - \frac{E_w I}{K} \Lambda \omega^2 U''(x) + \Lambda \omega^2 U(x) = 0. \quad (14.1)$$

Equation (14.1) generalizes the classical beam dynamics. In addition to the common bending, it includes at the leading order shear and inner bending. Usual descriptions (as Timoshenko, Euler–Bernoulli, Shear) can be recovered by vanishing one or two of these mechanisms. Note that for this category, the rotation inertia of the section do not appear, while the shear may governs the behavior of slender structures.

The dominating effect(s) that drive(s) the effective behavior of a given structure can be identified through a dimensional analysis performed on the generic beam. Introducing the characteristic size of vibration L the change of variables $\boldsymbol{x} = x/L$ transforms the governing equation (14.1) into

$$C\gamma U^{*(6)}(\boldsymbol{x}) - (1+\gamma) U^{*(4)}(\boldsymbol{x}) - \Omega^2 U^{*(2)}(\boldsymbol{x}) + (\Omega^2/C) U^*(\boldsymbol{x}) = 0, \quad (14.2)$$

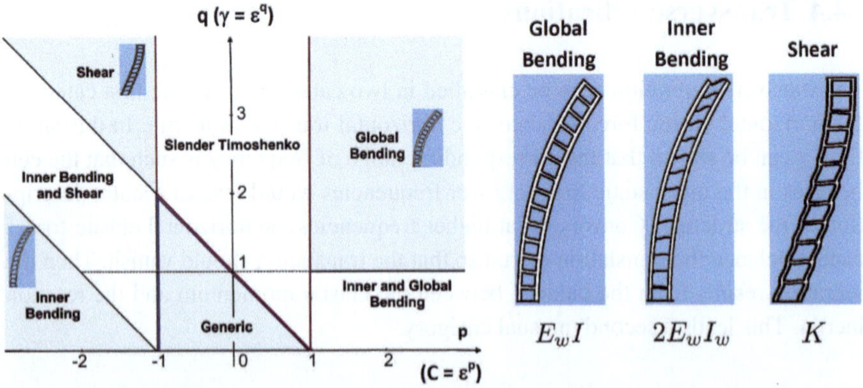

Fig. 14.4 (*Left*) Map of transverse behaviors in function of the dimensionless parameters $C = \tilde{\varepsilon}^p$ and $\gamma = \tilde{\varepsilon}^q$. (*Right*) The three transverse mechanisms

where, by construction, the dimensionless terms denoted by a * are $O(1)$ and using for L the actual vibration's size $\tilde{L} = 2H/\pi$ (for a clamped-free beam):

$$C = \frac{E_w I}{K\tilde{L}^2}, \qquad \gamma = \frac{2E_w I_w}{E_w I} = \frac{2I_w}{I}, \qquad \Omega^2 = \frac{\Lambda\omega^2\tilde{L}^2}{K}. \tag{14.3}$$

Here C evaluates the contrast between global bending and shear, and γ the contrast between inner and global bending. C and γ supply identification criteria of behavior: according to their order of magnitude with respect to $\tilde{\varepsilon} = \ell/\tilde{L} = \pi/(2N)$, (14.2) degenerates in simplified forms. For instance, if $C = O(1)$ and $\gamma = O(\tilde{\varepsilon})$, the terms related to $C\gamma$ and γ disappear and the resulting model is:

$$U^{*(4)} + \Omega^2 U^{*(2)} - (\Omega^2/C)U^* = 0$$

that corresponds to a slender Timoshenko beam, etc. Doing so, seven behaviors are obtained depending on the value of C, $C\gamma$ and γ compared to $\tilde{\varepsilon}$ powers. A synthetic representation (Fig. 14.4) is deduced by mapping the domain of validity of each behavior according to the two parameters p and q defined by: $C = \tilde{\varepsilon}^p$ and $\gamma = \tilde{\varepsilon}^q$.

14.4.2 Higher Frequency Range—Gyration Beam Model

At higher frequencies the translation U must vanish at the leading order (when seeking for a macroscopic description). A new dynamic equilibrium between elastic momentum and rotation inertia leads to the second category of macroscopic transverse vibrations. The corresponding atypical gyration beam model involves only the two kinematic variables α, θ, driven by the mechanisms of opposite traction-compression of vertical elements (global bending without translation (!), see Fig. 14.1(b)) and the frame shear. Because of the higher frequency domain, the condition of local quasi statism is not necessarily respected and thus two models are obtained.

Inner Quasi-Statism This situation corresponds to bending (then compression) wavelength much longer than the elements length. The governing equation is of the second degree whose parameters are the elastic stiffnesses already determined and the usual rotation inertia J. It reads

$$E_w I \alpha''(x) - K\alpha(x) + J\omega^2\alpha(x) = 0. \tag{14.4}$$

It can be interpreted as the balance of the global momentum with a classical inertial source term and an inner elastic source of momentum.

Inner Dynamics In this case, the bending wavelength in the element is of the order of their length, meanwhile the compression wavelength remains much larger. This enables expanding the compression forces and deriving a macro behavior. The governing equation of the second degree presents the same global momentum parameters as for local quasi-statism but differs fundamentally by the inertia term and the inner elastic source of momentum, both depending on frequency

$$E_w I \alpha''(x) - K(\omega)\alpha(x) + J(\omega)\omega^2\alpha(x) = 0. \tag{14.5}$$

The reason of these modifications lies in the non-expanded forces associated to the local dynamic bending that strongly depends on the frequency and gives rise to apparent inertia $J(\omega)$ and momentum source. This effect also appears in longitudinal vibrations and is discussed in the next section.

14.5 Longitudinal Vibrations

The longitudinal vibrations, described by (V, Φ, Δ), present a lesser complexity because the main mechanism is the vertical compression. The difference between the identified models only relies in the possible presence of inner dynamics.

Inner Quasi-Statism This case leads to the classical description of beam characterized by the compression modulus $2E_w A_w$ and the line mass Λ

$$2E_w A_w V''(x) + \Lambda\omega^2 V(x) = 0. \tag{14.6}$$

The domain of validity of this model is derived by expressing that the orders of magnitude of the fundamental frequency of the whole structure (described by (14.6)) is much smaller than the one of the elements in bending. This leads to the following lower bound of the number of cells: $N \geq$ slender ratio of the elements. In other words, the validity of usual model requires a sufficiently large number of cells.

Inner Dynamics Similarly to gyration modes, the inner dynamics introduces an frequency depending apparent mass, that can be expressed analytically [4]

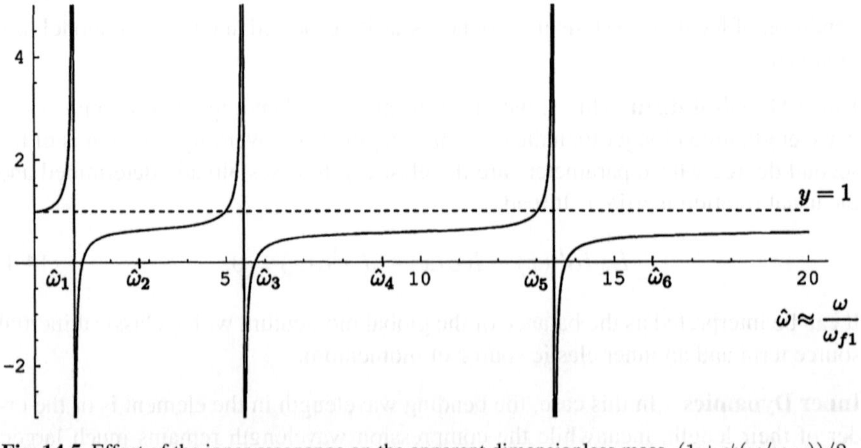

Fig. 14.5 Effect of the inner resonance on the apparent dimensionless mass, $(1+\psi(\omega/\omega_{f1}))/2$

$$2E_w A_w V''(x) + \Lambda(\omega)\omega^2 V(x) = 0; \quad \Lambda(\omega) = \Lambda_w + \Lambda_f \psi\left(\omega/\omega_{f_1}\right), \tag{14.7}$$

$$\psi(\tilde{\omega}) = \frac{8\sin(\frac{3}{4}\pi\sqrt{\tilde{\omega}})\mathrm{sh}(\frac{3}{4}\pi\sqrt{\tilde{\omega}})}{3\pi\sqrt{\tilde{\omega}}(\sin(\frac{3}{4}\pi\sqrt{\tilde{\omega}})\mathrm{ch}(\frac{3}{4}\pi\sqrt{\tilde{\omega}})) + \mathrm{sh}(\frac{3}{4}\pi\sqrt{\tilde{\omega}})\cos(\frac{3}{4}\pi\sqrt{\tilde{\omega}}))}. \tag{14.8}$$

The study of $\Lambda(\omega)$ (cf. Fig. 14.5), shows that (i) $\Lambda(\omega) \to \Lambda$ when $\omega \to 0$, and (ii) $|\Lambda(\omega)| \to \infty$ when $\omega \to \omega_{f_{2k+1}}$, where $\omega_{f_{2k+1}}$ are the odd eigenmodes of horizontal elements in bending. This induces abnormal response in the vicinity of the $\omega_{f_{2k+1}}$ that results in discrete spectrum of cut-off frequencies. The domain of validity of this model is derived by equating the orders of magnitude of the fundamental frequencies of (i) the whole structure and (ii) the elements in bending. This leads to bounding of the number of cells by $N \leq$ slender ratio of the elements.

14.6 Analogy with Micromorphic Media

This section points out an analogy between the several beam behaviors and those of micromorphic materials [6]. Conveniently, we focus on the first (low frequency) mode of transverse vibration polarized in direction e_2, propagating in direction e_1.

14.6.1 From 1D to 2D Structures

Consider a cell $\mathscr{F}$ of shear stiffness K, inner bending stiffness EI_w and global bending stiffness EI, corresponding to the intrinsic coefficients $\beta = EI/Kl_w^2$ and $\gamma = I_w/I$. From this frame, build a periodic beam $\mathscr{B} = N_{e_1} \times \mathscr{F}$, made of N_{e_1} frame cells repeated in the direction e_1, as in Fig. 14.2. The transverse behavior of

$\mathscr{B}$ is driven by the two dimensionless parameters and the actual scale ratio

$$C = \beta\left(\frac{\pi}{2N_{e_1}}\right)^2 = \tilde{\varepsilon}^p, \qquad \gamma = \tilde{\varepsilon}^q, \quad \tilde{\varepsilon} = \frac{\pi}{2N_{e_1}}.$$

Build the 2D periodic structure $\mathscr{S}$ by N_{e_2} lateral repetition in the transverse direction e_2 of $\mathscr{B}$, that is, $\mathscr{S} = N_{e_2} \times \mathscr{B} = N_{e_1} \times (N_{e_2} \times \mathscr{F})$. The properties of the $\mathscr{S}$'s cell made of $N_{e_2} \times \mathscr{F}$ frames can be estimated from those of $\mathscr{F}$; the stiffnesses are $O(N_{e_2}K)$ in shear, $O(N_{e_2}EI_w)$ in inner bending, and $O(N_{e_2}^3 EI)$ in global bending. Thus, the transverse behavior of $\mathscr{S}$ is driven by C_S and γ_S

$$C_S = \left(\frac{\pi N_{e_2}}{2N_{e_1}}\right)^2 \beta > C, \qquad \gamma_S = \left(\frac{1}{N_{e_2}}\right)^2 \gamma < \gamma$$

$$\text{then } C_S\gamma_S = C\gamma = \left(\frac{\pi}{2N_{e_1}}\right)^2 \beta\gamma. \tag{14.9}$$

Moreover, since $\mathscr{B}$ and $\mathscr{S}$ have the same number N_{e_1} of cells, $\tilde{\varepsilon}_S = \tilde{\varepsilon}$. Introducing (p_S, q_S) such that $C_S = \tilde{\varepsilon}^{p_S}$ and $\gamma_S = \tilde{\varepsilon}^{q_S}$, relations (14.9) lead to

$$p_S < p, \qquad q_S > q,$$
$$p_S + q_S = p + q.$$

Consequently, according to the previous results (cf. Fig. 14.4), the nature of $\mathscr{B}$ and $\mathscr{S}$ behavior would be the same if $p \leq -1$, but may differ if $p > -1$.

14.6.2 Micromorphic Media

Consider now an infinite media made of the cell $\mathscr{F}$ and assume that—in a domain $\mathscr{D}$ restricted to $N_{e_1} \times N_{e_2} \times \mathscr{F}$—appropriate boundary conditions impose normally to the direction e_1 a shear polarized in the direction e_2. The equivalent behavior in $\mathscr{D}$ is identified by comparing the power of $\tilde{\varepsilon} = \pi/(2N_{e_1})$ with the two parameters:

$$C_{\mathscr{D}} = \left(\frac{\pi N_{e_2}}{2N_{e_1}}\right)^2 \beta, \qquad \gamma_{\mathscr{D}} = \left(\frac{1}{N_{e_2}}\right)^2 \gamma.$$

Four behaviors may arise according to the independent kinematic variable(s).

Cauchy Elastic Continuum Behavior This will be observed when the kinematic involves the solely translation U, which occurs when $C_{\mathscr{D}}\gamma_D \leq \tilde{\varepsilon}$ and $C_{\mathscr{D}} \geq \tilde{\varepsilon}^{-1}$. These conditions require that the $\mathscr{D}$ geometry follows the constraints: in the direction normal to the shear motion, $N_{e_1} \geq (\pi/2)\beta\gamma$, i.e., a number of cells higher than the intrinsic critical number $N_c = (\pi/2)\beta\gamma = EI_w/Kl_w^2$; in the direction of the shear motion, $N_{e_2} \geq (2N_{e_1}/\pi)^{3/2}/\sqrt{\beta} = M_2$, i.e., a weak slenderness aspect ratio of $\mathscr{D}$.

Micromorphic Elastic Media with Inner Deformation This appears when the kinematic involves the translation U and the inner rotation θ, i.e., when $C_{\mathscr{D}}\gamma_{\mathscr{D}} \geq \tilde{\varepsilon}$ and $C_{\mathscr{D}} \geq \tilde{\varepsilon}^{-1}$. In terms of $\mathscr{D}$ geometry, this requires that $N_{e_1} \leq N_c$ and $N_{e_2} \geq M_2$.

Cosserat Like Behavior The kinematic variables are the translation U and the rotation α. This appears when $C_{\mathscr{D}}\gamma_{\mathscr{D}} \leq \tilde{\varepsilon}$ and $C_{\mathscr{D}} \leq \tilde{\varepsilon}^{-1}$, i.e., when $N_{e_1} \geq N_c$ and $N_{e_2} \leq M_2$. In this case, the cell rotation breaks the lateral periodicity, whereas the periodicity in the perpendicular direction is kept. For this reason, such a phenomenon is not be described by the usual 2D homogenization applied the basic $\mathscr{F}$ (that would impose the periodicity in both directions). Nevertheless, the deformation presents a scale separation in a single direction that enables the treatment by a 1D homogenization. In such a domain, the effects of global bending and shear are of same order.

Complex Micromorphic Media with Inner Deformation and Rotation involving the three variables U, α and θ. This situation is obtained when $C_{\mathscr{D}}\gamma_{\mathscr{D}} \geq \tilde{\varepsilon}$ and $C_{\mathscr{D}} \leq \tilde{\varepsilon}^{-1}$ that is for rather restricted domain geometry defined by $N_{e_1} \leq N_c$ and $N_{e_2} \leq M_2$.

To respect the scale separation, the micromorphic behaviors with inner deformation can only be reached if $N_c = (\pi/2)\beta\gamma$ is significantly larger than 1.

14.7 Conclusion

Reticulated media present a much larger diversity of behavior than the usual "massive" media. In particular, the enriched local kinematics introduces modes of different nature (as gyration modes), based on different kind of equilibrium (as inner bending modes). The atypical inner resonance effect is also demonstrated to be possible for those reticulated structure, while they cannot develop in massive beams.

The comparisons of these theoretical results with numerical modeling are satisfactory [7], and dynamics experiments on real regular buildings [2] also demonstrate the reliability of this approach and its interest in engineering domain. Finally, these results may be extended to the rheology of reticulated materials such as foam, glass wool, vegetal, bones, etc. and presents strong analogies with generalized continua.

References

1. Boutin, C., Hans, S., Homogenisation of periodic discrete medium: application to dynamics of framed structures. Comput. Geotech. **30**(4), 303–320 (2003)
2. Boutin, C., Hans, S., Ibraim, E., Roussillon, P.: In situ experiments and seismic analysis of existing buildings—Part II. Earthquake Eng. Struct. Dyn. **34**, 1531–1546 (2005)
3. Caillerie, D., Trompette, P., Verna, P.: Homogenisation of periodic trusses. In: Congres IASS, Madrid, pp. 7139–7180 (1989)

4. Chesnais, C., Hans, S., Boutin, C.: Wave propagation and diffraction in discrete structures—effect of anisotropy and internal resonance. In: ICIAM, Zurich, 16–20 July 2007
5. Cioranescu, D., Saint Jean Paulin, J.: Homogenization of Reticulated Structures. Applied Mathematical Sciences, vol. 136. Springer, Berlin (1999)
6. Eringen, A.C.: Mechanics of micromorphic continua. In: IUTAM Symposium on the Generalized Cosserat Continuum and the Continuum Theory of Dislocations with Applications, pp. 18–35. Springer, Berlin (1968)
7. Hans, S., Boutin, C.: Dynamics of discrete framed structures: a unified homogenized description. J. Mech. Mater. Struct. **3**(9), 1709–1739 (2008)
8. Kerr, A.D., Accorsi, M.L.: Generalization of the equations for frame-type structures—a variational approach. Acta Mech. **56**(1–2), 55–73 (1985)
9. Moreau, G. and Caillerie, D.: Continuum modeling of lattice structures in large displacement applications to buckling analysis. Comput. Struct. **68**(1–3), 181–189 (1998)
10. Noor, A.K.: Continuum modeling for repetitive lattice structures. Appl. Mech. Rev. **41**(7), 285–296 (1988)
11. Sanchez-Palencia E.: Non-Homogeneous Media and Vibration Theory. Lecture Note in Physics, vol. 127. Springer, Berlin (1980)
12. Tollenaere, H., Caillerie, D.: Continuum modeling of lattice structures by homogenization. Adv. Eng. Softw. **29**(7–9), 699–705 (1998)

4. Chesnais, C., Hans, S., Boutin, C.: Wave propagation and diffraction in discrete structures—effect of anisotropy and internal resonance. In: ICTAM, Zurich, 16–20 July 2007
5. Cioranescu, D., Saint Jean Paulin, J.: Homogenization of Reticulated Structures. Applied Mathematical Sciences, vol. 136. Springer, Berlin (1999)
6. Eringen, A.C.: Mechanics of micromorphic continua. In: IUTAM Symposium on the Generalized Cosserat Continuum and the Continuum Theory of Dislocations with Applications, pp. 18–35. Springer, Berlin (1968)
7. Hans, S., Boutin, C.: Dynamics of discrete framed structures: a unified homogenized description. J. Mech. Mater. Struct. 3(9), 1709–1739 (2008)
8. Kerr, A.D., Accorsi, M.L.: Generalization of the equations for frame-type structures—a variational approach. Acta Mech. 56(1–2), 55–73 (1985)
9. Moreau, G. and Caillerie, D.: Continuum modeling of lattice structures in large displacement applications to buckling analysis. Comput. Struct. 68(1–3), 181–189 (1998)
10. Noor, A.K.: Continuum modeling for repetitive lattice structures. Appl. Mech. Rev. 41(7), 285–296 (1988)
11. Sanchez-Palencia, E.: Non Homogeneous Media and Vibration Theory. Lecture Note in Physics, vol. 127. Springer, Berlin (1980)
12. Tollenaere, H., Caillerie, D.: Continuum modeling of lattice structures by homogenization. Adv. Eng. Softw. 29(7–9), 699–705 (1998)

Chapter 15
Wave Propagation in Damaged Materials Using a New Generalized Continuum Model

Vladimir I. Erofeyev, Elena A. Nikitina, and Alla V. Sharabanova

Abstract An approach is proposed that allows formulating self-consistent problem that includes equations of the material's dynamics and conditions of its damage. It is shown that damage in the material introduces a frequency-dependent damping and dispersion of the phase velocity of ultrasonic acoustic waves that allows estimating damage using the acoustic method. Applied field of deformations leads to the accumulation of damage. A kinetic equation is obtained, whose analysis shows that damage grows exponentially. The parameters of the system for which accumulation of damage can be considered as linear are estimated.

15.1 Introduction

Nowadays, ensuring the safety of mechanical engineering structures is performed with the help of non-destructive control methods. The acoustic method is currently considered as the most promising one. The precision of parameter measurement and further interpretation of the condition of structural material depend on the numerous factors, including the exploitation conditions. The durability and longevity of a structure depend on strength characteristics of the material in local zones that experience highest loading. During the exploitation, the material experiences structural changes, and the degree of degradation depends on the exploitation conditions of loading. Clearly, during the diagnostics of structures exploited for a long time period, such structural changes lead to a significant change of data compared to the original calibration data.

The goal of this work is to develop methods of deciphering the data of acoustic measurement tools to account for the damage of structural material. An approach

V.I. Erofeyev (✉) · E.A. Nikitina · A.V. Sharabanova
A.A. Blagonravov Mechanical Engineering Institute RAS, Nizhny Novgorod Branch, Belinskogo str., 85, Nizhny Novgorod 603024, Russia
e-mail: erf04@sinn.ru

G.A. Maugin, A.V. Metrikine (eds.), *Mechanics of Generalized Continua*, Advances in Mechanics and Mathematics 21,
DOI 10.1007/978-1-4419-5695-8_15, © Springer Science+Business Media, LLC 2010

is proposed that allows formulating a self-consistent problem that includes equations of the material's dynamics and conditions of its damage. It is shown that damage in the material introduces a frequency-dependent damping and dispersion of the phase velocity of ultrasonic acoustic wave that allows estimating damage using the acoustic method. Applied field of deformations leads to the accumulation of damage. A kinetic equation is obtained, whose analysis shows that damage grows exponentially. Parameters of the system for which accumulation of damage can be considered as linear are estimated.

15.2 Mathematical Model

The mechanics of damaged continuum is being actively developed, starting with the fundamental works of L.M. Kachanov [1] and Y.N. Rabotnov [2].

What is usually meant by damage is the shortening of elastic response of the body due to a reduction of effective area that transmits internal stress from one part of the body to the other. This reduction, in turn, is caused by appearance and development of a disperse field of micro-defects (micro-cracks in elasticity, dislocations in plasticity, micro-pores for creepage and surface micro-cracks for fatigue).

In traditional approaches to damageability computations, a measure of damageability during deformation development is represented by a scalar parameter $\psi(x,t)$ which characterizes the relative density of equally-spread micro-defects in the unit volume. This parameter is equal to zero when there is no damage, and is close to 1 at the moment of destruction.

We study a sample, shaped as a rod, through which a longitudinal wave can propagate. We denote middle-line particle's dislocation by $u(x,t)$. Let us consider that the rod undergoes static and cyclic testing and its material can accumulate damage. To measure damage, we introduce a function $\psi(x,t)$.

As a rule, in mechanics of deformable solid body, dynamic problems are studied separately from damage accumulation problems. During these method developments, it is common to postulate that velocity of an elastic wave is a function of damageability and then determine the proportional coefficients experimentally.

Phase velocity (υ_{ph}) of the wave and its dissipation (frequency-dependent damping) are considered to be power functions of frequency (ω) and linear functions of damageability (ψ):

$$\upsilon_{\text{ph}}(\omega) = c_0\big(1 - h_1\psi - h_2\psi\omega^2\big), \tag{15.1}$$

$$\alpha(\omega) = (h_3 + h_4\psi)\omega^4, \tag{15.2}$$

where $C_0 = \sqrt{\mathrm{E}/\rho}$ is the velocity of a longitudinal wave in the main medium if there were no damage; E Young modulus; ρ material density, h_i, $i = 1,\ldots,4$, coefficients that need experimental determination.

Damageability evolution can be described with the following kinetic equation:

$$\frac{d\psi}{dt} = f(\sigma, \psi), \tag{15.3}$$

where σ is the external stress.

The function $f(\sigma, \psi)$ is most often approximated as a linear function. Sometimes as a polynomial.

While this method, based on (15.1) and (15.2), has an important advantage, namely, it is rather simple, it also has drawbacks like all methods that are not based on mathematical models of processes and systems.

Let us consider that the studied problem is self-consistent and includes, besides an equation for damageability development which we rewrite as

$$\frac{\partial \psi}{\partial t} + \alpha\psi = \beta_2 E \frac{\partial u}{\partial x}, \tag{15.4}$$

an equation for the rod dynamics

$$\frac{\partial^2 u}{\partial t^2} - C_0^2 \frac{\partial^2 u}{\partial x^2} = \beta_1 \frac{\partial \psi}{\partial x}. \tag{15.5}$$

Here α, β_1, β_2 are constants that characterize material damageability and a connection between cyclic testing and damage accumulation.

15.3 Damping and Dispersion of Elastic Waves

We look for a solution of the system (15.4) and (15.5) in the form of harmonic waves $u, \psi \approx e^{i(\omega t - kx)}$, where ω is the circular frequency, $K = 2\pi/\lambda$ the wave number (λ the wave length), and arrive at the dispersion equation

$$\omega^2 - \left(C_0^2 + \frac{E\beta_1\beta_2}{\alpha}\right)K^2 + \frac{i}{\alpha}\omega^3 - \frac{iC_0^2}{\alpha}\omega K^2 = 0. \tag{15.6}$$

Please note that (15.6), which connects special and temporal scales of the longitudinal wave, also contains complex coefficient, which means that the wave will not only propagate through medium, it will also dissipate.

Let us consider the wave number to be complex $K = K^1 + iK^{11}$, where K^1 characterizes the propagation constant ($v_{\text{ph}} = \omega/K^1$ is the phase velocity of the wave), and $K^{11} = \alpha(\omega)$ characterizes wave dissipation.

Solving algebraic equation (15.6) allows the determination of both components of the complex wave number

$$K^1 = \pm\sqrt{\frac{a\omega^2 + C_0^2\omega^4/\alpha^2 \pm \sqrt{a^2\omega^4 + C_0^4\omega^8/\alpha^4 + (a^2 + C_0^4)\omega^6/\alpha^2}}{2[a^2 + C_0^4\omega^2/\alpha^2]}}, \tag{15.7}$$

$$K^{11}=\pm\sqrt{\frac{(a-C_0^2)\omega^3/\alpha}{a\omega^2+C_0^2\omega^4/\alpha^2\pm\sqrt{a^2\omega^4+C_0^4\omega^8/\alpha^4+(a^2+C_0^4)\omega^6/\alpha^2}}}. \tag{15.8}$$

Here a denotes $a=C_0^2+E\beta_1\beta_2/\alpha$.

From (15.7) and (15.8), it follows that the existence of damage causes dispersion, i.e., the phase velocity of the wave depends on the wave frequency $\upsilon_{\text{ph}}=\upsilon_{\text{ph}}(\omega)$ (Fig. 15.1) and frequency-dependent dissipation $K^{11}=K^{11}(\omega)$ (Fig. 15.2).

In the lower frequency range ($\omega\to 0$), the wave velocity works for

$$\upsilon_{\text{ph}}(0)\approx\sqrt{C_0^2+E\beta_1\beta_2/\alpha}.$$

The wave dissipation is proportional to the square of wave frequency:

$$K^{11}(0)\approx\frac{E\beta_1\beta_2\omega^2}{\alpha^2\sqrt{C_0^2+E\beta_1\beta_2/\alpha}}.$$

In the higher-frequency range ($\omega\to\infty$), the phase velocity works for C_0: $\upsilon_{\text{ph}}(\infty)\approx C_0$, and dissipation is proportional to the frequency:

$$K^{11}(\infty)\approx\frac{E\beta_1\beta_2\omega}{\alpha C_0}.$$

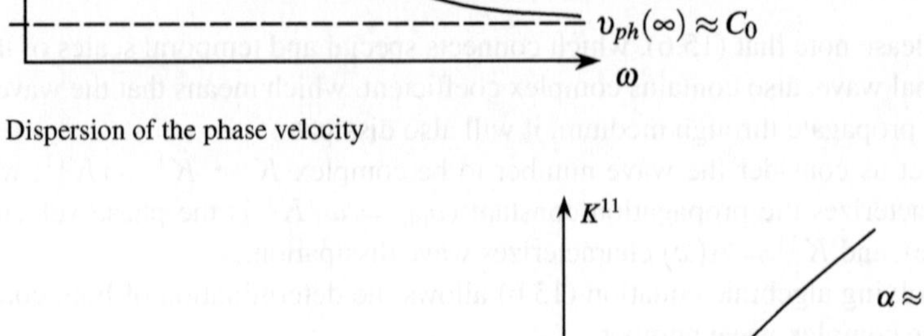

Fig. 15.1 Dispersion of the phase velocity

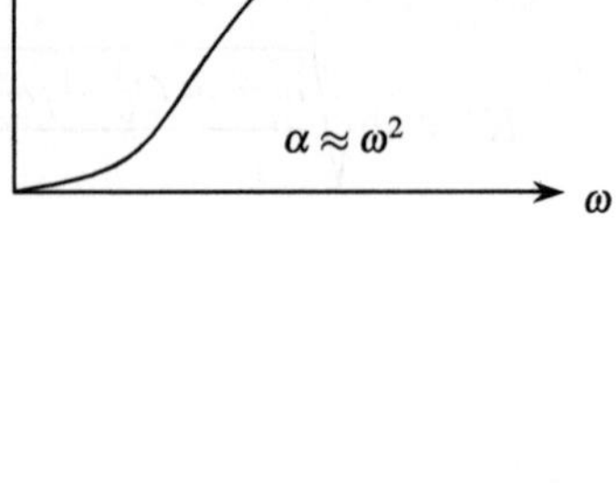

Fig. 15.2 Frequency-dependent damping

In the lower frequency range,

$$\frac{K^{1}(0)}{K^{11}(0)} = \left(1 + \frac{C_0^2\alpha}{\beta_1\beta_2 E}\right)\frac{\alpha}{\omega^2} \to \infty,$$

which means that the wave propagates almost without dissipation.

In the higher frequency range,

$$\frac{K^{1}(\infty)}{K^{11}(\infty)} = \frac{C_0^2\alpha}{\beta_1\beta_2 E\omega} \to 0,$$

which means that in this case dissipation is a dominant factor.

Note that the imaginary part of the wave number K^{11} can be measured in both the low-frequency and higher-frequency ranges. Therefore, the constants introduced in (15.4) and (15.5), namely α, β_1, β_2, can be computed from measured parameters:

$$\alpha = \frac{K^{11}(\infty)\omega}{K^{11}(0)\sqrt{1 + K^{11}(\infty)/(C_0\omega)}}, \tag{15.9}$$

$$\beta_1\beta_2 = \frac{C_0(K^{11}(\infty))^2}{EK^{11}(0)\sqrt{1 + K^{11}(\infty)/(C_0\omega)}}. \tag{15.10}$$

The system (15.4)–(15.5) can be rewritten as one equation in terms of ψ, characterizing damageability:

$$\frac{\partial^2\psi}{\partial t^2} - \left(C_0^2 + \frac{E\beta_1\beta_2}{\alpha}\right)\frac{\partial^2\psi}{\partial x^2} + \frac{1}{\alpha}\frac{\partial^3\psi}{\partial t^3} - \frac{C_0^2}{\alpha}\frac{\partial^3\psi}{\partial x^2\partial t} = 0. \tag{15.11}$$

Equation (15.11) represents a kinetic equation of damage accumulation. From its analysis it follows that damage grows exponentially. The exponent factor is determined from expression (15.8), and only with a few values of these parameters the process can be approximated as a linear function.

15.4 Conclusions

The suggested approach has allowed obtaining the new dependencies relating the dynamical equations of a material and the kinetics of its damage. This fact enables one to consider a problem about a deformation of a structural material and its damage as a unified self-consistent process.

Acknowledgements This study was supported in part by the RFBR, project no. 09-08-00827.

References

1. Kachanov, L.M.: Introduction to Damage Mechanics. Martinus Nijhoff, Dordrecht (1986)
2. Rabotnov, Y.N.: Elements of Hereditary Solid Mechanics. Mir, Moscow (1980)

Chapter 16
On the Uniqueness of the Lagrangian of Gradient Elastic Continua

Andrei V. Metrikine and Julia M. Prokhorova

Abstract In this paper, the volumetric density of the Lagrangian of a second-order isotropic gradient continuum is critically examined. This density is first derived from a cubic lattice using an implicit continualization procedure. Then, using the derived Lagrangian, an equation of motion of the continuum as well as expressions for the standard and second-order stresses are obtained. It is shown that these expressions are contradictory. To obtain non-contradictory expressions for the stresses, the Lagrangian density is trimmed using the so-called null-Lagrangians that do not affect the equation of motion but influence the expressions for the stresses. This results in a unique non-contradictory expression for the Lagrangian of the continuum.

16.1 Introduction

This paper contributes to the theory of the gradient continua [22]. Such continua are a quasi-local case of the non-local continua and the continua with memory. The integral relations between the field variables and the kinematic variables in the latter continua are replaced by differential equations in the gradient continua. The gradient continua with invariant material properties are capable of describing the weak wave dispersion in inhomogeneous materials [7, 11–13, 16, 20, 19, 21, 24, 29, 30, 33] and are instrumental for regularization of solutions of the elasticity, plasticity and damage problems of mechanics, see, e.g., [1, 2, 6, 3, 8–10, 15, 14, 17, 23, 25, 27, 28, 34, 35]. The gradient continua are applicable for modeling the processes whose wavelength is comparable with but a few times larger than the characteristic length of inhomogeneity.

A.V. Metrikine (✉)
Faculty of Civil Engineering and Geosciences, Delft University of Technology, 2600 GA Delft, The Netherlands
e-mail: A.Metrikine@tudelft.nl

G.A. Maugin, A.V. Metrikine (eds.), *Mechanics of Generalized Continua*, Advances in Mechanics and Mathematics 21,
DOI 10.1007/978-1-4419-5695-8_16, © Springer Science+Business Media, LLC 2010

The governing equations for gradient continua can be obtained phenomenologically [11, 22], by homogenization of an inhomogeneous continuum [7, 12, 13] or by continualization of a lattice [4, 20, 19, 21, 26, 31, 32]. The two latter approaches seem to be more useful as they allow the parameters of the gradient continuum to be explicitly expressed through parameters of the underlying micro-structural or inhomogeneous material.

Conventionally, in the homogenization and continualization techniques, attention has been paid to derivation of the equation of motion in dynamics and the force-balance equation in statics. The boundary-value problems have been addressed, but the natural boundary conditions have been retrieved from the equation of motion without noting that this procedure is not unique.

In this paper, the Lagrangian density of a gradient continuum is addressed. It seems to be important to derive this density as it should allow both the equation of motion and the natural boundary conditions to be defined consistently. It is shown, however, that a Lagrangian density derived by direct continualization of the Lagrangian of a cubic lattice results in contradictory expressions for the stresses in the continuum. It is argued then that the Lagrangian density should be trimmed by using the null-Lagrangians [21] such as to obtain non-contradictory expressions for the stresses. This is shown to be possible and a non-contradictory Lagrangian density is obtained for an isotropic second-order continuum, which constitutes the main novel result of this paper.

16.2 The Governing Equations and the Lagrangian of the Lattice

A cubic lattice of period l is considered that consists of identical particles of mass M and linear-elastic translational springs which connect these particles. Each particle of the lattice is connected to 50 neighbors which belong to the four smallest concentric spheres around this particle. This number of connections is the minimal necessary allowing an *isotropic* second-order gradient continuum to be derived.

Consider a central particle (m,n,k) whose undisturbed position is defined by $x_1 = nl$, $x_2 = ml$, $x_3 = kl$. The closest six neighbors of this particle belong to the sphere of radius l and are connected to the central particle by springs of stiffness K_1. The numbers of particles on the next spheres, the radii of which are $\sqrt{2}l$, $\sqrt{3}l$ and $\sqrt{5}l$, are 12, 8 and 24, respectively. The stiffnesses of the springs that connect the particles on these spheres to the central particle are K_2, K_3 and K_4, respectively. All springs are not pre-stressed.

The kinetic energy $\mathscr{U}_{\text{kin}}^{(m,n,k)}$ and the potential energy $\mathscr{U}_{\text{pot}}^{(m,n,k)}$, which are associated with the central particle (m,n,k), can be written as

$$\begin{aligned}
\mathscr{U}_{\text{kin}}^{(m,n,k)} &= \frac{M}{2}\sum_{i=1}^{3}\left(\dot{x}_i^{(m,n,k)}\right)^2,\\
\mathscr{U}_{\text{pot}}^{(m,n,k)} &= \frac{K_1}{2}\sum_{j=1}^{6}(\Delta l_j)^2 + \frac{K_2}{2}\sum_{j=7}^{18}(\Delta l_j)^2 + \frac{K_3}{2}\sum_{j=19}^{26}(\Delta l_j)^2 + \frac{K_4}{2}\sum_{j=27}^{50}(\Delta l_j)^2,
\end{aligned} \tag{16.1}$$

where $x_i^{(m,n,k)}(t)$, $i = 1, 2, 3$ are the displacements of the particle (m, n, k) from its undisturbed position, Δl_j are the elongations of the springs attached to the particle (m, n, k) that are obtained assuming small displacements of the particles.

To obtain the equations of motion of the particle $\Delta l_j^{(m,n,k)}$, the Lagrangian $\mathscr{L}^{(m,n)}$ is composed as the difference of $\mathscr{U}_{\text{kin}}^{(m,n,k)}$ and $\mathscr{U}_{\text{pot}}^{(m,n,k)}$, and then substituted into the Euler–Lagrange equation.

16.3 Continualization of the Lattice

To continualize the lattice, a procedure proposed by Metrikine and Askes [20] is used. According to this procedure, a non-local relation is introduced between the displacements $x_i^{(m,n,k)}$ of the particles and the displacements $u_i(x_1, x_2, x_3, t)$ of the to-be-derived continuum. The latter displacement is assumed to be a weighted average of the displacements of the particles surrounding the central particle (m, n, k):

$$u_i(\mathbf{x}, t) = \frac{a_1 \sum_1^6 x_i^{(\text{sphere 1})} + a_2 \sum_1^{12} x_i^{(\text{sphere 2})} + a_3 \sum_1^8 x_i^{(\text{sphere 3})} + a_4 \sum_1^{24} x_i^{(\text{sphere 4})}}{1 + 6a_1 + 12a_2 + 8a_3 + 24a_4}, \tag{16.2}$$

where a_1, a_2, a_3, a_4 are the dimensionless weighting coefficients for the particles belonging to the corresponding spheres. These coefficients may not be negative and should be smaller than unity.

To continualize the lattice, the kinematic relation (16.2) is to be inverted to express the displacements of the particles $x_i^{(m,n,k)}(t)$ through the continuum displacements. This inversion is carried out by assuming that the deviation of $x_i^{(m,n,k)}(t)$ from $u_i(\mathbf{x}, t)$ is small, so that the following relation holds

$$x_i^{(m,n,k)}(t) = u_i(\mathbf{x}, t) + \sum_{j=1}^{2N} l^j f_i^{(j)}(\mathbf{x}, t) + \mathscr{O}(L_i^{2N+1}), \tag{16.3}$$

where l is the period of the lattice, N is the order of the to-be-derived continuum and $f_i^{(j)}(\mathbf{x}, t)$ are unknown deviation functions, which have to be defined in correspondence with (16.2). The differential operators $L_i = l\partial/\partial x_i$ in (16.3) should be much smaller than unity in the sense that the result of application of these operators to a function should be much smaller that the function itself. Physically, this means that the wavelength of the processes to be considered should be sufficiently large relative to the period of the lattice. To find explicit expressions for the deviation functions, the displacements $x_i^{(m,n,k)}(t)$ and the displacements of the surrounding particles are to be substituted into the kinematic relation (16.2). The latter displacements can be derived expanding equation (16.3) into the Taylor series.

Using the above-outlined procedure the following equation of motion for a second-order ($N = 2$) gradient continuum can be derived:

$$\rho\ddot{u}_i - \rho l^2 A\ddot{u}_{i,jj} = \frac{2}{5}E(u_{i,jj} + 2u_{j,ij}) + \frac{1}{18}El^2(u_{i,jj} + 4u_{j,ij})_{,kk}$$
$$- \frac{2}{5}El^2 A(u_{i,jj} + 2u_{j,ij})_{,kk}, \tag{16.4}$$

where

$$A = \frac{a_1 + 4a_2 + 4a_3 + 20a_4}{1 + 6a_1 + 12a_2 + 8a_3 + 24a_4},$$
$$\rho = \frac{m}{l^3}, \qquad E = \frac{6K_3}{l}. \tag{16.5}$$

To achieve isotropy of the continuum in the bulk, when deriving (16.2), the following constraints were imposed on the spring constants:

$$K_1 = 4K_3/3, \qquad K_2 = 0, \qquad K_4 = K_3/3. \tag{16.6}$$

It is of great significance to note that all four weighting coefficients have been automatically accumulated in only one constant in the equation of motion.

To assure that both shear and compressional waves in the derived continuum have real speeds and taking into account that all weighting coefficients must be smaller than unity, it can be found that parameter A may vary within the following limits

$$\frac{25}{108} < A < \frac{4}{5}. \tag{16.7}$$

The exact value of this parameter for a physical material should be determined experimentally.

Equation (16.4) governs the linear dynamics of a second-order unbounded isotropic gradient continuum with Poisson's ration equal to 0.25. This continuum is unconditionally stable provided that constraint (16.7) is satisfied.

16.4 Volumetric Density of the Lagrangian

In order to formulate the natural boundary conditions that are in one-to-one correspondence with the derived equation of motion (16.4), it is customary to derive the volumetric density of the Lagrangian of the continuum. This can be done via direct continualization of the Lagrangian of the lattice. The continualization procedure is the same as proposed by Metrikine and Askes. Employing this procedure, the following expression for the volumetric density of the Lagrangian $\lambda(\mathbf{x}, t)$ can be derived:

$$\lambda = \frac{1}{2}\rho\dot{u}_i\dot{u}_i - \frac{1}{2}\rho l^2 A\dot{u}_i\dot{u}_{i,jj} - \frac{1}{5}E(u_{i,j}u_{j,i} + u_{i,j}u_{i,j} + u_{i,i}u_{j,j})$$
$$- \frac{1}{9}l^2 E\Big[u_{i,jk}u_{j,ik} + u_{i,ik}u_{j,jk} + u_{i,kk}u_{j,ij} + \frac{1}{4}u_{i,jj}u_{i,kk}$$

$$+ \frac{1}{2}u_{i,jk}u_{i,kj} + u_{i,j}u_{i,jkk} + u_{i,j}u_{j,ikk} + u_{i,i}u_{j,jkk} + 2u_{i,j}u_{k,ijk}\Big]$$
$$+ \frac{1}{5}l^2 AE(u_{i,j}u_{i,jkk} + u_{i,j}u_{j,ikk} + u_{i,i}u_{j,jkk}), \tag{16.8}$$

where the mass density and the Young's modulus are defined by (16.6).

Application of the Hamilton–Ostrogradsky variational principle allows the equation of motion and the natural boundary conditions (the standard and higher-order stresses) to be expressed in terms of the density of the Lagrangian. As shown in [21], for any Lagrangian density whose dependence of the continuum displacements can be expressed as

$$\lambda = \lambda(\dot{u}_i, u_{i,j}, u_{i,jk}, \dot{u}_{i,jk}, u_{i,jkl}) \tag{16.9}$$

the equation of motion, the standard stresses, and the second-order stresses are given as

$$\frac{\partial}{\partial t}\frac{\partial\lambda}{\partial\dot{u}_i} + \frac{\partial}{\partial x_j}\frac{\partial\lambda}{\partial u_{i,j}} - \frac{\partial^2}{\partial x_j\partial x_k}\frac{\partial\lambda}{\partial u_{i,jk}}$$
$$+ \frac{\partial^3}{\partial t\partial x_j\partial x_k}\frac{\partial\lambda}{\partial\dot{u}_{i,jk}} + \frac{\partial^3}{\partial x_j\partial x_k\partial x_l}\frac{\partial\lambda}{\partial u_{i,jkl}} = 0, \tag{16.10}$$
$$\sigma_{ij} = -\frac{\partial\lambda}{\partial u_{i,j}} + \frac{\partial}{\partial x_k}\frac{\partial\lambda}{\partial u_{i,jk}} - \frac{\partial^2}{\partial t\partial x_k}\frac{\partial\lambda}{\partial\dot{u}_{i,jk}} - \frac{\partial^2}{\partial x_k\partial x_l}\frac{\partial\lambda}{\partial u_{i,jkl}}, \tag{16.11}$$
$$\tau_{ijk} = -\frac{\partial\lambda}{\partial u_{i,jk}} + \frac{\partial}{\partial t}\frac{\partial\lambda}{\partial\dot{u}_{i,jk}} + \frac{\partial}{\partial x_l}\frac{\partial\lambda}{\partial u_{i,jkl}}. \tag{16.12}$$

Substituting (16.8) into (16.10) results in the equation of motion (16.4). The stresses resulting from (16.11) and (16.12) read

$$\sigma_{ij} = \frac{2}{5}E(u_{i,j} + u_{j,i} + u_{k,k}\delta_{ij}) + \frac{l^2 E}{18}(u_{i,jkk} + u_{j,ikk} + 2u_{k,ijk} + u_{k,kll}\delta_{ij})$$
$$+ \frac{1}{2}\rho l^2 A\ddot{u}_{i,j} + \frac{2}{15}l^2 AE(3u_{i,jkk} + 2u_{j,ikk} + 2u_{k,ijk} + 2u_{k,kll}\delta_{ij}), \tag{16.13}$$
$$\tau_{ijk} = -\frac{\rho l^2}{2}A\ddot{u}_i\delta_{jk} + \frac{l^2 E}{54}\big[2u_{i,jk} + 2u_{k,ij} + 2u_{j,ik} + (2u_{l,kl} + u_{k,ll})\delta_{ij}$$
$$+ (2u_{l,jl} + u_{j,ll})\delta_{ik} + (2u_{l,il} + u_{i,ll})\delta_{jk}\big]$$
$$+ \frac{l^2 A}{15}E\big[2u_{i,jk} + u_{j,ik} + u_{k,ij} + u_{l,kl}\delta_{ij} + u_{l,jl}\delta_{ik} + (u_{i,ll} + 2u_{l,il})\delta_{jk}\big]. \tag{16.14}$$

Let us check the symmetry of the standard stress tensor. From (16.13), the following holds:

$$\sigma_{ij} - \sigma_{ji} = \frac{1}{2}\rho l^2 A(\ddot{u}_{i,j} - \ddot{u}_{j,i}) + \frac{2}{15}l^2 AE(u_{i,jkk} - u_{j,ikk}). \tag{16.15}$$

Obviously, the standard stress tensor is asymmetric. The asymmetry associated with the time derivatives can be expected as the non-local relation (16.2) introduces an effective mass moment of inertia to every differential element of the continuum. In statics, however, the asymmetry of the standard stress tensor must disappear and the following equation must hold:

$$\sigma_{ij}^{\text{static}} = \sigma_{ji}^{\text{static}}. \tag{16.16}$$

Yet, (16.15) obviously shows that σ_{ij} remains asymmetric also in statics. Thus, we have come to an inconsistency that the density of the Lagrangian given by (16.8) results in a seemingly non-contradictory equation of motion (16.2) and in a contradictory non-symmetric stress tensor. This inconsistency will be addressed in the next section.

Before that, however, let us illuminate a few other controversies contained in (16.13) and (16.14) which are associated with the second-order stress. As argued in [4], the standard stresses and the second-order stresses have to satisfy the following equation that reflects the balance of forces at the lower observation scale:

$$\sigma_{ij}^{(1)} = -\tau_{ijk,k}, \tag{16.17}$$

where $\sigma_{ij}^{(1)}$ is the higher-order part of the stress tensor that is proportional to l^2. From (16.13) and (16.15), it follows that

$$\begin{aligned}\sigma_{ij}^{(1)} + \tau_{ijk,k} = &\frac{1}{9} l^2 E(u_{i,jkk} + u_{j,ikk} + 2u_{k,ijk} + u_{k,kll}\delta_{ij}) \\ &+ \frac{1}{15} l^2 AE(9u_{i,jkk} + 5u_{j,ikk} + 8u_{k,ijk} + 5u_{k,kll}\delta_{ij}).\end{aligned} \tag{16.18}$$

Equations (16.17) and (16.18) are obviously incompatible, which constitutes an inconsistency of the standard and second-order stresses obtained from the Lagrangian (16.8).

The second-order stresses given by (16.14) are also self-inconsistent. The physical meaning of these stresses was suggested in [22] and elaborated upon in [4]. Based on the above papers, it can be concluded that in order to ensure equilibrium of the moment of momentum at the lower scale and to have non-contradicting expressions at the corners of a considered lower-scale volume, τ_{ijk} must be invariant with respect to all possible permutations of the indices, i.e., the following identities must hold:

$$\tau_{ijk}^{\text{static}} = \tau_{ikj}^{\text{static}} = \tau_{jik}^{\text{static}} = \tau_{jki}^{\text{static}} = \tau_{kij}^{\text{static}} = \tau_{kji}^{\text{static}}. \tag{16.19}$$

Obviously, the stress-tensor defined by (16.14) does not satisfy the above identities.

Thus, the Lagrangian (16.8) results in a handful of contradictions. In the next section, this function is modified to a unique form that allows the above-formulated contradictions to be removed.

16.5 Non-Contradictory Lagrangian and Stresses

Strictly speaking, the Lagrangian derived in the previous section cannot be expected to automatically satisfy the symmetries of the standard and the higher-order stress-tensors. Indeed, (16.8) was derived from the Lagrangian of an inner particle of the lattice. Consequently, the derived density of the Lagrangian of the continuum must provide a non-contradictory equation of motion, whereas the natural boundary conditions that result from this function can be just expected to be correct as they are obtained from the same variational procedure. As shown in the previous section, this expectation is false and the obtained Lagrangian density results in the contradictory stress-tensors. Thus, one and the same Lagrangian gives a "correct" equation of motion and "incorrect" natural boundary conditions. Given that, it seems reasonable to raise the following question: Is it possible to modify the Lagrangian (16.8) so that, upon insertion into (16.10)–(16.12), it would give the "correct" equation of motion (16.2) and non-contradictory expressions for the stresses? Such a question would be unthinkable for a local non-gradient continuum as the equation of motion of such a continuum is in one-to-one correspondence with a Lagrangian defined to within an additive total time derivative of any function of coordinates and time. The gradient continua, however, do not possess this property. It is known, see, e.g., [21], that there exists more than one Lagrangian that leads to the same equation of motion. Indeed, it is obvious that the volumetric density of the Lagrangian

$$\tilde{\lambda} = \lambda(\dot{u}_i, u_{i,j}, u_{i,jk}, \dot{u}_{i,jk}, u_{i,jkl}) + \lambda_0(u_{i,j}, u_{i,jk}, u_{i,jkl}) \tag{16.20}$$

results in the same equation of motion as the density λ given by (16.9), provided that λ_0 satisfies the following equation:

$$\frac{\partial}{\partial x_j}\frac{\partial \lambda_0}{\partial u_{i,j}} - \frac{\partial^2}{\partial x_j \partial x_k}\frac{\partial \lambda_0}{\partial u_{i,jk}} + \frac{\partial^3}{\partial x_j \partial x_k \partial x_l}\frac{\partial \lambda_0}{\partial u_{i,jkl}} = 0. \tag{16.21}$$

The following expressions for λ_0 can be given as examples:

$$\lambda_0 = u_{i,j}u_{i,jkk} + u_{i,jk}u_{i,jk}. \tag{16.22}$$

Additionally, a number of quadratic forms in the Lagrangian density (that include derivatives of the order higher than one) result in the same term in the equation of motion. For example, $u_{i,jk}u_{j,ik}$, $u_{i,ik}u_{j,jk}$, and $u_{i,j}u_{j,ikk}$ upon substitution into (16.10) result in $-2u_{j,ijkk}$. Thus, it is definitely possible to modify the Lagrangian density in (16.8) without changing the corresponding equation of motion. The question now is: What modifications are to be undertaken? To answer this question, it is customary to classify the terms in the Lagrangian density as shown in Table 16.1. The second column in Table 16.1 shows the higher-order quadratic forms that are contained in the Lagrangian (16.8). The next columns contain the corresponding terms in the equation of motion, in the standard stress and in the second-order stress obtained according to (16.10), (16.11) and (16.12), respectively. The last two columns show if the symmetry of the standard stress tensor and the relation between the standard and the second-order stresses are respected.

The most important conclusion that can be drawn from Table 16.1 is that the relation $\sigma_{ij}^{(1)} = -\tau_{ijk,k}$ is respected by none of the quadratic forms that contain the third derivatives of the displacement (the last four terms in the λ-column). Therefore, to respect the above relation these terms must be removed from the Lagrangian. This is a far-reaching conclusion that means, actually, that the Lagrangian of a consistent second-order continuum *may not contain derivatives of the displacement of the order higher than two*. Most likely, this conclusion can be generalized to an Nth order gradient continuum: the highest derivative of the displacement in the Lagrangian must be the same as the order of the continuum.

Thus, the question now is: How to remove the terms containing the third displacement derivatives from the Lagrangian without changing the resulting equation of motion? As the third column of Table 16.1 shows, it can be done quite easily, as all quadratic terms in the second column of Table 16.1 produce in the equation of motion either $\pm 2u_{j,ijkk}$ or $\pm 2u_{i,jjkk}$. Straightforward evaluations show that the originally derived Lagrangian (16.8) has to be replaced by the following trimmed Lagrangian, the higher-order part of which contains only second derivatives of the displacement:

$$\begin{aligned}\lambda = &\frac{1}{2}\rho\dot{u}_i\dot{u}_i - \frac{1}{2}\rho l^2 A\dot{u}_{i}\dot{u}_{i,jj} - \frac{1}{5}E(u_{i,j}u_{j,i} + u_{i,j}u_{i,j} + u_{i,i}u_{j,j})\\ &+ \frac{l^2 E}{27}\left[u_{i,jk}u_{j,ik} + u_{i,ik}u_{j,jk} + u_{i,kk}u_{j,ij} + \frac{1}{4}u_{i,jj}u_{i,kk} + \frac{1}{2}u_{i,jk}u_{i,kj}\right]\\ &- \frac{l^2 A}{5}E(u_{i,jk}u_{i,kj} + 2u_{i,jk}u_{j,ik}). \end{aligned} \tag{16.23}$$

The insertion of the above Lagrangian into (16.13) and (16.14) results in the following expressions for the standard and the second-order stresses:

$$\begin{aligned}\sigma_{ij} = &\frac{2}{5}E(u_{i,j} + u_{j,i} + u_{k,k}\delta_{ij}) + \frac{l^2 E}{18}(u_{i,jkk} + u_{j,ikk} + 2u_{k,ijk} + u_{k,kll}\delta_{ij})\\ &+\frac{1}{2}\rho l^2 A\ddot{u}_{i,j} - \frac{2}{5}l^2 AE(u_{i,jkk} + u_{j,ikk} + u_{k,ijk}), \end{aligned} \tag{16.24}$$

$$\begin{aligned}\tau_{ijk} = &-\frac{l^2 E}{54}\left[2u_{i,jk} + 2u_{j,ik} + 2u_{k,ij} + (2u_{l,kl} + u_{k,ll})\delta_{ij}\right.\\ &\left.+ (2u_{l,jl} + u_{j,ll})\delta_{ik} + (2u_{l,il} + u_{i,ll})\delta_{jk}\right]\\ &- \frac{1}{2}\rho l^2 A\ddot{u}_i\delta_{jk} + \frac{2}{15}l^2 AE(u_{i,jk} + u_{j,ik} + u_{k,ij}). \end{aligned} \tag{16.25}$$

It can be checked straightforwardly that the insertion of the Lagrangian (16.23) into the Euler–Lagrange equation gives the equation of motion (16.4) that was derived by the direct continualization of the lattice. The symmetry of the standard stress tensor in statics, (16.16), is obviously observed by (16.24). The second order stress is invariant in statics with respect to the permutation of indices, which means that

Table 16.1

	λ	EM	σ_{ij}	τ_{ijk}	$\sigma_{ij} = \sigma_{ji}$	$\sigma_{ij}^{(1)} = -\tau_{ijk,k}$
1	$u_{i,jk}u_{j,ik}$	$-2u_{j,ijkk}$	$u_{j,ikk} + u_{k,ijk}$	$-u_{j,ik} - u_{k,ij}$	no	yes
2	$u_{i,ik}u_{j,jk}$	$-2u_{j,ijkk}$	$u_{l,kkl}\delta_{ij} + u_{l,ijl}$	$-u_{l,kl}\delta_{ij} - u_{l,jl}\delta_{ik}$	yes	yes
3	$u_{i,kk}u_{j,ij}$	$-2u_{j,ijkk}$	$(u_{j,ill} + 2u_{l,ijl} + u_{k,kll}\delta_{ij})/2$	$-(u_{j,ll}\delta_{ik} + u_{k,ll}\delta_{ij} + 2u_{l,il}\delta_{jk})/2$	no	yes
4	$u_{i,jj}u_{i,kk}$	$-2u_{i,jjkk}$	$2u_{i,jkk}$	$-2u_{i,ll}\delta_{jk}$	no	yes
5	$u_{i,jk}u_{i,kj}$	$-2u_{i,jjkk}$	$2u_{i,jkk}$	$-2u_{i,jk}$	no	yes
6	$u_{i,j}u_{i,jkk}$	$2u_{i,jjkk}$	$-2u_{i,jkk}$	$(2u_{i,jk} + u_{i,ll}\delta_{jk})/3$	no	no
7	$u_{i,j}u_{j,ikk}$	$2u_{j,ijkk}$	$-2(2u_{j,ikk} + u_{k,ijk})/3$	$(u_{j,ik} + u_{k,ij} + u_{l,il}\delta_{jk})/3$	no	no
8	$u_{i,i}u_{j,jkk}$	$2u_{j,ijkk}$	$-2(2u_{k,kll}\delta_{ij} + u_{k,ijk})/3$	$(u_{l,kl}\delta_{ij} + u_{l,jl}\delta_{ik} + u_{l,il}\delta_{jk})/3$	yes	no
9	$u_{i,j}u_{k,ijk}$	$2u_{j,ijkk}$	$-(4u_{k,ijk} + u_{j,ikk} + u_{k,kll}\delta_{ij})/3$	$(u_{j,ll}\delta_{ik} + u_{l,jl}\delta_{ik} + u_{k,ll}\delta_{ij}$ $+ u_{l,kl}\delta_{ij} + u_{j,ik} + u_{k,ij})/6$	no	no

the requirement (16.19) is satisfied. So is (16.17), as can be checked by direct substitution of (16.24) and (16.25). Thus, the Lagrangian given by (16.23) provides a non-controversial description of a second-order isotropic gradient elastic continuum.

16.6 Discussion and Conclusions

It is well-known that the Lagrangian of a system is defined to within an additive total time derivative of any function of coordinates and time [18]. In this paper, a less-known non-uniqueness of the Lagrangian that is associated with gradient continua has been addressed. For such continua, the Lagrangian is defined to within the so-called null-Lagrangians whose contribution to the Euler–Lagrange equation is zero. It is important to note that the null-Lagrangians are not total time derivatives of a function.

Whereas the classical non-uniqueness of the Lagrangian does not influence the prediction of the dynamics of systems, the null-Lagrangians do affect that of gradient continua. The reason for the latter is that the null-Lagrangians do contribute to the natural boundary conditions (expressions for the stresses). This means that the non-uniqueness associated with the null-Lagrangians is unacceptable, as every physical system must correspond to one or a set of the Lagrangians that result in one and the same boundary value problem.

The main aim of this paper has been to show that a unique Lagrangian can be identified for an isotropic gradient continuum. To achieve this, first, by means of continualization of the Lagrangian of a cubic lattice, a density of the Lagrangian of a second-order gradient isotropic continuum has been derived. Then, it has been shown that the derived density corresponds to a non-contradictory equation of motion and, at the same time, results in contradictory expressions for both the classical and higher-order stresses. To resolve this, a procedure has been proposed that, making use of the null-Lagrangians, allows the Lagrangian density to be trimmed so that the stresses become not contradictory while the equation of motion remains unaffected. The presented trimming procedure has clearly shown that the Lagrangian of a second-order continuum may not contain the displacement derivatives of the order higher than second. It is tempting to generalize this statement to an Nth order gradient continuum by stating that the order of the displacement derivatives in the Lagrange function of such a continuum may not be higher than N. The latter statement, however, has not been proven in this paper.

It should be noted that whenever the Lagrangian density of an Nth order gradient continuum is derived (not postulated phenomenologically), either by homogenization of an inhomogeneous continuum or continualization of a lattice, this density seems to contain the displacement derivatives of the order higher than N [5, 21, 24]. The results of this paper suggest that these densities should not be used in formulating boundary value problems which require formulation of the natural boundary conditions. Another indirect but useful conclusion of this paper can be formulated as follows. When the equation of motion is taken as the starting point in the finite element formulation of a boundary-value problem [9, 12, 13], one has to keep in mind that the natural boundary conditions cannot be derived uniquely unless certain symmetry requirements are employed to obtain a non-contradictory set of the boundary conditions.

References

1. Aifantis, E.C.: On the role of gradients in the localization of deformation and fracture. Int. J. Eng. Sci. **30**, 1279–1299 (1992)
2. Altan, B.S., Aifantis, E.C.: On some aspects in the special theory of gradient elasticity. J. Mech. Behav. Mater. **8**, 231–282 (1997)
3. Askes, H., Aifantis, E.C.: Gradient elasticity theories in statics and dynamics a unification of approaches. Int. J. Fract. **139**, 297–304 (2006)
4. Askes, H., Metrikine, A.V.: One-dimensional dynamically consistent gradient elasticity models derived from a discrete microstructure. Part 1: Static and dynamic response. Eur. J. Mech. A, Solids **21**, 573–588 (2002)
5. Askes, H., Metrikine, A.V.: Higher-order continua derived from discrete media: continualisation aspects and boundary conditions. Int. J. Solids Struct. **42**, 187–202 (2005)
6. Askes, H., Sluys, L.J.: Explicit and implicit gradient series in damage mechanics. Eur. J. Mech. A, Solids **21**, 379–390 (2002)
7. Chen, W., Fish, J.: A dispersive model for wave propagation in periodic heterogeneous media based on homogenization with multiple spatial and temporal scales. J. Appl. Mech., Trans. ASME **68**, 153–161 (2001)
8. Chen, J.Y., Huang, Y., Ortiz, M.: Fracture analysis of cellular materials: a strain gradient model. J. Mech. Phys. Solids **46**, 789–828 (1998)
9. de Borst, R., Mühlhaus, H.-B.: Gradient-dependent plasticity—formulation and algorithmic aspects. Int. J. Numer. Methods Eng. **35**, 521–539 (1992)
10. de Borst, R., Pamin, J., Peerlings, R.H.J., Sluys, L.J.: On gradient-enhanced damage and plasticity models for failure in quasi-brittle and frictional materials. Comput. Mech. **17**, 130–141 (1995)
11. Erofeyev, V.I.: Wave Processes in Solids with Microstructure. World Scientific, Singapore (2003)
12. Fish, J., Chen, W., Nagai, G.: Non-local dispersive model for wave propagation in heterogeneous media: one-dimensional case. Int. J. Numer. Methods Eng. **54**, 331–346 (2002)
13. Fish, J., Chen, W., Nagai, G.: Non-local dispersive model for wave propagation in heterogeneous media: multi-dimensional case. Int. J. Numer. Methods Eng. **54**, 347–363 (2002)
14. Fleck, N.A., Hutchinson, J.W.: A reformulation of strain gradient plasticity. J. Mech. Phys. Solids **49**, 2245–2271 (2001)
15. Fleck, N.A., Muller, G.M., Ashby, M.F., Hutchinson, J.W.: Strain gradient plasticity—theory and experiment. Acta Metall. Sin. **42**, 475–487 (1994)
16. Georgiadis, H.G., Vardoulakis, I., Lykortafitis, G.: Torsional surface waves in a gradient-elastic half-space. Wave Motion **31**, 333–348 (2000)
17. Kuroda, M., Tvergaard, V.: A finite deformation theory of higher-order gradient crystal plasticity. J. Mech. Phys. Solids **56**, 2573–2584 (2008)
18. Landau, L.D., Lifshitz, E.M.: Mechanics, 3rd edn. Butterworth-Heinemann, Oxford (2003)
19. Metrikine, A.V.: On causality of the gradient elasticity models. J. Sound Vib. **297**, 727–742 (2006)
20. Metrikine, A.V., Askes, H.: One-dimensional dynamically consistent gradient elasticity models derived from a discrete microstructure, Part 1: Generic formulation. Eur. J. Mech. A, Solids **21**, 555–572 (2002)

21. Metrikine, A.V., Askes, H.: An isotropic dynamically consistent gradient elasticity model derived from a 2D lattice. Philos. Mag. **86**, 3259–3286 (2006)
22. Mindlin, R.D., Tiersten, H.F.: Effects of couple-stresses in linear elasticity. Arch. Ration. Mech. Anal. **11**, 415–448 (1962)
23. Mühlhaus, H.-B., Aifantis, E.C.: A variational principle for gradient plasticity. Int. J. Solids Struct. **28**, 845–857 (1991)
24. Mühlhaus, H.-B., Oka, F.: Dispersion and wave propagation in discrete and continuous models for granular materials. Int. J. Solids Struct. **33**, 2841–2858 (1996)
25. Pamin J.: Gradient-dependent plasticity in numerical simulation of localization phenomena. PhD thesis, Delft University of Technology (1996)
26. Pasternak, E., Mühlhaus, H.-B.: Generalised homogenisation procedures for granular materials. J. Eng. Math. **52**, 199–229 (2005)
27. Peerlings, R.H.J., de Borst, R., Brekelmans, W.A.M., de Vree, J.H.P.: Gradient enhanced damage for quasi-brittle materials. Int. J. Numer. Methods Eng. **39**, 3391–3403 (1996)
28. Peerlings, R.H.J., de Borst, R., Brekelmans, W.A.M., Geers, M.G.D.: Gradient-enhanced damage modelling of concrete fracture. Mech. Cohesive-Frict. Mater. **3**, 323–342 (1998)
29. Rubin, M.B., Rosenau, P., Gottlieb, O.: Continuum model of dispersion caused by an inherent material characteristic length. J. Appl. Phys. **77**, 4054–4063 (1995)
30. Sluys, L.J.: Wave propagation, localisation and dispersion in softening solids. PhD thesis, Delft University of Technology (1992)
31. Suiker, A.S.J., de Borst, R., Chang, C.S.: Micro-mechanical modelling of granular material. Part 1: Derivation of a second-gradient micro-polar constitutive theory. Acta Mech. **149**, 161–180 (2001)
32. Suiker, A.S.J., de Borst, R., Chang, C.S.: Micro-mechanical modelling of granular material. Part 2: Plane wave propagation in infinite media. Acta Mech. **149**, 181–200 (2001)
33. Suiker, A.S.J., Metrikine, A.V., de Borst, R.: Comparison of wave propagation characteristics of the Cosserat continuum model and corresponding discrete lattice models. Int. J. Solids Struct. **38**, 1563–1583 (2001)
34. Triantafyllidis, N., Aifantis, E.C.: A gradient approach to localization of deformation. 1. Hyperelastic materials. J. Elast. **16**, 225–237 (1986)
35. Wang, W.M., Askes, H., Sluys, L.J.: Gradient viscoplastic modelling of material instabilities in metals. Met. Mater. Int. **4**, 537–542 (1998)

Chapter 17
Dynamic Properties of Essentially Nonlinear Generalized Continua

Alexey V. Porubov, Eron L. Aero, and B.R. Andrievsky

Abstract Essentially nonlinear proper structural model for continua with internal structure is analyzed using the localized strain wave solutions of the governing equations for macro- and micro-strains. The intervals of the velocity are found when either compression or tensile localized strain waves exist. It is obtained that simultaneous existence of compression and tensile macro-strain waves is impossible in contrast to the micro-strains responsible for the lattice defects. Also, it is shown that similar profiles of the macro-strain solitary waves may be accompanied by distinct profiles of the micro-strain waves. Generation of the bell-shaped defects in the lattice is studied numerically that allows us to describe structural deviations caused by the dynamical loading due to the localized macro-strain wave propagation.

17.1 Modeling of Essentially Nonlinear Processes in Media with Internal Structures

Usually, classic elastic materials possess weakly nonlinear features that allow us to employ truncated power series in strains to model nonlinearity. However, the presence of an internal structure gives rise to an essential nonlinearity [1, 2]. Thus, it was found in [3, 4] that the presence of the components with contrasting elastic features in some rocks and soils gives rise to the essential nonlinearity, and the contribution of the quadratic and cubic nonlinearities in the stress-strain relationship

A.V. Porubov (✉) · E.L. Aero · B.R. Andrievsky
Institute of Problems in Mechanical Engineering, Bolshoy av. 61, V.O., St. Petersburg 199178, Russia
e-mail: porubov.math@mail.ioffe.ru

E.L. Aero
e-mail: 16aero@mail.ru

B.R. Andrievsky
e-mail: bandri@yandex.ru

G.A. Maugin, A.V. Metrikine (eds.), *Mechanics of Generalized Continua*,
Advances in Mechanics and Mathematics 21,
DOI 10.1007/978-1-4419-5695-8_17, © Springer Science+Business Media, LLC 2010

turns out to be of the same order. Similar effect has been found in [5, 6] while studying an influence of an external magnetic field on paramagnetic crystals. In both cases, the weakly nonlinear models in the form of power series truncation cannot be applied in a strict sense. Nevertheless, they are used already as the exact expressions in the framework of a *phenomenological* approach. Some of such phenomenological models are considered in [8] where it is shown that they are usually transformed into a single equation for macro-strains, while micro- or internal field is introduced via variations of the constants relative to those of the classic elastic materials. An advantage of these models is in the possibility to measure or estimate the values of these constants.

Essentially nonlinear processes may be described by a proper model for the internal structure of the material. Thus, the rotatory molecular groups were added to the usual one in atomic chain in [7], and large rotations were considered. A more complicated internal motion is modeled in [1, 2] where translational internal motion is considered together with rotations. In both models, this essential nonlinearity has not been modeled by any power series truncation.

Here we consider essentially proper structural nonlinear models that treats a continuum approach and a crystal translational symmetry without making a continuum limit of a discrete model [1, 2]. The equations are derived for the vectors of macro-displacement U and relative micro-displacement u for the pair of atoms with masses m_1, m_2,

$$\mathbf{U} = \frac{m_1\mathbf{U}_1 + m_2\mathbf{U}_2}{m_1 + m_2}, \qquad \mathbf{u} = \frac{\mathbf{U}_1 - \mathbf{U}_2}{a},$$

where a is a period of sub-lattice. The first variable allows describing macro-strains, while the second variable accounts for the reference displacement of the internal or the lattice structure. The following coupled governing equations are obtained in [1, 2] in the 1D case:

$$\rho U_{tt} - EU_{xx} = S\big(\cos(u) - 1\big)_x, \tag{17.1}$$

$$\mu u_{tt} - \kappa u_{xx} = (SU_x - p)\sin(u). \tag{17.2}$$

The choice of the trigonometric function allows us to describe translational symmetry of the crystal lattice. A comparison between phenomenological models and the structural model (17.1), (17.2) has been done in [8]. Here we continue this comparison emphasizing an advantage of the structural modeling in description of the lattice defects using exact solitary wave solutions and numerical solutions of (17.1) and (17.2).

17.2 Localized Macro-Strain Waves and Corresponding Deviations in Crystalline Lattice

It is possible to obtain exact traveling wave solutions to (17.1) and (17.2) depending on the phase variable $\theta = x - Vt$. Indeed, (17.1) is resolved for the micro-field by

$$\cos(u) = 1 - \frac{(E - \rho V^2)U_\theta - \sigma}{S}, \tag{17.3}$$

where σ is a constant of integration. Equation (17.2) is integrated once, multiplied by u_θ and integrated again. Then (17.3) is substituted into this equation, finally giving an ordinary differential equation for the macro-strain $v = U_\theta$:

$$v_\theta^2 = a_0 + a_1 v + a_2 v^2 + a_3 v^3 + a_4 v^4, \tag{17.4}$$

with the coefficients found in [8] where the solutions of (17.4) vanishing at infinities were studied in detail. Here we only briefly mention the main results. When $a_0 = 0$, $a_1 = 0$, the ODE (17.4) possesses known exact localized solutions of two kinds that may be obtained by direct integration

$$v_1 = \frac{A}{Q\cosh(k\theta) + 1}, \tag{17.5}$$

$$v_2 = -\frac{A}{Q\cosh(k\theta) - 1}. \tag{17.6}$$

The parameters for $\sigma = 0$ read

$$A = \frac{4S}{\rho(c_0^2 + c_L^2 - V^2)}, \qquad Q_\pm = \pm\frac{c_L^2 - V^2 - c_0^2}{c_L^2 - V^2 + c_0^2}, \qquad k = 2\sqrt{\frac{p}{\mu(c_l^2 - V^2)}}, \tag{17.7}$$

and for $\sigma = -2S$ we have

$$A = \frac{4S}{\rho(c_0^2 + V^2 - c_L^2)}, \qquad Q_\pm = \pm\frac{V^2 - c_L^2 - c_0^2}{V^2 - c_L^2 + c_0^2}, \qquad k = 2\sqrt{\frac{p}{\mu(V^2 - c_l^2)}}, \tag{17.8}$$

where $c_L^2 = E/\rho$, $c_l^2 = \kappa/\mu$, $c_0^2 = S^2/(p\rho)$.

An analysis of the shapes of the exact solutions also should take into account the coupling governed by (17.3). Depending upon the value of the first derivative at $\theta = 0$, one can express it in a different way. Reversing the cosine function for derivation of the expression for u, one has to avoid the point where the first derivative does not exist. This breaking happens for $\theta = 0$ at $\sigma = 0$ and for $Q = Q_+$. Therefore, the solution for u obtained using both (17.5) and (17.6) should be written as

$$u = \pm\arccos\left(\frac{(\rho V^2 - E)U_x}{S} + 1\right) \quad \text{for } \theta \le 0, \tag{17.9}$$

$$u = \pm\pi \mp \arccos\left(\frac{(\rho V^2 - E)U_x}{S} + 1\right) \quad \text{for } \theta > 0, \tag{17.10}$$

Table 17.1 Wave shapes for $\sigma = 0$

V^2	$(0; c_L^2 - c_0^2)$	$(c_L^2 - c_0^2; c_L^2)$	$(c_L^2; c_L^2 + c_0^2)$	$> c_L^2 + c_0^2$
Shape of v	Tensile (17.5)	Tensile (17.5)	Compression (17.6)	Compression (17.5)
Shape of u	Kink	Bell-shaped	Kink	Kink
Choice of $Q_\pm$	Q_+	Q_-	Q_+	Q_+

Table 17.2 Wave shapes for $\sigma = -2S$

V^2	$(0; c_L^2 - c_0^2)$	$(c_L^2 - c_0^2; c_L^2)$	$(c_L^2; c_L^2 + c_0^2)$	$> c_L^2 + c_0^2$
Shape of v	Compression (17.5)	Compression (17.6)	Tensile (17.5)	Tensile (17.5)
Shape of u	Kink	Kink	Bell-shaped	Kink
Choice of $Q_\pm$	Q_+	Q_+	Q_-	Q_+

However, the first derivative is zero for $Q = Q_-$ at $\theta = 0$, and the expression for u reads

$$u = \pm \arccos\left(\frac{(\rho V^2 - E)U_x}{S} + 1\right). \tag{17.11}$$

The solution (17.9)–(17.10) accounts for the kink-shaped profile of the wave, while solution (17.11) describes bell-shaped localized wave. The velocity intervals when one or another profile exists are shown in Table 17.1. Similar results for $\sigma = -2S$ are summarized in Table 17.2.

It is known [8] that simultaneous existence of macro-strain waves of tensile and compression is impossible. This is not true for the waves u describing deviations in the lattice. In particular, the localized bell-shaped profile exists in the interval $(c_L^2; c_L^2 + c_0^2)$ for $\sigma = 0$. Shown in Fig. 17.1 is the propagation of macro-strain tensile wave and the tensile wave of u. However, the same macro-strain wave may be accompanied by the micro-strain compression wave, this is governed by the initial condition for u. An important problem is to know which wave is generated when an initial condition for u differs from the exact solution (17.11) at $t = 0$. It will be studied numerically in the next section.

17.3 Generation of Localized Defects in Crystalline Lattice

We consider only the interval $(c_L^2; c_L^2 + c_0^2)$ for $\sigma = 0$. According to the exact solution, see Table 17.1, here only bell-shaped localized waves may propagate. Numerical simulations with initial conditions coinciding with the exact solution confirm this prediction of the exact solution as shown in Fig. 17.1.

What happens when an initial profile contains both the tensile and compression parts, and how its initial position relative to the initial position of the macro-strain

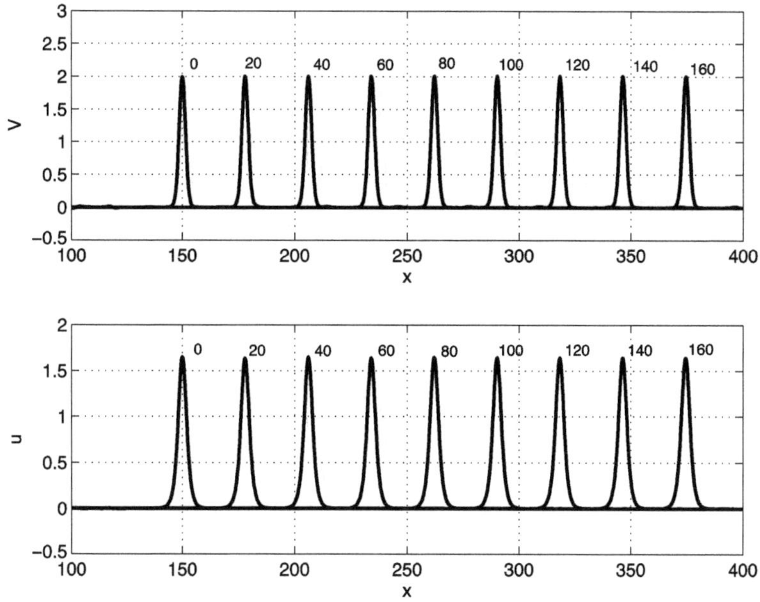

Fig. 17.1 Simultaneous propagation of macro- and micro-strain waves according to the exact solution. Points of time correspond to the neighboring peaks

wave v affect generation of the localized waves? To address this question, we keep the initial condition for v in the form of the exact solution while only a small disturbance of u is used.

One can see in Fig. 17.2 that tensile localized wave u is generated, and its amplitude is modulated in time below the value of the exact solution shown in Fig. 17.1. The value of the amplitude of the macro-strain wave v is close to that of the exact solution shown in Fig. 17.1 but both the macro-strain wave and the wave u accelerate. Also additional disturbances arise around the bell-shaped waves. Note that an initial position of the peak of small disturbance u coincides with that of the macro-strain wave v.

One can see in Fig. 17.3 that a compression bell-shaped wave is generated if the initial position of the negative part of the input of u coincides with that of the peak of the macro-strain wave v. The positive part does not give rise to the formation of the tensile waves. Profiles of the localized compression wave is mirror to the tensile one shown in Fig. 17.2.

However, the coincidence of the initial positions is not necessary for the formation of localized waves. One can see in Fig. 17.4 that the compression wave is generated as soon as the macro-strain wave v overtakes the micro-strain wave u. It meets first the compression part of u that results in generation of the compression localized wave while the tensile wave is not generated. Similarly, a tensile wave is generated if the macro-wave first achieves the tensile part of the micro-strain wave u.

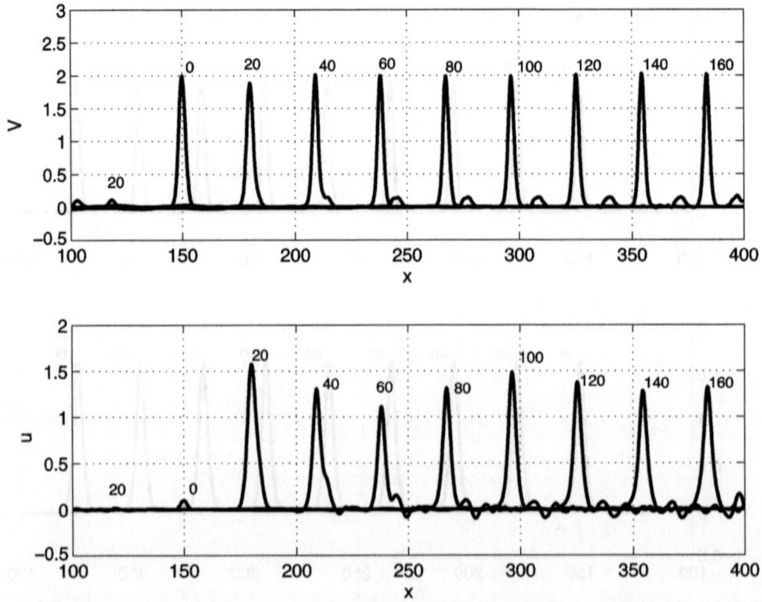

Fig. 17.2 Generation of moving tensile localized defect of lattice. Points of time correspond to the neighboring peaks

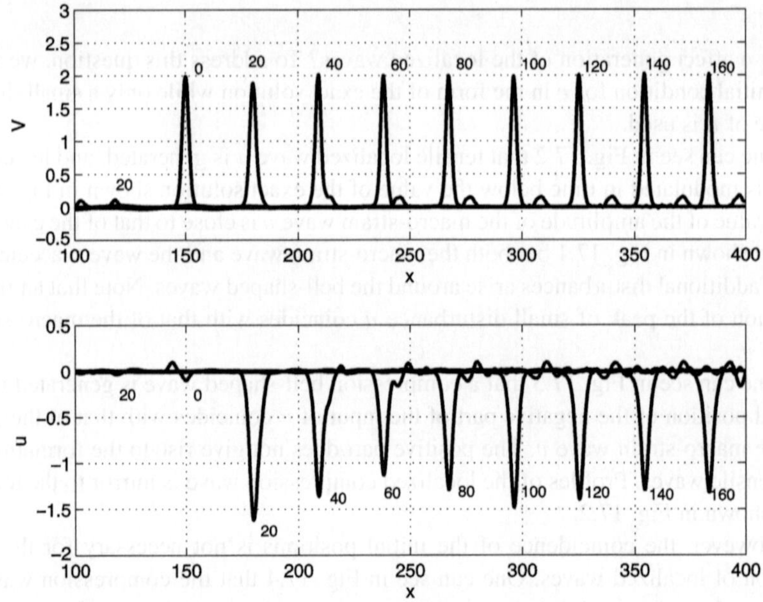

Fig. 17.3 Influence of an initial position of a lattice disturbance on generation of moving compression localized defect in a lattice. Points of time correspond to the neighboring peaks

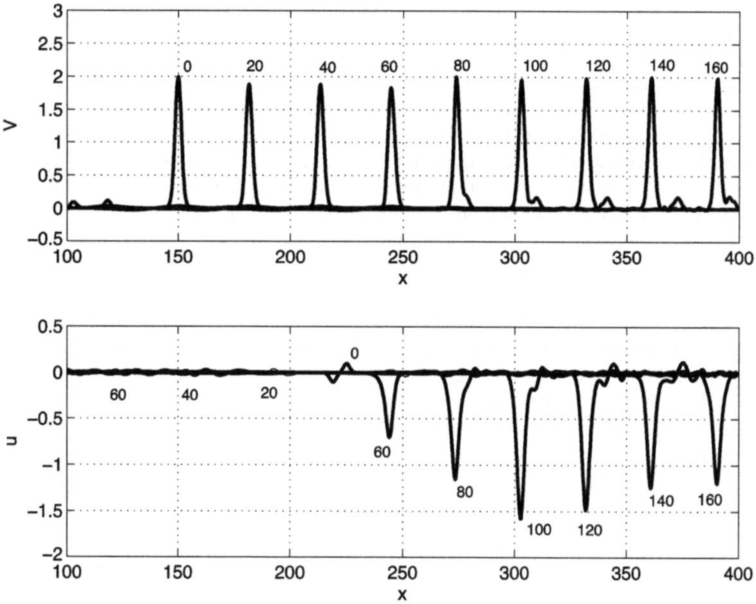

Fig. 17.4 Influence of an initial position of a lattice disturbance on generation of moving compression localized defect in a lattice. Points of time correspond to the neighboring peaks

17.4 Conclusions

To sum up, the microstructural crystalline model (17.1)–(17.2) allows us to describe deviations in the crystalline lattice depending on the velocity of the bell-shaped localized macro-strain wave. It is shown that predictions of the traveling wave exact solution are realized in a more general numerical simulation. Some important features of the lattice defects generation are revealed in numerics. Therefore, an analysis based on the use of the strain solitary waves demonstrates an advantage of the proper structural approach relative to the phenomenological ones studied in [8].

Acknowledgements The work of AVP has been supported by a grant of the Russian Science Support Foundation. The work also has been supported by the Russian Foundation for Basic Researches, grants Nos. 07-01-00213 and 09-01-00469a.

References

1. Aero, E.L.: Micromechanics of a double continuum in a model of a medium with variable periodic structure. J. Eng. Math. **55**, 81–95 (2002)
2. Aero, E.L., Bulygin, A.N.: Strongly nonlinear theory of nanostructure formation owing to elastic and nonelastic strains in crystalline solids. Mech. Solids **42**, 807–822 (2007)

3. Belyaeva, I.Yu., Zaitsev, V.Yu., Ostrovsky, L.A.: Nonlinear acoustical properties of granular media. Acoust. Phys. **39**, 11–16 (1993)
4. Belyaeva, I.Yu., Zaitsev, V.Yu., Ostrovsky, L.A., Sutin, A.M.: Elastic nonlinear parameter as an informative characteristic in problems of prospecting seismology. Izv. Acad. Sci. USSR, Phys. Solid Earth **30**, 890–894 (1994)
5. Bugai, A.N., Sazonov, S.V.: The influence of transverse perturbations on the propagation of picosecond acoustic pulses in a paramagnetic crystal. Phys. Solid State **47**, 1914–1921 (2005)
6. Bugai, A.N., Sazonov, S.V.: Soliton-like propagation modes of picosecond acoustic pulses in a paramagnetic crystal. Phys. Solid State **49**, 118–125 (2007)
7. Maugin, G.A., Pouget, J., Drouot, R., Collet, B.: Nonlinear Electromechanical Couplings. Wiley, New York (1992)
8. Porubov, A.V., Aero, E.L., Maugin, G.A.: Two approaches to study essentially nonlinear and dispersive properties of the internal structure of materials. Phys. Rev. E **79**, 046608 (2009)
9. Porubov, A.V., Maugin, G.A.: Longitudinal strain solitary waves in presence of cubic nonlinearity. Int. J. Nonlinear Mech. **40**, 1041–1048 (2005)
10. Porubov, A.V., Maugin, G.A.: Propagation of localized longitudinal strain waves in a plate in presence of cubic nonlinearity. Phys. Rev. E **74**, 046617–046624 (2006)
11. Porubov, A.V., Maugin, G.A.: Improved description of longitudinal strain solitary waves. J. Sound Vib. **310**, 694–701 (2008)
12. Porubov, A.V., Maugin, G.A.: Cubic nonlinearity and longitudinal surface solitary waves. Int. J. Nonlinear Mech. **44**, 552–559 (2009)

Chapter 18
Reissner–Mindlin Shear Moduli of a Sandwich Panel with Periodic Core Material

Arthur Lebée and Karam Sab

Abstract Cecchi and Sab homogenization method (Cecchi and Sab in Int. J. Solids Struct. 44(18–19):6055–6079, 2007) for the derivation of the effective Reissner–Mindlin shear moduli of a periodic plate is applied to sandwich panels including chevron pattern. Comparison with existing bounds (Lebée and Sab in Int. J. Solids Struct., 2010) and full 3D finite element computation validates the method. Finally, the skins effect on transverse shear stiffness is put forward.

18.1 Introduction

Sandwich panels made of two thin skins separated by a thick periodic core such as honeycomb are commonly used in many fields. They offer good compromise between strength and weight especially for aeronautics applications.

When bending the sandwich panel, the skins are subjected to in-plane traction and compression whereas the core is subjected to transverse shear. Recently, new types of promising cores have emerged. Especially, folded cores are promising because of new production means [1, 11, 7, 9]. Among them, chevron folded cores were probably the first to be machined (Fig. 18.1).

Chevron folded cores transverse shear behavior has been experimentally investigated by Kintscher et al. [9] and Fisher et al. [5]. In a previous work, Lebée and Sab [10] applied Kelsey et al. [8] method to the chevron pattern and derived analytical and numerical bounds for transverse shear moduli. Kelsey et al. method is a First order Shear Deformation Theory (FSDT) approximation according to [3]: the core material is homogenized first with respect to out-of-plane shearing. It yields

A. Lebée (✉) · K. Sab
Université Paris-Est, UR Navier, École des Ponts ParisTech, 6 et 8 avenue Blaise Pascal, 77455 Marne-la-Vallée Cedex 2, France
e-mail: arthur.lebee@lami.enpc.fr

K. Sab
e-mail: sab@enpc.fr

G.A. Maugin, A.V. Metrikine (eds.), *Mechanics of Generalized Continua*, Advances in Mechanics and Mathematics 21,
DOI 10.1007/978-1-4419-5695-8_18, © Springer Science+Business Media, LLC 2010

Fig. 18.1 Chevron folded paper

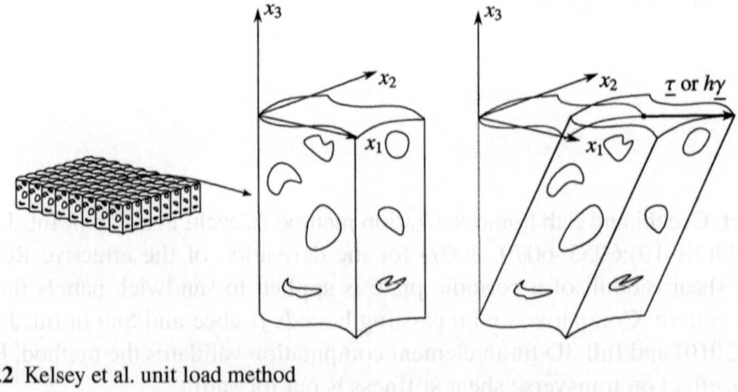

Fig. 18.2 Kelsey et al. unit load method

transverse shear constitutive equation $\tau_{\alpha 3} = G_{\alpha 3 \beta 3} \gamma_{\beta 3}$, where $\alpha, \beta, \gamma, \delta = 1, 2$. Then, the overall Reissner–Mindlin (R–M) shear moduli of the panel are evaluated as $F_{\alpha\beta} = hG_{\alpha 3 \beta 3}$, where h is the panel thickness. $F_{\alpha\beta}$ links the shear forces to the dual R–M strains through $Q_\alpha = F_{\alpha\beta} d_\beta$. The upper bound $G^+_{\alpha 3 \beta 3}$ is derived by submitting the upper face of the core to a unit displacement load $h\gamma_{\beta 3}$ while the lower face remains clamped (Fig. 18.2). The average upper-skin stress $\tau_{\alpha 3}$ is equated to $G^+_{\alpha 3 \beta 3} \gamma_{\beta 3}$. Likewise, the lower bound $G^-_{\alpha 3 \beta 3}$ is derived by submitting the upper face of the core to a uniform stress load $\tau_{\beta 3}$ and the lower face to $-\tau_{\beta 3}$. The average relative displacement of the skins $h\gamma_{\alpha 3}$ is computed and $G^-_{\alpha 3 \beta 3}$ is identified. See also Gibson and Ashby [6].

The main conclusion of [10] is that the upper and lower bounds for the already produced chevron pattern geometries are very loose (100% discrepancy). The interpretation proposed is that the interaction between the skins and the core is not taken into account. In this paper, Cecchi and Sab [4] Higher Order Homogenization (HOH) method is recalled and applied to sandwich panels including the chevron pattern. Its ability to capture the influence of the skins on R–M shear moduli will be demonstrated.

18.2 The Chevron Pattern

Like the honeycomb, the chevron pattern is periodic in the in-plane directions. Four identical parallelogram faces are necessary to generate the whole pattern by

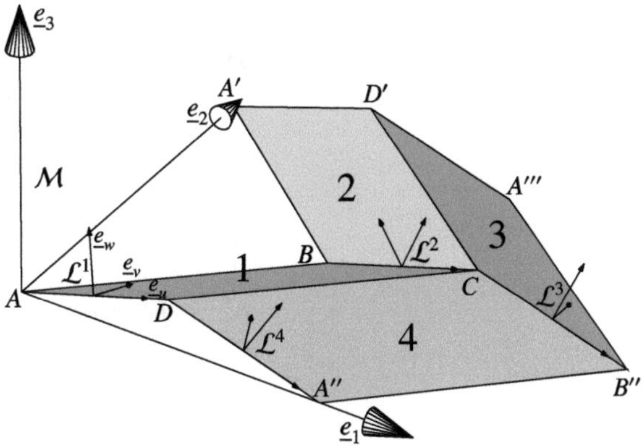

Fig. 18.3 The four elementary faces of the pattern

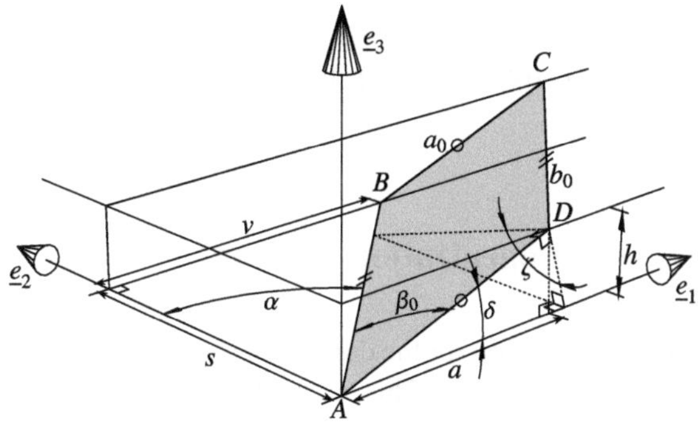

Fig. 18.4 Geometric parameters for the chevron pattern

Table 18.1 Vertices coordinates

Vertex	A	B	C	D	A'	D'	A''	B''	A'''
x_1	0	v	$a+v$	a	0	a	$2a$	$2a+v$	$2a$
x_2	0	s	s	0	$2s$	$2s$	0	s	$2s$
x_3	0	0	h	h	0	h	0	0	0

periodicity along the $\mathbf{e}_1$ vector (period $2a$) and the $\mathbf{e}_2$ vector (period $2s$) where $\mathscr{M} = (A, \mathbf{e}_1, \mathbf{e}_2, \mathbf{e}_3)$ is the main coordinate system (Fig. 18.3). Table 18.1 gives the coordinates of the vertices in terms of four geometric parameters: a, s, v, and h, where v is a horizontal offset parameter ($v = 0$ when B is aligned with A and A') and h is the pattern height. The full parametrization is given in Fig. 18.4. The face thickness is t_f.

The chevron pattern is glued or welded between two skins along AB, BA' edge for the lower skin and along DC, CD' edge for the upper skin and then becomes a sandwich panel (Fig. 18.5).

The sandwich panel including the chevron pattern has three major symmetries. These symmetries lead to several simplifications in the Reissner–Mindlin plate constitutive law [12]. The central symmetry with respect to the intersection point of line AC and BD uncouples in-plane (membrane) stresses and out-of-plane (flexural) stresses. The symmetry with respect to the (B, C, B'') plane uncouples trans-

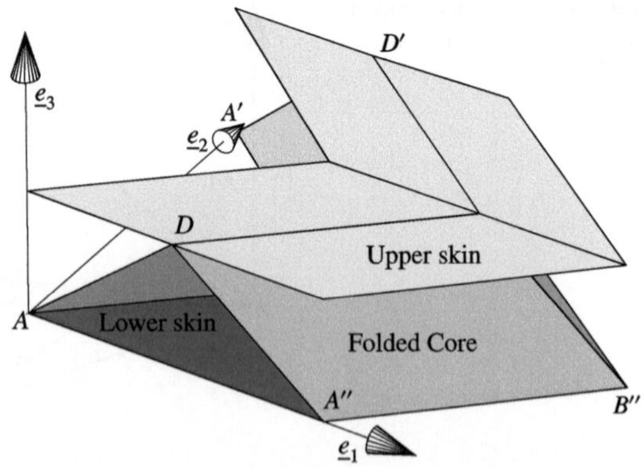

Fig. 18.5 The unit cell of a sandwich panel including the chevron pattern

verse shear stresses. Finally, there are only two R–M shear moduli to be determined: F_{11} and F_{22}.

18.3 Homogenization Method

The HOH method for the derivation of the effective R–M shear moduli of a periodic plate [4] is recalled. This method is implemented in two steps. First, the Love–Kirchhoff homogenization auxiliary problem proposed by Caillerie [2] is solved on a unit cell using finite elements. It yields the effective Love–Kirchhoff membrane and flexural moduli of the periodic plate, as well as the local 3D stresses. Then, a second unit cell problem is solved using a linear combination of the above mentioned local stress fields as external loading. The overall elastic energy is used to identify the R–M shear moduli. For more details, see [4]. Y is the unit cell, and $|Y|$ is its volume. The upper face ∂Y_3^+ and the lower face ∂Y_3^- are free. The lateral faces ∂Y_l must fulfill periodicity conditions (Fig. 18.6).

18.3.1 Homogenized Love–Kirchhoff Model

The unit cell auxiliary problem is:

$$\begin{cases} \operatorname{div} \boldsymbol{\sigma} = 0, \\ \boldsymbol{\sigma} = \mathbf{a}(\mathbf{y}) : \boldsymbol{\varepsilon}, \\ \boldsymbol{\varepsilon} = \mathbf{e} + y_3 \boldsymbol{\chi} + \boldsymbol{\nabla}^s(\mathbf{u}^{\text{per}}), \\ \boldsymbol{\sigma} \cdot \mathbf{e}_3 = 0 \quad \text{on free faces } \partial Y_3{}^{\pm}, \\ \boldsymbol{\sigma} \cdot \mathbf{n} \quad \text{skew periodic on } \partial Y_l, \\ \mathbf{u}^{\text{per}}(y_1, y_2, y_3) \quad (y_1, y_2)\text{-periodic on } \partial Y_l, \end{cases} \tag{18.1}$$

Fig. 18.6 A unit cell

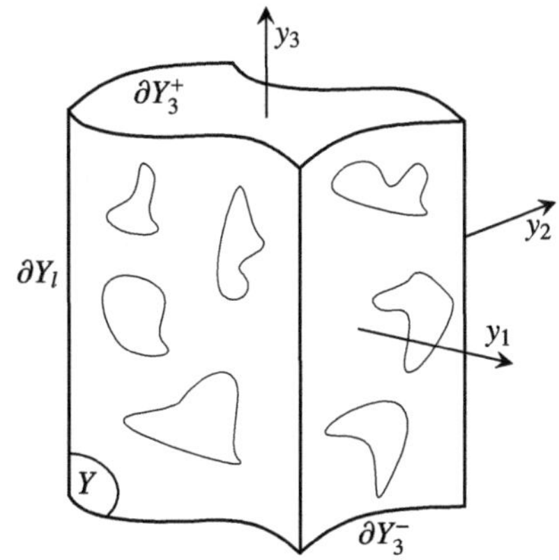

where $e_{\alpha\beta}$ are the membrane strains and $\chi_{\alpha\beta}$ are the curvatures. Solving the problem for each individual component of $\mathbf{e}$ and χ leads to the local 3D stress fields $\boldsymbol{\sigma}^{e_{\alpha\beta}}(\mathbf{y})$ and $\boldsymbol{\sigma}^{\chi_{\alpha\beta}}(\mathbf{y})$. The complete local stress field can be reconstructed by linear combination:

$$\boldsymbol{\sigma}^{LK} = e_{\alpha\beta}\boldsymbol{\sigma}^{e_{\alpha\beta}}(\mathbf{y}) + \chi_{\alpha\beta}\boldsymbol{\sigma}^{\chi_{\alpha\beta}}(\mathbf{y}). \tag{18.2}$$

The Love–Kirchhoff plate moduli are evaluated as follows:

$$\begin{aligned} A_{\alpha\beta\gamma\delta} &= \frac{h}{|Y|}\int_Y \boldsymbol{\sigma}^{e_{\alpha\beta}} : \boldsymbol{\varepsilon}^{e_{\gamma\delta}}\mathrm{d}Y, \\ B_{\alpha\beta\gamma\delta} &= \frac{h}{|Y|}\int_Y \boldsymbol{\sigma}^{e_{\alpha\beta}} : \boldsymbol{\varepsilon}^{\chi_{\gamma\delta}}\mathrm{d}Y, \\ D_{\alpha\beta\gamma\delta} &= \frac{h}{|Y|}\int_Y \boldsymbol{\sigma}^{\chi_{\alpha\beta}} : \boldsymbol{\varepsilon}^{\chi_{\gamma\delta}}\mathrm{d}Y. \end{aligned} \tag{18.3}$$

The Love–Kirchhoff plate constitutive law writes:

$$N_{\alpha\beta} = A_{\alpha\beta\gamma\delta}e_{\gamma\delta} + B_{\alpha\beta\gamma\delta}\chi_{\gamma\delta}, \qquad M_{\alpha\beta} = B_{\gamma\delta\beta\alpha}e_{\gamma\delta} + D_{\alpha\beta\gamma\delta}\chi_{\gamma\delta}, \tag{18.4}$$

where $N_{\alpha\beta}$ are the membrane stress and $M_{\alpha\beta}$ are the bending moments.

Chevron pattern symmetries uncouple membrane and flexural stiffnesses which enforce $B_{\alpha\beta\gamma\delta} = 0$. This is a prescribed condition for applying the second homogenization step.

18.3.2 Homogenized Reissner–Mindlin Model

The second auxiliary problem to be solved on the unit cell is:

$$\begin{cases} \operatorname{div} \boldsymbol{\sigma}^Q + \mathbf{f}^Q(\mathbf{y}) = 0, \\ \boldsymbol{\sigma}^Q = \mathbf{a}(\mathbf{y}) : \boldsymbol{\nabla}^s(\mathbf{u}^Q), \\ \boldsymbol{\sigma}^Q \cdot \mathbf{e}_3 = 0 \quad \text{on free faces } (\partial Y_3^{\pm}), \\ \boldsymbol{\sigma}^Q \cdot \mathbf{n} \quad \text{skew periodic on } (\partial Y_l), \\ \mathbf{u}^Q(y_1, y_2, y_3) \quad (y_1, y_2)\text{-periodic on } \partial Y_l, \end{cases} \tag{18.5}$$

where

$$f_i^Q = d_{\alpha\alpha\beta\beta}\sigma_{\alpha i}^{\chi_{\beta\beta}} Q_\alpha, \tag{18.6}$$

Q_α is the shear force and $d_{\alpha\beta\gamma\delta}$ is the inverse tensor of the flexural stiffness $D_{\alpha\beta\gamma\delta}$. As for (18.2), this leads to the local 3D stress fields associated to $\mathbf{Q}$ which are obtained by the linear combination:

$$\boldsymbol{\sigma}^Q = Q_\alpha \boldsymbol{\sigma}^{Q_\alpha}(\mathbf{y}). \tag{18.7}$$

It is then possible to identify the R–M shear moduli with:

$$\frac{1}{F_{\alpha\beta}} = \frac{h}{|Y|}\int_Y \boldsymbol{\sigma}^{Q_\alpha} : \boldsymbol{\varepsilon}^{Q_\beta} \mathrm{d}Y. \tag{18.8}$$

18.4 Application to the Chevron Pattern

The considered sandwich panel is fully made of aluminium sheets with $E = 73$ GPa and $\nu = 0.3$. Similarly to Nguyen et al. [11], the following geometric parameters are investigated: $a_0 = 30$ mm, $b_0 \in [20, 60]$ mm, $t_f = 0.1$ mm, $\delta = 72°$, $\zeta = 34°$. Figure 18.7 shows configurations for several investigated shape ratios a_0/b_0. Three skin thicknesses are considered in order to point out t_s effect on shear constants, $t_s = [0.5, 1, 2]$ mm.

The computation of the R–M shear modulus for each direction has been performed within the linear elasticity framework. Since faces are very thin, Kirchhoff shell elements were used with ABAQUS finite element software (Linear quadrangular S4R element). The unit cell of Fig. 18.5 was chosen. The R–M shear modulus normalized with $\rho G h$ was compared to FSDT bounds derived in Lebée and Sab [10] (ρ is the volume fraction of the core, G is the shear modulus of the constitutive material of the core).

In order to validate the HOH method, full 3D simulation of a 3-point bending of a sandwich panel including the chevron pattern has been performed. Results are not detailed here; however, less than 5% discrepancy has been found between HOH and the 3D estimated R–M shear moduli.

Figure 18.8 shows von Mises stress and the deformed unit cell under Q_1 loading for $a_0 = b_0$ and $t_s = 1$ mm. The overall deformation consists of a horizontal relative displacement of the skins which is commonly associated to transverse shear deformation in thick sandwich panels. Moreover, it is clear that the skins do not remain planar in Fig. 18.8.

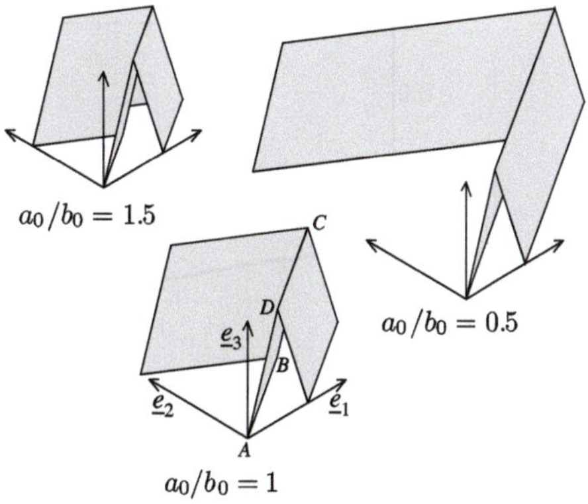

Fig. 18.7 Chevron pattern investigated configurations

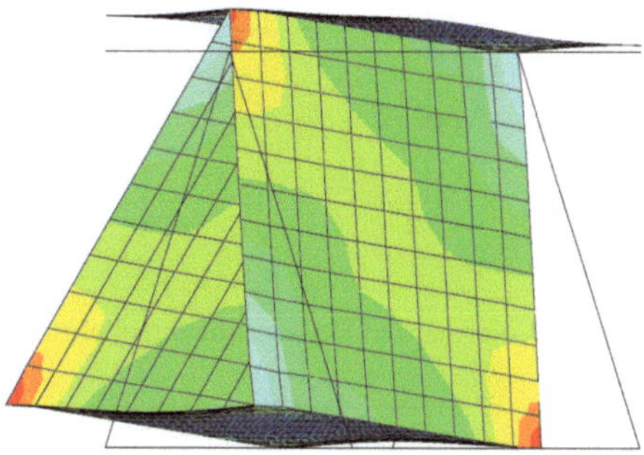

Fig. 18.8 Mesh deformation under Q_1 loading for $t_s = 1$ mm, $t_f = 0.1$ mm. View from $\mathbf{e}_2$

Figures 18.9 and 18.10 show the normalized R–M shear moduli for each loading direction versus shape ratio. FSDT finite element bounds are compared with the HOH method. The skin thickness has clear influence on the actual shear stiffness of the sandwich plate. The reader may have noticed that FSDT bounds are not fully respected even if the estimation is good. Actually, these bounds are only valid under specific hypothesis which will be developed in a further work.

For both loading cases, the thicker the skins, the closer to the upper bound the R–M shear moduli. These observations are consistent with Kelsey et al. conclusion that the upper bound is relevant for thick skins while the lower bound is relevant for thin skins.

Moreover, for loading case Q_1, shear constants globally remain close to the lower bound, whereas for loading case Q_2 they are much closer to the upper bound. It has also been noticed that for loading case Q_1 the skins are much more distorted than for loading case Q_2. Thus, the core geometry has a key role in the emergence of skins distortion.

18.5 Conclusion

FSDT bounds have been widely used for honeycomb sandwich panels since they give quite good estimation of the actual R–M shear moduli. However, in the case

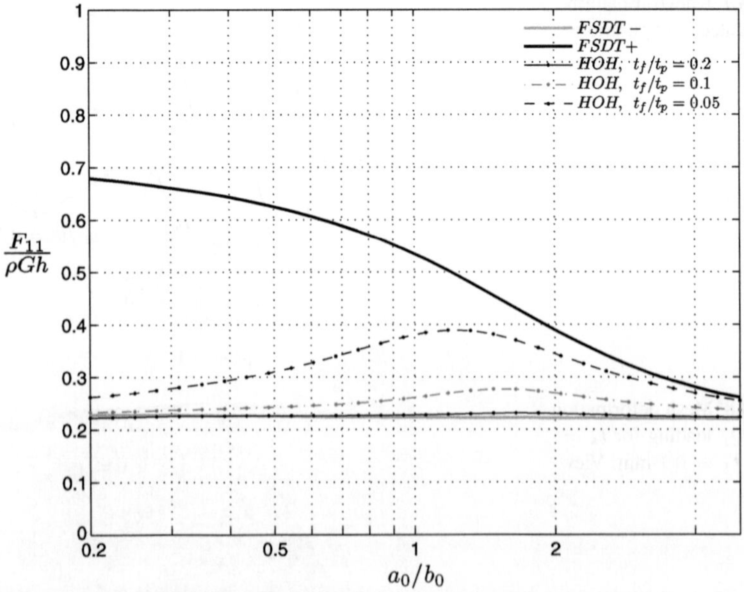

Fig. 18.9 Normalized Reissner shear modulus for Q_1 loading versus shape ratio

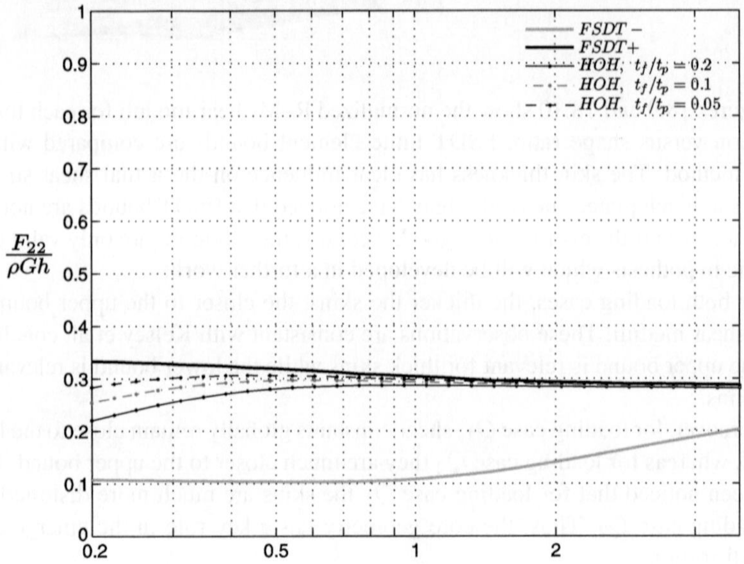

Fig. 18.10 Normalized Reissner shear modulus for Q_2 loading versus shape ratio

of the chevron pattern, the gap between the upper and lower bounds is much larger than in the case of honeycombs, and requires more refined estimation procedures. In this paper, we applied Cecchi and Sab higher order homogenization method for estimating the R–M shear moduli and confronted results both to 3D full computation and to FSDT bounds. From this, we conclude that the proposed method gives accurate estimation of shear behavior and enables pointing out the critical effect of skins on transverse shear stiffness, especially for the chevron pattern.

References

1. Basily, B., Elsayed, E.: Dynamic axial crushing of multi-layer core structures of folded chevron patterns. Int. J. Mater. Prod. Technol. **21**, 169–185 (2004)
2. Caillerie, D.: Thin elastic and periodic plates. Math. Methods Appl. Sci. **6**(2), 159–191 (1984)
3. Carrera, E.: Theories and finite elements for multilayered, anisotropic, composite plates and shells, Arch. Comput. Methods Eng. **9**(2), 87–140 (2002)
4. Cecchi, A., Sab, K.: A homogenized Reissner–Mindlin model for orthotropic periodic plates: Application to brickwork panels. Int. J. Solids Struct. **44**(18–19), 6055–6079 (2007)
5. Fischer, S., Drechsler, K., Kilchert, S., Johnson, A.: Mechanical tests for foldcore base material properties. Compos. A, Appl. Sci. Manuf. **40**, 1941–1952 (2009)
6. Gibson, L.J., Ashby, M.F.: Cellular Solids Pergamon, Elmsford (1988)
7. Heimbs, S., Mehrens, T., Middendorf, P., Maier, A., Schumacher, M.: Numerical determination of the nonlinear effective mechanical properties of folded core structures for aircraft sandwich panels. In: 6th European LS DYNA User's Conference, 2006.
8. Kelsey, S., Gellatly, R., Clark, B.: The shmar Modulus of foil honeycomb cores: a theoretical and experimental investigation on cores used in sandwich construction. Aircr. Eng. Aerospace Technol. **30**(10), 294–302 (1958)
9. Kintscher, M., Karger, L., Wetzel, A., Hartung, D.: Stiffness and failure behaviour of folded sandwich cores under combined transverse shear and compression. Compos. A, Appl. Sci. Manuf. **38**(5), 1288–1295 (2007)
10. Lebée, A., Sab, K.: Transverse shear stiffness of a chevron folded core used in sandwich construction. Int. J. Solids Struct. (2010, submitted)
11. Nguyen, M., Jacombs, S., Thomson, R., Hachenberg, D., Scott, M.: Simulation of impact on sandwich structures. Compos. Struct. **67**(2), 217–227 (2005)
12. Reissner, E.: Reflections on the theory of elastic plates. Appl. Mech. Rev. **38**(11), 1453–1464 (1985)

Chapter 19
Waves in Residual-Saturated Porous Media

Holger Steeb, Marcel Frehner, and Stefan Schmalholz

Abstract We present a three-phase model describing wave propagation phenomena in residual-saturated porous media. The model consists of a continuous non-wetting phase and a discontinuous wetting phase and is an extension of classical biphasic (Biot-type) models. The model includes resonance effects of single liquid bridges or liquid clusters with miscellaneous eigenfrequencies taking into account a visco-elastic restoring force (pinned oscillations and/or sliding motion of the contact line). For the quasi-static limit case, i.e., $\omega \mapsto 0$, the results of the model are identical with the phase velocity obtained with the well-known Gassmann–Wood limit.

19.1 Introduction and Motivation

Understanding the dynamical and acoustical behavior of porous rocks is of great importance in geophysics, e.g., earthquakes, and for various seismic applications, e.g., hydrocarbon exploration. While many studies investigated wave propagation in fully-saturated porous rocks analytically and numerically, cf. [13, 5–7, 27], studies for partially-saturated porous rocks are rare, cf. [8, 20, 14, 4, 23, 24, 26, 29, 31, 32, 9, 18, 19]. However, several physical processes relevant at low frequencies take place only in partially saturated rocks but not in fully saturated rocks such as, for

H. Steeb (✉)
Ruhr-University Bochum, Mechanics-Continuum Mechanics, Universitätsstrasse 150, 44780 Bochum, Germany
e-mail: Holger.Steeb@rub.de

M. Frehner
University of Vienna, Department of Geodynamics and Sedimentology, Althanstrasse 14, 1090 Vienna, Austria
e-mail: marcel.frehner@univie.ac.at

S. Schmalholz
ETH Zurich, Geological Institute, Sonneggstrasse 5, 8092 Zurich, Switzerland
e-mail: schmalholz@erdw.ethz.ch

G.A. Maugin, A.V. Metrikine (eds.), *Mechanics of Generalized Continua*, Advances in Mechanics and Mathematics 21,
DOI 10.1007/978-1-4419-5695-8_19, © Springer Science+Business Media, LLC 2010

example, capillarity-induced resonance of oil blobs [16] or wave-induced fluid flow (cf. [33, 11]). Capillarity-induced resonance of trapped oil blobs at residual saturation can be excited by sound waves and is caused by the contact-line dynamics [16, 15]. The resonant oscillations of fluid blobs can also cause considerable attenuation at the resonance frequency [2]. However, in models describing such resonance effects the porous skeleton is often assumed rigid and only relative movements between the discontinuous fluid blobs and the solid skeleton are considered [3, 15, 16]. Reference [12] presented a first attempt to combine the dynamical behavior of wave propagation and rock-internal oscillations caused by capillarity effects.

Wave-induced fluid flow in partially-saturated rocks is considered to be an important wave attenuation mechanism at low frequencies where the viscous (i.e., dissipative) flow is caused by the differences in pore fluid pressures, cf. [33, 11, 22, 28]. Wave-induced flow caused by partial saturation is frequently studied with the so-called patchy-saturation models that apply Biot's theory (i.e., a two-phase model for full saturation) with spatially varying material parameters, for example, fluid compressibilities, to model a partial saturation on the meso-scale, i.e., scale larger than the pore size but smaller than the wavelength, cf. [33, 11]. Such models have attracted considerable attention because they predict significant attenuation and dispersion for low frequencies in wide agreement with laboratory measurements [21] and field observations [10]. However, applying the theory for fully-saturated rocks to study partially-saturated rocks causes some disadvantages such as: (i) the saturation is not a primary model parameter, (ii) partial saturation can only be modeled on the meso-scale but not on the micro-scale, (iii) the calculation of the effective meso-scale patch size distribution from the real continuous saturation distribution is not obvious, and (iv) the impact of capillary pressure is neglected which may be especially important for residual saturations (e.g., [7] speculated that capillary pressure could have a significant influence on phase velocities and attenuation).

In the present investigation, our aim is to study attenuation due to fluid oscillations and attenuation due to wave-induced flow with the macroscopic three-phase model, i.e., a mixture consisting of one solid constituent building the elastic skeleton and two immiscible fluid constituents. Thus we briefly describe the governing field equations, especially the momentum interaction between the inherent constituents. Finally, we study monochromatic waves in transversal and longitudinal direction and discuss the resulting dispersion relations for a typical reservoir sandstone equivalent (Berea sandstone).

19.2 Field Equations of the Multi-Phase Model

In a certain Representative Volume Element (RVE) with volume $\mathrm{d}v$, the volume fractions of the constituents φ^α are defined as $n^\alpha = \mathrm{d}v^\alpha/\mathrm{d}v$. The volume occupied by a single constituent φ^α is defined as $\mathrm{d}v^\alpha$. Furthermore, the partial densities are introduced as the mass of the constituent $\mathrm{d}m^\alpha$ related to the volume of the RVE $\mathrm{d}v$, i.e., $\rho^\alpha := \mathrm{d}m^\alpha/\mathrm{d}v$. The true densities are given by $\rho^{\alpha R} := \mathrm{d}m^\alpha/\mathrm{d}v^\alpha$. Note

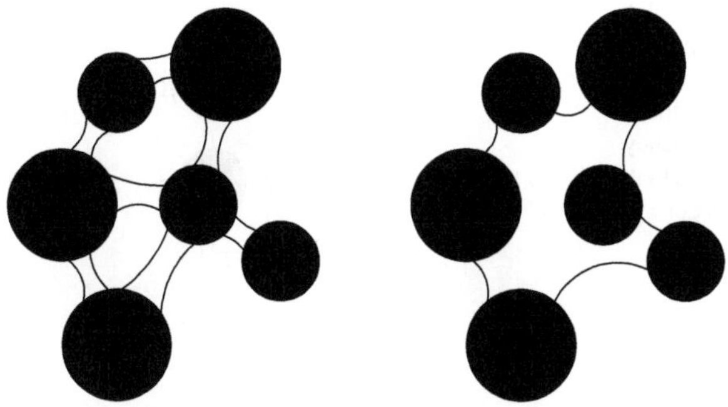

Fig. 19.1 A residual-saturated porous medium. (*Left*) Pendular state with individual liquid bridges. (*Right*) Funicular state with a liquid cluster, cf. [1, 25]

that both densities are related by the volume fractions, i.e., $\rho^\alpha = n^\alpha \rho^{\alpha R}$. Let us denote the continuous phases as φ^β with $\beta \in \{\mathfrak{s}, \mathfrak{n}\}$ ($\mathfrak{s}$ denoting the solid phase and $\mathfrak{n}$ denoting the non-wetting phase) while the discontinuous wetting fluid phase is denoted as $\varphi^{\mathfrak{w}}$. On the one hand, the geometry and mass distribution of the discontinuous phase is rather complex in realistic porous media. On the other hand, this information, obtained, e.g., by modern non-destructive imaging techniques, in principle allows calculating the eigenmodes ω_i of the liquid blobs numerically, cf. Fig. 19.1. This topic will be investigated in detail in the future. An obvious result from the realistic, i.e., inhomogeneous distribution of the discontinuous liquid in the pore space of the material is that liquid bridges or clusters with varying eigenfrequencies ω_i have to be taken into account. From a modeling perspective, this could be, for instance, captured with a probability distribution function in the RVE. We begin by formulating the partial density $\rho_i^{\mathfrak{w}}$ of a single blob $\varphi_i^{\mathfrak{w}}$ with a certain eigenfrequency ω_i. Furthermore, we can introduce the partial density of the total wetting phase $\rho^{\mathfrak{w}}$ in the RVE

$$\rho_i^{\mathfrak{w}} = \frac{z_i \, \mathrm{d}m_i^{\mathfrak{w}}}{\mathrm{d}v}, \quad \text{and} \quad \rho^{\mathfrak{w}} = \sum_{i=1}^{z} \rho_i^{\mathfrak{w}} = \rho^{\mathfrak{w}R} \sum_{i=1}^{z} n_i^{\mathfrak{w}}. \tag{19.1}$$

In contrast to a continuous distribution of eigenfrequencies in the RVE, we focus ourselves on a set of discrete eigenfrequencies. Note that the number of blobs in the RVE with the eigenfrequency ω_i is given by z_i, and z is the total number of (discrete) eigenfrequencies.

19.2.1 Balance Equations of Momentum of Continuous Phases

The balance equations of momentum in local form for a biphasic mixture consisting of a solid skeleton $\varphi^{\mathfrak{s}}$ and a continuous non-wetting phase $\varphi^{\mathfrak{n}}$ are written in the

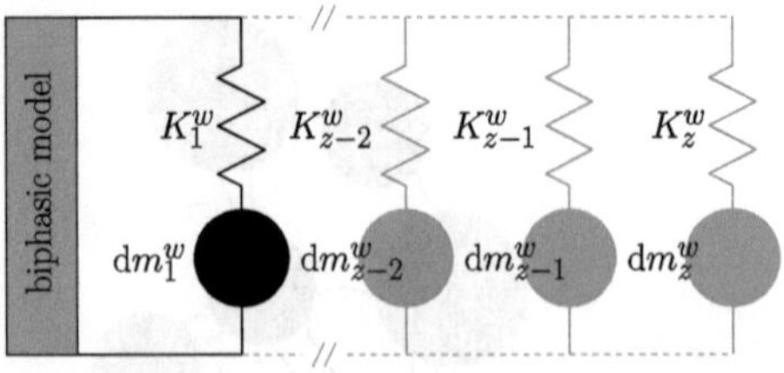

Fig. 19.2 Rheology of a residual-saturated porous model with a coupled set of z undamped oscillators with eigenfrequency $\omega_i(K_i^z, \mathrm{d}m_i^{\mathfrak{w}})$. Note that each oscillator could occur z_i-time

following form

$$\rho^{\mathfrak{s}}\ddot{\mathbf{u}}_{\mathfrak{s}} - \operatorname{div}\boldsymbol{\sigma}^{\mathfrak{s}} = \hat{\mathbf{p}}^{\mathfrak{s}}, \quad \text{and} \quad \rho^{\mathfrak{n}}\ddot{\mathbf{u}}_{\mathfrak{n}} - \operatorname{div}\boldsymbol{\sigma}^{\mathfrak{n}} = \hat{\mathbf{p}}^{\mathfrak{n}}. \tag{19.2}$$

Furthermore, we assume that the wetting phase $\varphi^{\mathfrak{w}}$ is in a discontinuous state of residual saturation. Thus, the momentum interaction of the solid constituent consists of a viscous (Darcy-like) term and an viscoelastic restoring force caused by the oscillations of the wetting phase. The viscous momentum term describes the exchange of (non-equilibrium) momentum between the solid and the non-wetting phase while the elastic momentum term captures the exchange of equilibrium momentum between the solid and the pinned discontinuous wetting phase. With respect to the large differences in compressibilities between an (incompressible) wetting fluid (oil, water) and a (compressible) non-wetting fluid (gas), we are able to neglect the exchange of momentum between the fluid phases.

$$\sum_\alpha \hat{\mathbf{p}}^\alpha = \mathbf{0}, \qquad \hat{\mathbf{p}}^{\mathfrak{s}} = -(\hat{\mathbf{p}}^{\mathfrak{n}} + \hat{\mathbf{p}}^{\mathfrak{w}}), \quad \text{and} \quad \hat{\mathbf{p}}^{\mathfrak{w}} = \sum_i \hat{\mathbf{p}}_i^{\mathfrak{w}}. \tag{19.3}$$

19.2.2 Balance Equation of Momentum of Discontinuous Wetting Phase

As we would like to describe pinned oscillations as well as sliding motion of the wetting constituent, the balance of momentum for one single liquid patch is given by a damped oscillator-like equation (Fig. 19.2)

$$\mathrm{d}m_i^{\mathfrak{w}}\ddot{\mathbf{u}}_{\mathfrak{w}}^i = -\mathrm{d}m_i^{\mathfrak{w}}\omega_i^2(\mathbf{u}_{\mathfrak{w}}^i - \mathbf{u}_{\mathfrak{s}}) - \mathrm{d}c_i^{\mathfrak{w}}(\dot{\mathbf{u}}_{\mathfrak{w}}^i - \dot{\mathbf{u}}_{\mathfrak{s}}) = \mathbf{f}_i^{\mathfrak{w}}. \tag{19.4}$$

This equation can be easily averaged in the volumetrical sense within the RVE. In the RVE, we obtain the coarse-grained equation for one eigenfrequency ω_i

$$\rho_i^{\mathfrak{w}}\ddot{\mathbf{u}}_{\mathfrak{w}}^i = -\rho_i^{\mathfrak{w}}\omega_i^2(\mathbf{u}_{\mathfrak{w}}^i - \mathbf{u}_{\mathfrak{s}}) - c_i(\dot{\mathbf{u}}_{\mathfrak{w}}^i - \dot{\mathbf{u}}_{\mathfrak{s}}) = \hat{\mathbf{p}}_i^{\mathfrak{w}}. \tag{19.5}$$

An equation equivalent to (19.5) exists for each set of oscillators with a certain eigenfrequency ω_i. According to classical results obtained from equilibrium and non-equilibrium evaluation of the balance of entropy, we obtain results for the momentum interaction terms $\hat{\mathbf{p}}^{\alpha} = \hat{\mathbf{p}}^{\alpha}_{\text{eq}} + \hat{\mathbf{p}}^{\alpha}_{\text{neq}}$ of the constituents:

$$
\begin{aligned}
\hat{\mathbf{p}}^{\mathfrak{n}}_{\text{eq}} &= p\,\text{grad}\,n^{\mathfrak{n}} \quad \text{and} \quad \hat{\mathbf{p}}^{\mathfrak{n}}_{\text{neq}} = -b_0\big(\dot{\mathbf{u}}_{\mathfrak{n}} - \dot{\mathbf{u}}_{\mathfrak{s}}\big),\\
\hat{\mathbf{p}}^{\mathfrak{w}}_{\text{eq}} &= p\,\text{grad}\,n^{\mathfrak{w}} \quad \text{and}\\
\hat{\mathbf{p}}^{\mathfrak{w}}_{\text{neq}} &= -\sum_i \big[\rho^{\mathfrak{w}}_i \omega_i^2 \big(\mathbf{u}^i_{\mathfrak{w}} - \mathbf{u}_{\mathfrak{s}}\big)\big] - \sum_i \big[c_i\big(\dot{\mathbf{u}}^i_{\mathfrak{w}} - \dot{\mathbf{u}}_{\mathfrak{s}}\big)\big],\\
\hat{\mathbf{p}}^{\mathfrak{s}}_{\text{eq}} &= p\,\text{grad}\,n^{\mathfrak{s}} \quad \text{and} \quad \hat{\mathbf{p}}^{\mathfrak{s}}_{\text{neq}} = \hat{\mathbf{p}}^{\mathfrak{n}}_{\text{neq}} + \hat{\mathbf{p}}^{\mathfrak{w}}_{\text{neq}},
\end{aligned}
\tag{19.6}
$$

with $b_0 = [(n^{\mathfrak{n}})^2 \gamma^{\mathfrak{g}R}]/k^{\mathfrak{n}}$. Note that Eq. (19.6)b refers to the wetting constituent, i.e., the sum of all damped oscillators. Furthermore, we have introduced the effective weight of the non-wetting fluid $\gamma^{\mathfrak{n}R}$ and the Darcy permeability (hydraulic conductivity) $k^{\mathfrak{n}}$. Inserting stress-strain relations (Hooke's law) and neglecting the convective part of the material time derivatives (linearized model), we obtain the linear set of field equations exemplarily written down for the special case of a set of oscillators with two distinct eigenfrequencies ω_1 and ω_2

$$
\begin{aligned}
N\,\text{div grad}\,\mathbf{u}_{\mathfrak{s}} + (A+N)\,\text{grad div}\,\mathbf{u}_{\mathfrak{s}} + Q\,\text{grad div}\,\mathbf{u}_{\mathfrak{n}} &= \rho^{\mathfrak{s}}\ddot{\mathbf{u}}_{\mathfrak{s}} - \hat{\mathbf{p}}^{\mathfrak{s}}_{\text{neq}},\\
Q\,\text{grad div}\,\mathbf{u}_{\mathfrak{s}} + R\,\text{grad div}\,\mathbf{u}_{\mathfrak{n}} &= \rho^{\mathfrak{n}}\ddot{\mathbf{u}}_{\mathfrak{n}} - \hat{\mathbf{p}}^{\mathfrak{n}}_{\text{neq}},\\
0 &= \rho^{\mathfrak{w}}_1 \ddot{\mathbf{u}}^1_{\mathfrak{w}} - \hat{\mathbf{p}}^{\mathfrak{w}}_{1,\text{neq}},\\
0 &= \rho^{\mathfrak{w}}_2 \ddot{\mathbf{u}}^2_{\mathfrak{w}} - \hat{\mathbf{p}}^{\mathfrak{w}}_{2,\text{neq}}.
\end{aligned}
\tag{19.7}
$$

The set of field equations (19.7) can be regarded is an extension of the classical Biot equations, cf. [5–7]. The bi-phasic limit case given by $n^{\mathfrak{w}} = 0$ reduces (19.7) to two the Biot equations (19.7)a,b. The parameters $P := 2N + A$, Q and R are the classical Biot parameters. They can be related to the physical-based shear modulus of the solid skeleton (G) and the bulk moduli of the grains ($K^{\mathfrak{s}}$), the skeleton (K) and the non-wetting fluid ($K^{\mathfrak{n}}$), cf. [26].

19.2.3 Monochromatic Waves

With the help of a standard ansatz for harmonic waves and usual splitting techniques, we study monochromatic waves for the developed three-phase model. For notational convenience, we restrict ourselves again to the special case of a set of oscillators with ω_1 and ω_2.

19.2.3.1 Transversal Mode

The generalized eigenvalue problem for shear waves can be written in matrix notation as $\det(\mathbf{A} - k^2\mathbf{B}_T) = 0$ with

$$
\mathbf{A} = \begin{bmatrix} \tilde{\rho}^{\mathfrak{s}} - \sum(\frac{\omega_i^2}{\omega^2}) - \sum(\frac{ic_i}{\omega}) & \tilde{\rho}^{\mathfrak{s}\,\mathfrak{n}} & \frac{\omega_1^2}{\omega^2}\tilde{\rho}_1^{\mathfrak{w}} + \frac{ic_1}{\omega} & \frac{\omega_2^2}{\omega^2}\tilde{\rho}_2^{\mathfrak{w}} + \frac{ic_2}{\omega} \\ \tilde{\rho}^{\mathfrak{s}\,\mathfrak{n}} & \tilde{\rho}^{\mathfrak{n}} & 0 & 0 \\ \frac{\omega_1^2}{\omega^2}\rho_1^{\mathfrak{w}} + \frac{ic_1}{\omega} & 0 & \rho_1^{\mathfrak{w}}(1 - \frac{\omega_1^2}{\omega^2} - \frac{ic_1}{\omega\rho_1^{\mathfrak{w}}}) & 0 \\ \frac{\omega_2^2}{\omega^2}\rho_2^{\mathfrak{w}} + \frac{ic_2}{\omega} & 0 & 0 & \rho_2^{\mathfrak{w}}(1 - \frac{\omega_2^2}{\omega^2} - \frac{ic_2}{\omega\rho_2^{\mathfrak{w}}}) \end{bmatrix} \omega^2 \qquad (19.8)
$$

and

$$
\mathbf{B}_T = \begin{bmatrix} N & 0 & 0 & 0 \\ 0 & 0 & 0 & 0 \\ 0 & 0 & 0 & 0 \\ 0 & 0 & 0 & 0 \end{bmatrix} \quad \text{and} \quad \begin{aligned} \tilde{\rho}^{\mathfrak{s}} &= \rho^{\mathfrak{s}} - ib_0/\omega, \\ \tilde{\rho}^{\mathfrak{n}} &= \rho^{\mathfrak{n}} - ib_0/\omega, \\ \tilde{\rho}^{\mathfrak{s}\mathfrak{n}} &= ib_0/\omega. \end{aligned} \qquad (19.9)
$$

The structure of the eigenvalue problems for both cases (transversal and shear mode) clearly shows that the dimension of the characteristic polynomial is of the same order as for the Biot case. Therefore, additional and higher order wave modes cannot be expected.

19.2.3.2 Longitudinal Mode

The generalized eigenvalue problem for compressional waves can be written in matrix notation as $\det(\mathbf{A} - k^2\mathbf{B}_L) = 0$. Note that the matrix $\mathbf{A}$ is equal to the case of shear waves.

$$
\mathbf{B}_L = \begin{bmatrix} P & Q & 0 & 0 \\ Q & R & 0 & 0 \\ 0 & 0 & 0 & 0 \\ 0 & 0 & 0 & 0 \end{bmatrix} \qquad (19.10)
$$

In Fig. 19.3 we plot the dispersion relations for a residual-saturated Berea sandstone, i.e., a typical reservoir sandstone equivalent.

19.3 Discussion and Conclusion

We propose a new model capturing resonance effects in partially-saturated porous media. The basic rheology of the material consists of the classical Biot mechanism (or physics introduced by equivalent mixture theory-based models) and a set of undamped oscillators with particular eigenfrequencies. The model contains all the effects described by the Biot model, e.g., the existence of a second compressional wave (i.e., Biot slow wave) or the characteristic velocity dispersion of the different waves. Additionally, the presented model shows a distinct dispersive effect governed by the discrete oscillating fluid-blobs or liquid clusters.

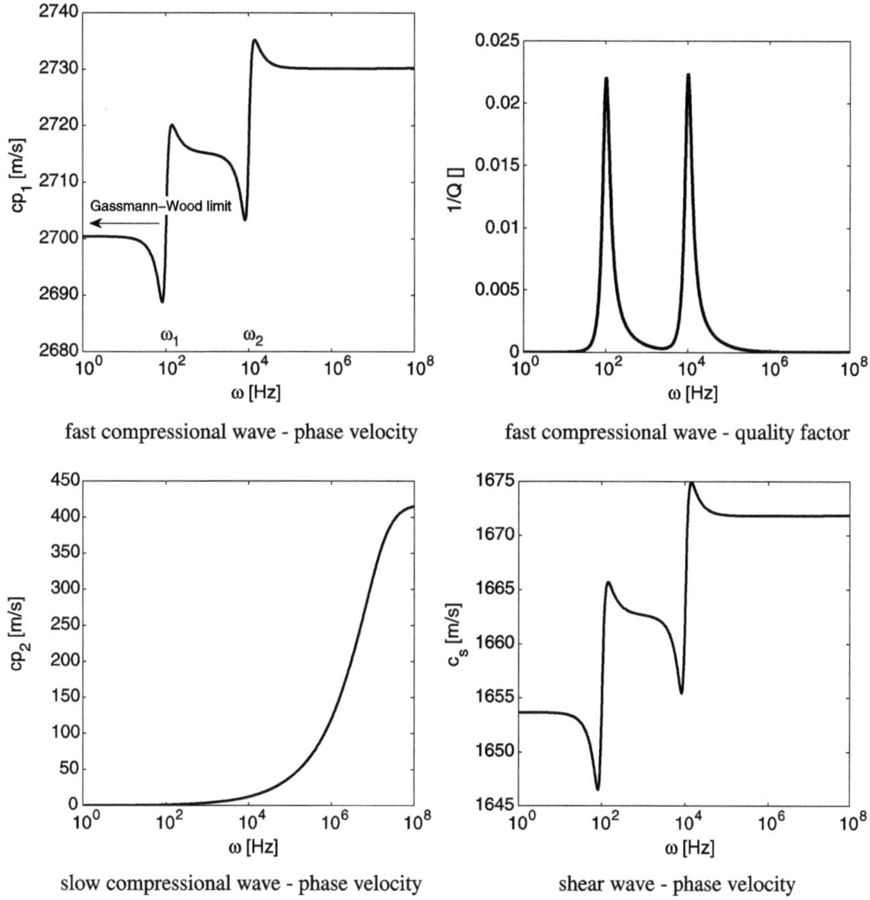

Fig. 19.3 Dispersion relations for a residual-saturated Berea sandstone. According to [30], we assume the following material properties: $G = 6.0$ GPa, $K = 8.0$ GPa, $K^{\mathfrak{s}} = 36.0$ GPa, $k^{\mathfrak{s}} = 190$ mD, $K^{\mathfrak{n}} = 131$ kPa, $\phi_0 = 1 - n^{\mathfrak{s}} = 0.19$ and the saturation of the wetting fluid is $s^{\mathfrak{w}} = 0.25$. Furthermore, the non-wetting fluid has the properties of water and the wetting fluid has the properties of air. The wetting fluid consists of two distinct patch sizes with eigenfrequencies of $\omega_1 = 100$ Hz and $\omega_s = 10$ kHz. The viscous damping parameters c_1 and c_2 are chosen arbitrarily

The proposed model, in its current form, is valid for a discontinuous wetting phase and a continuous non-wetting phase with a much smaller compressibility. One example for such a system is residual water in an otherwise air-filled porous rock or soil, as it occurs, for example, in the vadose zone above the groundwater table. In future studies, we will seek for a more general model that can handle two immiscible fluids with similar compressibilities. With such a model we will be able to study, for example, oscillation effects in hydrocarbon reservoir rocks partially saturated by oil and water. This will be particularly interesting for the case of residual oil saturation in a water-flooded reservoir.

One yet unsolved problem of the current model is the determination of the resonance frequency of individual fluid blobs, fluid bridges or fluid patches. References [16] or [17] give analytical solutions for the resonance frequency of very particular geometries of the fluid blobs. However, it is questionable if these geometries are realistic in real partially saturated rocks. In the presented study, we assume only two discrete values for the eigenfrequency, representing two different sizes of fluid blobs or patches. Eigenfrequencies of naturally occurring fluid blobs or patches will need to be measured in the laboratory or determined by numerical analysis techniques. We expect a certain range of eigenfrequencies for natural geometries, rather than discrete values. However, such eigenfrequency distributions will be relatively easy to implement in the current model.

A second, yet unsolved, problem is the determination of the viscous damping force acting on the individual oscillators. For the presented study, we arbitrarily choose two values, one for each eigenfrequency. However, for future studies, we will try to present a physically-based derivation of the viscous damping force.

Despite the discussed problems, we believe that the newly presented model is an important extension of the Biot model. We include a micro-scale effect, i.e., oscillation of the residual fluid phase, that was not considered in these types of models until now. The presented model contains all the effect described by the Biot model, matches the high- and low-frequency limits of the Biot model and reduces to the original Biot model when the newly introduced parameters are set accordingly. The new model therefore represents the possibility to study micro-scale oscillation effects using a well-known and accepted theory.

References

1. Bear, J.: Dynamics of Fluids in Porous Media. Dover, New York (1972)
2. Bedford, A., Stern, M.: A model for wave propagation in gassy sediments. J. Acoust. Soc. Am. **73**, 409–417 (1983)
3. Beresnev, I.A.: Theory of vibratory mobilization on nonwetting fluids entrapped in pore constrictions. Geophysics **71**, N47–N56 (2006)
4. Berryman, J.G., Thigpen, L., Chin, R.C.Y.: Bulk elastic wave propagation in partially saturated porous solids. J. Acoust. Soc. Am. **84**, 360–373 (1988)
5. Biot, M.A.: Theory of propagation of elastic waves in a fluid-saturated porous solid. I. Lowfrequency range. J. Acoust. Soc. Am. **29**, 168–178 (1956)
6. Biot, M.A.: Mechanics of deformation and acoustic propagation in porous media. J. Appl. Phys. **33**, 1482–1498 (1962)
7. Bourbié, T., Coussy, O., Zinszner, B.: Acoustics of Porous Media. Editions Technip, Paris (1987)
8. Brutsaert: The propagation of elastic waves in unconsolidated unsaturated granular mediums. J. Geophys. Res. **69**, 243–257 (1964)
9. Carcione, J.M., Cavallini, F., Santos, J.E., Ravazzoli, C.L., Gauzellino, P.M.: Wave propagation in partially saturated porous media: simulation of a second slow wave. Wave Motion **39**, 227–240 (2003)
10. Chapman, M., Liu, E., Li, X.Y.: The influence of fluid-sensitive dispersion and attenuation on AVO analysis. Geophys. J. Int. **167**, 89–105 (2006)

11. Dutta, N.C., Odé, H.: Attenuation and dispersion of compressional waves in fluid-filled porous rocks with partial gas saturation (White model)—Part I: Biot theory, Part II: Results. Geophysics **44**, 1777–1805 (1979)
12. Frehner, M., Schmalholz, S.M., Podladchikov, Y.: Spectral modification of seismic waves propagating through solids exhibiting a resonance frequency: a 1-D coupled wave propagation–oscillation model. Geophys. J. Int. **176**, 589–600 (2009)
13. Frenkel, J.: On the theory of seismic and seismoelectric phenomena in a moist soil. J. Phys. **3**, 230–241 (1944)
14. Garg, S.K., Nayfeh, A.H.: Compressional wave propagation in liquid and/or gas saturated elastic porous media. J. Appl. Phys. **60**, 3045–3055 (1986)
15. Hilpert, M.: Capillarity-induced resonance of blobs in porous media: Analytical solutions, Lattice–Boltzmann modeling, and blob mobilization. J. Colloid Interface Sci. **309**, 493–504 (2007)
16. Hilpert, M., Jirka, G.H., Plate, E.J.: Capillarity-induced resonance of oil blobs in capillary tubes and porous media. Geophysics **65**, 874–883 (2000)
17. Holzner, R., Eschle, P., Dangel, S., Frehner, M., Narayanan, C., Lakehal, D.: Hydrocarbon microtremors interpreted as nonlinear oscillations driven by oceanic background waves. Commun. Nonlinear Sci. Numer. Simul. **14**, 160–173 (2009)
18. Lo, W.C., Sposito, G.: Wave propagation through elastic porous media containing two immiscible fluids. Water Resour. Res. **41**, W02,025 (2005)
19. Lo, W.C., Sposito, G., Majer, E.: Low-frequency dilatational wave propagation through unsaturated porous media containing two immiscible fluids. Transp. Porous Media **68**, 91–105 (2007)
20. Mavko, G.M., Nur, A.: Wave attenuation in partially saturated rocks. Geophysics **44**, 161–178 (1979)
21. Murphy, W.F.: Effects of water saturation on attenuation of Massilon sandstone and Vycor porous plate. J. Acoust. Soc. Am. **71**, 1458–1468 (1982)
22. Pride, S.R., Berryman, J.G., Harris, J.M.: Seismic attenuation due to wave-induced flow. J. Geophys. Res. **109**, B01,201 (2004)
23. Santos, J.E., Corbero, J.M., Douglas, J.: Static and dynamic behavior of a porous solid saturated by a two-phase fluid. J. Acoust. Soc. Am. **87**, 1428–1438 (1990)
24. Santos, J.E., Douglas, J., Corbero, J.M., Love, O.M.: A model for wave propagation in a porous medium saturated by a two-phase fluid. J. Acoust. Soc. Am. **87**, 1439–1448 (1990)
25. Scheidegger, A.E.: The Physics of Flow Through Porous Media. MacMillan, New York (1957)
26. Smeulders, D.M.J., de la Rosette, J.P.M., Dongen, M.E.H.V.: Waves in partially saturated porous media. Transp. Porous Media **9**, 25–37 (1992)
27. Stoll, R.D.: Sediment Acoustics. Lecture Notes in Earth Sciences. Springer, Berlin (1989)
28. Toms, J., Müller, T.M., Cizc, R., Gurevich, B.: Comparative review of theoretical models for elastic wave attenuation and dispersion in partially saturated rocks. Soil Dyn. Earthq. Eng. **26**, 548–565 (2006)
29. Tuncay, K., Corapcioglu, M.Y.: Body waves in poroelastic media saturated by two immiscible fluids. Geophys. Res. Lett. **101**, 25,149-25,159 (1996)
30. Wang, H.F.: Theory of Linear Poroelasticity. Princeton University Press, Princeton (2000)
31. Wei, C., Muraleetharan, K.K.: A continuum theory of porous media saturated by multiple immiscible fluids: I. Linear poroelasticity. Int. J. Eng. Sci. **40**, 1807–1833 (2002)
32. Wei, C., Muraleetharan, K.K.: A continuum theory of porous media saturated by multiple immiscible fluids: II. Lagrangian description and variational structureinear poroelasticity. Int. J. Eng. Sci. **40**, 1835–1854 (2002)
33. White, J.E., Mikhaylova, N.G., Lyakhovitskiy, F.M.: Low-frequency seismic waves in fluid-saturated layered rocks. Earth Phys. **10**, 44–52 (1975)

Part V
Gradient Theory (Weakly Nonlocal Theories)

Chapter 20
A Personal View on Current Generalized Theories of Elasticity and Plastic Flow

Elias C. Aifantis

To the Memory of Ioannis Vardoulakis

Abstract A brief discussion of some current generalized continuum mechanics theories of elasticity and plasticity is provided. Attention is focused on works directly or indirectly motivated by the initial gradient models proposed by the author which, in turn, rest on ideas pioneered by Maxwell and van der Waals for fluid-like bodies but within a solid mechanics framework in the spirit of the celebrated monograph of brothers Cosserat published a quarter of a century later. The work of Cosserat, being dormant for half a century, ignited in the 1960s a plethora of generalized elasticity theories by the founders of modern continuum mechanics, as described in the treatises of Truesdell and Toupin and Truesdell and Noll. But it was not until another quarter of a century later that the interest in generalized continuum mechanics theories of elasticity and plasticity was revived, partly due to the aforementioned robust gradient models introduced and elaborated upon by the author and his co-workers in relation to some unresolved material mechanics and material physics issues; namely, the elimination of elastic singularities from dislocation lines and crack tips, the interpretation of size effects, and the description of dislocation patterns and spatial features of shear bands. This modest contribution is not aiming at a detailed account and/or critical review of the current state-of-the-art in the field. It only aims at a brief account of selected recent developments with some clarification on difficult points that have not been adequately considered or still remain somewhat obscure (origin and form of gradient terms, boundary conditions, thermodynamic potentials).

E.C. Aifantis (✉)
Laboratory of Mechanics and Materials, Polytechnic School, Aristotle University of Thessaloniki, Thessaloniki 54124, Greece
e-mail: mom@mom.gen.auth.gr

E.C. Aifantis
Center for Mechanics of Material Instabilities and Manufacturing Processes, College of Engineering, Michigan Technological University, Houghton, MI 49931, USA
e-mail: mom@mtu.edu

G.A. Maugin, A.V. Metrikine (eds.), *Mechanics of Generalized Continua*, Advances in Mechanics and Mathematics 21,
DOI 10.1007/978-1-4419-5695-8_20, © Springer Science+Business Media, LLC 2010

20.1 Introduction

Even though gradient type generalizations of classical elasticity may be traced back to original writings of Bernoulli, Euler and Cauchy and more explicitly discussed by Voigt and the brothers Cosserat, it was not until the early 1960s that an abundance of mathematical theories were developed in this area due to contributions by Rivlin, Toupin, Mindlin, Kroner, Eringen, and many others. After that the field remained dormant until the early 1990s when a robust gradient elasticity model proposed by the author was shown by him and his collaborators, as well as others, to eliminate elastic singularities and discontinuities and to also account for size effects. While the aforementioned model (involving the Laplacian of strain) may be viewed as a special case of Mindlin's strain gradient theory, there was no physical basis or experimental evidence for adopting such particular specialization. Moreover, various variants of the model (involving the Laplacian of stress) which were also subsequently elaborated upon, in connection with the interpretation of related experiments on size dependence, could not be obtained as specializations of any of the previously proposed generalized theories. One of the purposes of this modest article is to provide a more flexible, yet simple, framework for gradient elasticity models and discuss some salient aspects concerning the origin of gradient terms and associated boundary conditions, as well as thermodynamics.

About ten years earlier than the appearance of the aforementioned gradient elasticity model, a simple gradient plasticity model was proposed by the author to predict the thickness and spacing of shear bands, as well as to dispense with the mesh-size dependence of related finite element calculations in the material softening regime. At the same time, a gradient dislocation dynamics model was also advanced by the author to interpret dislocation patterning phenomena. Despite the fact that both of these early models were the predecessors of what has evolved to become today the disciplines of strain gradient plasticity and discrete/statistical dislocation dynamics, respectively, it is not always pointed out what the advantages and disadvantages of the original models are. Another modest purpose of the article is to pave the way for forthcoming work illustrating how the original gradient plasticity and gradient dislocation models can be improved and slightly generalized to account for some recent advances in the literature.

The present article is dedicated to the memory of an early close collaborator and profound contributor on current generalized continuum mechanics theories of both Cosserat and gradient type: Ioannis Vardoulakis, who had a fatal accident falling from his yard tree during the first weekend of this last September. Vardoulakis was one of the first soil mechanics researchers to realize the importance of both Cosserat and gradient type theories in modeling shear bands and size effects in granular media. In the late 1980s—early 1990s, a series of discussions with him and our common friend Hans Muhlhaus have led to an adaptation of gradient plasticity for metals (as progressing at that time with the vigorous participation of my students Doug Bammann and Hussein Zbib) to gradient plasticity for soils and rocks. This resulted in several joint or separate publications where the participation and numerical expertise of René de Borst and his students have provided a long-lived effect in the field.

Later, in the mid 1990s, similar discussions with Ioannis and George Exadaktylos on generalized elasticity have led to an extension of the initial gradient elasticity model to include surface energy effects and to its usage for exploring these effects in fracture and wave propagation studies. This latter work has sparked a remarkable recent research activity on this field by several prominent faculty in Greece (e.g., Georgiadis, Beskos, Aravas, Giannakopoulos, Tsamasphyros, Lazopoulos, and their co-workers).

20.2 Gradient Elasticity

20.2.1 Simple Gradient Elasticity Models

A brief review of generalized theories of elasticity with the emphasis on gradient type may be found in [18]. The model that was elaborated upon in that article reads

$$\boldsymbol{\sigma} = \lambda(\operatorname{tr}\boldsymbol{\varepsilon})\mathbf{1} + 2\mu\boldsymbol{\varepsilon} - c\nabla^2\big[\lambda(\operatorname{tr}\boldsymbol{\varepsilon})\mathbf{1} + 2\mu\boldsymbol{\varepsilon}\big] = \boldsymbol{\sigma}^c - c\nabla^2\boldsymbol{\sigma}^c, \tag{20.1}$$

where $(\boldsymbol{\sigma}, \boldsymbol{\varepsilon})$ denote the stress and strain tensors; (λ, μ) are the Lamé constants; and c is a newly introduced gradient coefficient ($c = \ell^2$, with ℓ denoting an internal length. [The auxiliary quantity $\boldsymbol{\sigma}^c = \lambda(\operatorname{tr}\boldsymbol{\varepsilon})\mathbf{1} + 2\mu\boldsymbol{\varepsilon}$ denotes the classical Hookean stress.] This model first proposed in [9] could be derived from a strain energy density function of the form

$$\begin{aligned} \mathcal{W} &= \mathcal{W}^c + \frac{1}{2}c\sigma^c_{ij,k}\varepsilon_{ij,k} = \mathcal{W}^c + c\left(\frac{1}{2}\lambda\varepsilon_{ii,k}\varepsilon_{jj,k} + \mu\varepsilon_{ij,k}\varepsilon_{ij,k}\right), \\ \mathcal{W}^c &= \frac{1}{2}\sigma^c_{ij}\varepsilon_{ij}, \end{aligned} \tag{20.2}$$

where the first term $\mathcal{W}^c$ of the right hand side is the classical contribution and the second term multiplied by c is the gradient contribution. Equation (20.2) is a special form of Mindlin's expression (e.g., [41] and references quoted therein)

$$\begin{aligned} \mathcal{W} = \mathcal{W}^c &+ c_1\varepsilon_{ij,j}\varepsilon_{ik,k} + c_2\varepsilon_{ii,k}\varepsilon_{kj,j} + c_3\varepsilon_{ii,k}\varepsilon_{jj,k} \\ &+ c_4\varepsilon_{ij,k}\varepsilon_{ij,k} + c_5\varepsilon_{ij,k}\varepsilon_{kj,i} \end{aligned} \tag{20.3}$$

involving five gradient coefficients. A virtual work 'principle' of the form

$$\int_\Omega \delta\mathcal{W}\,\mathrm{d}\Omega = \int_{\partial\Omega}(t_i\delta u_i + \tau_i D\delta u_i)\,\mathrm{d}(\partial\Omega), \tag{20.4}$$

where t_i denotes traction, τ_i denotes hypertraction and $D\delta u_i = \delta u_{i,j} n_j$ ($\mathbf{n}$ is the unit outer normal of $\partial\Omega$ and $\mathbf{u}$ is the displacement), gives

$$\begin{aligned} &\sigma_{ij,j} = \sigma^c_{ij,j} - c\sigma^c_{ij,kkj} = 0, \\ &t_i = \sigma_{ij} n_j + c t^G_i, \qquad \tau_i = c\sigma^c_{ij,k} n_k n_j, \end{aligned} \tag{20.5}$$

with the extra 'gradient' traction $\mathbf{t}^G = n_j n_k (D_m n_m)\sigma^c_{ij,k} - (D_j n_k)\sigma^c_{ij,k} - n_k(D_j \sigma^c_{ij,k})$ and $D_j = (\delta_{jk} - n_j n_k)\partial_k$. Thus, the appropriate boundary conditions associated with this virtual work statement consist of the pairs (u_i or t_i) and ($u_{i,j} n_j$ or τ_i) being prescribed on $\partial\Omega$. It also turns out that uniqueness is ensured provided that $(3\lambda + 2\mu, \mu, c)$ are all positive constants. This was first established in [18] and rederived by others, e.g., [29], without reference to the original work; it was also later generalized to the theory of gradient elasticity with surface tension by Exadaktylos et al. [24]).

A yet simpler model than that given by (20.1) has been used by the author and co-workers [10] to interpret size effects in bore holes under plane strain conditions. This model reads $\boldsymbol{\sigma} = \lambda(\operatorname{tr}\boldsymbol{\varepsilon})\mathbf{1} + 2\mu\boldsymbol{\varepsilon} - c\nabla^2(\operatorname{tr}\boldsymbol{\varepsilon})\mathbf{1}$, while a conjugate model, $\boldsymbol{\varepsilon} = [(1+\nu)/E]\boldsymbol{\sigma} - (\nu/E)(\operatorname{tr}\boldsymbol{\sigma})\mathbf{1} - c(\nu/E)\nabla^2(\operatorname{tr}\boldsymbol{\sigma})\mathbf{1}$ (E and ν denote Young's and Poisson's moduli), was also used in [12] for interpreting size effects in perforated thin plates under plane stress conditions. Such variants of gradient elasticity cannot be obtained as special cases of (20.1) or Mindlin's general theory. The same is true for the model discussed in [11] involving both stress and strain gradients which reads $\boldsymbol{\sigma} - c\nabla^2\boldsymbol{\sigma} = \lambda(\operatorname{tr}\boldsymbol{\varepsilon})\mathbf{1} + 2\mu\boldsymbol{\varepsilon} - c\nabla^2[\lambda(\operatorname{tr}\boldsymbol{\varepsilon})\mathbf{1} + 2\mu\boldsymbol{\varepsilon}]$, where the gradient coefficients measuring the effect of the Laplacian of stress and the Laplacian of strain were set equal for simplicity. It follows that a more general framework than the one leading to (20.1) is desirable. Such a framework is provided below by making use of the concept of implicit constitutive equations.

20.2.2 Implicit Gradient Elasticity Models

The idea of implicit constitutive equations seems to have been first introduced by Morgan [42] in relation to plasticity and the formulation of yield conditions. It was later generalized and elaborated upon in a more general manner by Rajagopal and coworkers (e.g., [47]) in relation to constitutive equations for both fluids and solids. In the present case of gradient elasticity, it may most simply be pursued through an implicit constitutive equation of the form

$$\mathbf{f}(\boldsymbol{\sigma}, \boldsymbol{\varepsilon}, \nabla^2\boldsymbol{\varepsilon}, \nabla^2\boldsymbol{\sigma}) = 0, \tag{20.6}$$

where $\mathbf{f}$ is a general linear isotropic function of both stress and strain, as well as their Laplacians. In the special case when $\mathbf{f}$ is a linear isotropic tensor function of its arguments, the relevant representation theorem for $\mathbf{f}$ gives $\operatorname{tr}(\alpha_1\boldsymbol{\varepsilon} + \alpha_2\boldsymbol{\sigma})\mathbf{1} + \alpha_3\boldsymbol{\varepsilon} + \alpha_4\boldsymbol{\sigma} + \nabla^2[\operatorname{tr}(\alpha_5\boldsymbol{\varepsilon} + \alpha_6\boldsymbol{\sigma})\mathbf{1} + \alpha_7\boldsymbol{\varepsilon} + \alpha_8\boldsymbol{\sigma}] = 0$. It follows that the various gradient models mentioned in the previous section can be obtained by properly selecting the constants α_i. It is also pointed out that the models obtained by allowing α_6 and α_8 not to vanish identically are not contained in Mindlin's strain gradient theory. In fact, the models containing the Laplacian of strain may not be invertible to produce models containing the Laplacian of stress. The situation is reminiscent of the case of linear viscoelasticity where the viscoelastic model containing the rate of strain (Kelvin–Voigt solid) exhibiting creep is different than the viscoelastic model containing the rate of stress (Maxwell's fluid) exhibiting relaxation.

In concluding this section, it is noted that the following gradient models may be obtained on the basis of the implicit constitutive equation (20.6)

$$\boldsymbol{\sigma} = \mathbf{C}(1 - c\nabla^2)\boldsymbol{\varepsilon}, \quad \boldsymbol{\varepsilon} = \mathbf{C}^{-1}(1 - c\nabla^2)\boldsymbol{\sigma}, \tag{20.7}$$

$$(1 - c_1\nabla^2)\boldsymbol{\sigma} = \mathbf{C}(1 - c_2\nabla^2)\boldsymbol{\varepsilon}, \tag{20.8}$$

where $\mathbf{C}$ is the usual fourth-order elastic stiffness tensor (the inverse $\mathbf{C}^{-1}$ denotes elastic compliance) which in the isotropic case is expressed in terms of the Lamé constants (λ, μ) as $C_{ijkl} = \lambda\delta_{ij}\delta_{kl} + \mu(\delta_{ik}\delta_{jl} + \delta_{il}\delta_{jk})$. Next, we consider the problem of boundary conditions associated with the gradient constitutive equations (20.7)–(20.8).

20.2.3 Some Comments on Boundary Conditions

Solutions of boundary value problems based on (20.1) and the boundary conditions resulting from (20.4) are, in general, difficult to obtain. Besides, the physical meaning of the extra boundary conditions is not clear. These difficulties are alleviated for certain situations where the solution of gradient elasticity may be reduced to the solution of an inhomogeneous Helmholtz equation with the source term being the solution of the corresponding classical elasticity problem. This was first shown by Ru and Aifantis ([48]; see also [49]) for special traction-type boundary conditions for which the solution of the model given by $(20.7)_1$ is obtained from (R–A theorem)

$$(1 - c\nabla^2)\mathbf{u} = \mathbf{u}^{\mathrm{c}}, \qquad \boldsymbol{\sigma} = \boldsymbol{\sigma}^{\mathrm{c}}, \tag{20.9}$$

where $\mathbf{u}^{\mathrm{c}}$ is the classical elasticity solution satisfying the same form of traction boundary condition, i.e., $\mathbf{t} = \boldsymbol{\sigma}\mathbf{n} = \boldsymbol{\sigma}^c\mathbf{n}$ and an extra boundary condition of the form $\partial^2\mathbf{u}/\partial n^2 = 0$. Such type of 'extra' boundary condition for the displacement may be deduced, for example, on a boundary surface ($y =$ const) for zero hypertraction. In this case, $(20.5)_3$ gives $u_{,yy} + v_{,xy} = 0$; $\lambda(u_{,xy} + v_{,yy} + w_{,zy}) + 2\mu v_{,yy} = 0$; $w_{,yy} + v_{,yz} = 0$, where (u, v, w) denote the corresponding components of the displacement field in the (x, y, z) directions. It is then seen that these conditions are reduced to the previously mentioned one $(u_{,yy} = v_{,yy} = w_{,yy})$ in the sense of a one-dimensional (y-direction) approximation. In other words, when the variation of displacement in the normal direction to the boundary is much larger than the corresponding one in the tangential direction, this extra boundary condition is drastically simplified. An analogous but less restrictive type of such boundary-layer approximation ($x \to \varepsilon x$, $y \to y$, $z \to \varepsilon z$; $\varepsilon \ll 1$) may be assumed only for the difference $\mathbf{u} - \mathbf{u}^c$ where $\mathbf{u}^c$ is the known classical displacement field corresponding to the same classical traction boundary condition. Then, the above listed extra boundary conditions become $u_{,yy} + v^c_{xy} = 0$; $(\lambda + 2\mu)v_{,yy} + \lambda(u^c_{,xy} + w^c_{,zy}) = 0$; $w_{,yy} + v^c_{,yz} = 0$. If, instead of the hypertraction τ_i, the condition $Du_i = u_{i,j}n_j$ is adopted, then the extra boundary condition reads $\partial\mathbf{u}/\partial n = 0$.

For 2-D cases, the above boundary conditions for the hypertraction τ_i result in $u_{,yy} + v_{,xy} = 0$ and $\lambda u_{,xy} + (\lambda + 2\mu)v_{,yy} = 0$ for plane problems; and $w_{,yy} = 0$ for anti-plane problems. In this connection, it is also noted that for two-dimensional plane problems ($y =$ const) the sufficient conditions which ensure the vanishing of the extra traction $\mathbf{t}^G$ in $(20.5)_2$ are given by $(\lambda + 2\mu)u_{,xy} + \mu v_{,yy} = 0$ and $u_{,yy} + v_{,xy} = 0$; while for anti-plane problems the corresponding condition is $w_{,y} = 0$. Both types of displacement conditions involving the first and second derivatives were first used and explored for dislocation and crack problems in an unpublished MTU report co-authored with Ru [48]; see also [49]. This report provided the foundation for obtaining subsequent gradient elasticity results (e.g., Gutkin [33] (see also [34, 35]), Askes [19] (see also [21, 20]), Lazar [37] (see also [39, 38]) and their co-workers).

Similar results can be deduced for the gradient elasticity model given by $(20.7)_2$. The counterpart of the R–A theorem of (20.9) is now expressed in terms of an Airy's stress type function ϕ and, in 2-D, reads

$$(1 - c\nabla^2)\phi = \phi^c; \qquad \boldsymbol{\varepsilon} = \boldsymbol{\varepsilon}^c, \tag{20.10}$$

where $(\phi^c, \boldsymbol{\varepsilon}^c)$ denote the classical solutions corresponding to the same boundary values of the displacement field for both classical and gradient problems. The extra boundary conditions can be deduced from a variational statement for the complementary strain energy density function [48]; see also [49]. For anti-plane problems, it turns out that the extra free-boundary conditions read $\partial^2\phi/\partial n^2 = 0$ or $\partial\phi/\partial n = 0$, in analogy to the displacement 'extra' boundary conditions for the model of $(20.7)_1$.

Solution techniques for the gradient elasticity model given by (20.8) have also been developed and certain non-singular solutions for dislocation and crack problems have been derived for the above type of boundary conditions [48]; see also [49]. These solutions, which dispense with unwanted elastic singularities for both stress and strain fields, have a rather complex form. More appealing analytical solutions have been obtained by the author (e.g., [12]). These may be viewed as improvements of previous solutions obtained for crack problems by the author and his co-workers (e.g., [24, 17]) which provide finite strains but singular stresses, as well as those obtained earlier by Eringen (e.g., [23]) which provide singular strains but finite stresses. Moreover, they may be considered as improvements of recent solutions obtained by various authors (e.g., [50, 22, 36, 51]) which seem to predict negative singular stresses at the tip of a tensile crack.

20.2.4 An Alternative Derivation of the ∇^2-Stress/∇^2-Strain Model

Due to the success of the (stress/strain) gradient elasticity model given by (20.8), we provide an alternative derivation, without resorting to the implicit formulation of Sect. 20.2.1. First, we start with Eringen's non-local model, i.e.,

$$\sigma_{ij} = \int_{\Omega} k\big(|\mathbf{x}-\mathbf{x}'|\big)\big(\lambda\varepsilon^c_{kk}(\mathbf{x}')\delta_{ij} + 2\mu\varepsilon^c_{ij}(\mathbf{x}')\big)\,\mathrm{d}\Omega, \tag{20.11}$$

and assume [18] the following form for the non-local kernel

$$k(\mathbf{x}-\mathbf{x}') = c_1/(2\pi)\,K_0\big(c_1\sqrt{(x-x')^2+(y-y')^2}\big),$$

where K_0 denotes the modified Bessel function. It then follows, through a Taylor expansion that, up to second order terms, the following expression holds

$$\sigma_{ij} - c_1\nabla^2\sigma_{ij} = \sigma^c_{ij}; \qquad \sigma^c_{ij} = \lambda\varepsilon^c_{kk}\delta_{ij} + 2\mu\varepsilon^c_{ij}. \tag{20.12}$$

Next, we use the definition of a 'macroscopic' strain tensor ε_{ij} as a volume-type average of the local strain tensor ε^c_{ij}, e.g.,

$$\varepsilon_{ij} = \frac{1}{V_\Omega}\int_{\Omega}\varepsilon^c_{ij}\,\mathrm{d}\Omega,$$

which through a Taylor expansion gives $\varepsilon_{ij} = \varepsilon^c_{ij} + c_2\nabla^2\varepsilon^c_{ij}$. The positive constant c_2 depends on the size of the elementary volume ($\sim 1/R^2$) for a spherical volume V_Ω of radius R) which, in turn, depends on the underlying microstructure and the degree of existing heterogeneity. A direct consequence of the above relation, within a second spatial order approximation (e.g., [53]; see also [43]), reads $\varepsilon^c_{ij} = \varepsilon_{ij} - c_2\nabla^2\varepsilon_{ij}$ and, thus, (20.12) is reduced to the model of (20.8).

Yet another, atomistic-like derivation of (20.8) may be obtained by utilizing an argument recently explored by the author [13] for a two-dimensional lattice. In this case, the strain and stress tensors $(\boldsymbol{\sigma}, \boldsymbol{\varepsilon})$ for a linearized isotropic configuration turn out to be of the form

$$\boldsymbol{\varepsilon} = \alpha e\mathbf{1} + \beta e\mathbf{M}; \qquad \boldsymbol{\sigma} = as\mathbf{1} + bs\mathbf{M}, \tag{20.13}$$

where $(e,\, s)$ denote the one-dimensional scalar measures of strain and stress in a representative 'atomic chain', while $(\alpha,\, \beta)$ and $(a,\, b)$ are constants, and $2M_{ij} = n_i\nu_j + \nu_j n_i$ with $(\mathbf{n},\, \boldsymbol{\nu})$ being the orthonormal vectors defining the lattice directions. If the atomic chain s–e relation is of the form $s - c_1\nabla^2 s = k(e - c_2\nabla^2 e)$, then it turns out that $(\boldsymbol{\sigma},\, \boldsymbol{\varepsilon})$ are related through (20.8), where the Lamé constants λ and μ are easily expressed in terms of the elastic-like modulus k of the atomic chain and the (lattice orientation) constants $(\alpha,\, \beta)$ and $(a,\, b)$.

20.2.5 Some Comments on Continuum Thermodynamics

Continuum thermodynamic arguments for gradient-dependent elastic media or elastic media with internal variables obeying 'complete balance laws' or evolution equations containing both a rate and a flux term have been advanced, among others, by the author (e.g., [3, 7]), as well as others (e.g., [40]; see also [52, 44] and references quoted therein). The main relevant issues to the present case of gradient elasticity are the concepts of a non-classical continuum which may exchange mass, momentum, energy and entropy with its bounding surface ([3]; see also [7]) and the modification of the energy equation (1st law) to include an additional working term to account for the extra work due to higher order gradients [3, 4]; see also [7, 6]. Associated with the form of this extra term is the concept of a divergence-free extra stress (null Lagrangian) which does not affect the equilibrium equations but may directly influence the form of the boundary conditions in 'finite' problems and the validity of equal area Maxwell's rule in the 'infinite' liquid–vapor transition problem [4]; see also [6]. The relevance of all these concepts to the problem of gradient elasticity has been recently undertaken and thoroughly examined by Polizzotto [44]. By applying these arguments (nonlocal thermodynamics and null Lagrangians) to the gradient elasticity model of (20.1), it was concluded that, under certain hypotheses, (20.5) hold with $t_i = \sigma_{ij}n_j$ and $\tau_i = 0$. This was essentially established by allowing an extra surface energy term $\int_{\partial\Omega} \delta\psi^\star \, \mathrm{d}(\partial\Omega)$ to enter in the variational statement given by (20.4) such that $\partial\psi^\star/\partial u_i = \sigma^s_{ij}n_j$, where σ^s_{ij} with $\sigma^s_{ij}n_j = G_j(\sigma^c_{ij,k}n_k)$ and $G_j \equiv n_i n_k D_k - D_j$, is a surface stress which acts on the boundary of the body but not in its interior as it is taken to be divergence free.

Further discussions on the gradient elasticity model given by (20.1) and its variants (including dynamic and wave propagation aspects, conservation laws, gauge theories of dislocations, and nanotube behavior) can be found in recent contributions by H. Askes and M. Lazar and their co-workers (e.g., their papers in this volume and references quoted therein). In this connection, it is remarked that the higher-order stress $\partial w/\partial \varepsilon_{ij,k} \equiv \tau_{ijk}$ used by Mindlin in the development of his general gradient theory based on (20.3) is identified as $\tau_{ijk} = c\sigma^c_{ij,k}$ for the robust model based on (20.2). This amounts to choosing $c_1 = c_2 = c_5 = 0$, $c_3 = \lambda c/2$, $c_4 = \mu c$ in (20.3). In Feynmann's linear theory of gravity, an expression identical to that containing the c_i's in (20.3) also appears but he has chosen $c_1 = -c_2 = -2\mu c$, $c_3 = -c_4 = -\mu c$, $c_5 = 0$. Then, the quantity $(1/2\mu c)\tau_{ijk,k} = -(\mathrm{inc}\,\varepsilon)_{ij} = \varepsilon_{ikl}\varepsilon_{jmn}\varepsilon_{ln,km}$ is equivalent to the 3-D linear Einstein's tensor. Initially, gauge theories of dislocations were based on Feynmann's choice of the constants c_i, but they gave unphysical results; the situation was repaired afterwards by essentially employing another gauge transformation motivated by the choice of c_i's leading to (20.1).

20.3 Gradient Plasticity

Only a few comments will be given here concerning the evolution of this field since the original model of the author was proposed in early 1980s (e.g., [5]; see also [8]).

The next major contribution to this field was due to the theory of Fleck/Hutchinson and co-workers (e.g., [28]; see also [25, 26]), which was developed by adopting Mindlin's strain gradient elasticity theory to the case of plasticity. But as was recently pointed out by Gurtin and Anand [32] this theory was not consistent with thermodynamics, a problem not encountered in the original 1984/1987 model. The next major contribution was the inclusion of interfacial energy, as introduced independently by Gudmundson [30] and in a more robust manner by Aifantis and Willis ([14]; see also [15, 16]) who produced an analytical expression for the grain boundary yield stress and provided experimental support for it through nanoindentation. A most recent work by Fleck and Willis [27] modified the Fleck/Hutchinson's 2001 version of their theory to make it consistent with thermodynamics and also generalized the earlier Aifantis/Willis strain gradient plasticity theory with interfacial energy [27].

In this connection, it is pointed out that a large number of generalized theories including interfacial energy have been proposed recently by several authors. For example, Abu Al-Rub and co-workers ([2]; see also [1]) included in their proposal both a scalar and tensorial measure of plastic flow, but they started with a variational statement involving inter-dependent quantities. A more general question for introducing thermodynamic potentials and variational principles in plasticity has to do with Bridgman's remark on 'plasticity being an island in the sea of irreversibility' and the related problem about the existence of entropy in this case. Perhaps, micromechanically based constitutive equations and mechanical equilibrium may be a more safe avenue to pursue, thus making economy of axioms in modeling nonlinear far-from-equilibrium dissipative phenomena. The situation of a van der Waals fluid and Maxwell's rule considered by the author in [4] (see also [6]) is only an example of this class of problems. Nevertheless, the recent thermodynamics works using the concept of nonlocal energy (or entropy production) residual (e.g., [45, 54]), as also utilized by the author in [4] (see also [6]), certainly deserve attention and further examination, but always in connection with the underlying physical (dislocation) mechanisms and in relation to critical experiments.

Finally, it should be mentioned that the 'lower-order' gradient plasticity models have also been proposed which do not require higher order boundary conditions. Despite their difficulty in predicting efficiently boundary layers, they also need further examination. A unifying framework for lower and higher gradient plasticity models has recently been outlined by Gudmundson and Aifantis [31].

20.4 Gradient Dislocation Dynamics

At the same time that the aforementioned gradient plasticity model was proposed, a gradient dislocation dynamics model (sometimes referred to as W–A model) was also outlined [5]; see also [8]. The motivation for it was the prediction of the ladder structure of persistent slip bands (PSB's) and related dislocation patterning phenomena. The situation has recently been discussed by Pontes/Walgraef/Aifantis [46], where a brief historical account and more recent developments on discrete and statistical dislocation mechanics considerations are reviewed. The basic premise of the W–A model was to write 'complete' balance laws for dislocation populations, modeling both dislocation transport and production/annihilation processes. This was done for both positive and negative dislocations (and thus for their sum $\rho^+ + \rho^- = \rho_{\text{statistical}}$, and their difference $\rho^+ - \rho^- = \rho_{\text{GND}}$), a practice which is extensively used today (e.g., Peerlings in this volume and refs quoted therein), sometimes without reference to the original work. Under certain conditions, the final evolution equations for dislocation populations turn out to contain diffusion-like terms (where the diffusivities depend on strain rates), and these terms were essential—in competition with the nonlinearities in the production/annihilation terms—for obtaining 'stable' dislocation patterns. Subsequently to the W–A model, several rigorous but a lot more complex works—based on statistical mechanics, as well as discrete dislocation dynamics (DDD) simulations—were published, thus bringing the field to a next level of development. The W–A model, however, still remains as a robust method to model dislocation patterns. It was recently revisited to produce an integrated model by incorporating both the W–A and the Groma/Zaiser approach ([56] and references quoted therein; see also [57, 55]).

References

1. Abu Al-Rub, R.K.: Interfacial gradient plasticity governs scale-dependent yield strength and strain hardening rates in micro/nano structured metals. Int. J. Plast. **24**, 1277–1306 (2008)
2. Abu Al-Rub, R.K., Voyiadjis, G.Z., Bammann, D.J.: A thermodynamic based higher-order gradient theory for size dependent plasticity. Int. J. Solids Struct. **44**, 2888–2923 (2007)
3. Aifantis, E.C.: A proposal for continuum with microstructure. Mech. Res. Commun. **5**, 139–145 (1978)
4. Aifantis, E.C.: Maxwell and van der Waals revisited. IMA Report, Michigan Technological University (1983)
5. Aifantis, E.C.: On the microstructural origin of certain inelastic models. Trans. ASME, J. Eng. Matter. Technol. **106**, 326–330 (1984)
6. Aifantis, E.C.: Maxwell and van der Waals revisited. In: Tsakalakos, T. (ed.) Phase Transformations in Solids, pp. 37–49. North-Holland, Amsterdam (1984)
7. Aifantis, E.C.: Remarks on media with microstructures. Int. J. Eng. Sci. **22**, 961–968 (1984)
8. Aifantis, E.C.: The Physics of plastic deformation. Int. J. Plast. **3**, 211–247 (1987)
9. Aifantis, E.C.: On the role of gradients in the localization of deformation and fracture. Int. J. Eng. Sci. **30**, 1279–1299 (1992)
10. Aifantis, E.C.: Higher order gradients and size effects. In: Carpinteri, A. (ed.) Size-Scale Effects in the Failure Mechanisms of Materials and Structures, pp. 231–242. Chapman and Hall, New York (1995)
11. Aifantis, E.C.: Update on a class of gradient theories. Mech. Mater. **35**, 259–280 (2003)
12. Aifantis, E.C.: Exploring the applicability of gradient elasticity to certain micro/nano reliability problems. Microsyst. Technol. **15**, 109–115 (2009)
13. Aifantis, E.C.: On scale invariance in anisotropic plasticity, gradient plasticity and gradient elasticity. Int. J. Eng. Sci. **47**, 1089–1099 (2009)
14. Aifantis, K.E., Willis, J.R.: Interfacial jump conditions in strain-gradient plasticity and relations of Hall–Petch type. In: Kounadis, A., Providakis, C., Exadaktylos, G. (eds.) Proc. 7th Nat. Congr. Mech. (June 24–26, 2004, Chania/Crete), pp. 372–376 (2004)

15. Aifantis, K.E., Willis, J.R.: The role of interfaces in enhancing the yield strength of composites and polycrystals. J. Mech. Phys. Solids **53**, 1047–1070 (2005)
16. Aifantis, K.E., Soer, W.A., De Hosson, J.Th.M., Willis, J.R.: Interfaces within strain gradient plasticity: theory and experiments. Acta Mater. **54**, 5077–5085 (2006)
17. Altan, B., Aifantis, E.C.: On the structure of the mode III crack-tip in gradient elasticity. Scripta Metall Mater. **26**, 319–324 (1992)
18. Altan, B., Aifantis, E.C.: On some aspects in the special theory of gradient elasticity. J. Mech. Behav. Mater. **8**, 231–282 (1997)
19. Askes, H., Aifantis, E.C.: Numerical modeling of size effects with gradient elasticity. Part I: Formulation, meshless discretization and examples. Int. J. Fract. **117**, 347–358 (2002)
20. Askes, H., Bennett, T., Gitman, I.M., Aifantis, E.C.: A multi-scale formulation of gradient elasticity and its finite element implementation. In: Papadrakakis, M., Topping, B.H.V. (eds.) Trends in Engineering Computational Technology, pp. 189–208. Saxe-Coburg, Stirling (2008)
21. Askes, H., Morata, I., Aifantis, E.C.: Finite element analysis with staggered gradient elasticity. Comput. Struct. **86**, 1266–1279 (2008)
22. Chen, J.Y., Wei, Y., Huang, Y., Hutchinson, J.W., Hwang, K.C.: The crack tip fields in strain gradient plasticity: the asymptotic and numerical analyses. Eng. Fract. Mech. **64**, 625–648 (1999)
23. Eringen, A.C.: Theory of micropolar elasticity. In: Liebowitz, H. (ed.) Fracture: An Advanced Treatise, pp. 622–728. Academic Press, London (1968)
24. Exadaktylos, G., Vardoulakis, I., Aifantis, E.C.: Cracks in gradient elastic bodies with surface energy. Int. J. Fract. **79**, 107–119 (1996)
25. Fleck, N.A., Hutchinson, J.W.: Strain gradient plasticity. Adv. Appl. Mech. **33**, 295–361 (1997)
26. Fleck, N.A., Hutchinson, J.W.: A reformulation of strain gradient plasticity. J. Mech. Phys. Solids **49**, 2245–2271 (2001)
27. Fleck, N.A., Willis, J.R.: A mathematical basis for strain-gradient plasticity theory—Part I: Scalar plastic multiplier. J. Mech. Phys. Solids **57**, 161–177 (2009)
28. Fleck, N.A., Muller, G.M., Ashby, M.F., Hutchinson, J.W.: Strain gradient plasticity: theory and experiment. Acta Metall. Mater. **42**, 475–487 (1994)
29. Georgiadis, H.G.: The mode III crack problem in microstructured solids governed by dipolar gradient elasticity. J. Appl. Mech. **70**, 517–528 (2003)
30. Gudmundson, P.: A unified treatment of strain gradient plasticity. J. Mech. Phys. Solids **52**, 1379–1406 (2004)
31. Gudmundson, P., Aifantis, E.C.: Connections between higher and lower order strain gradient plasticity theories. Preprint (2009)
32. Gurtin, M.E., Anand, L.: Thermodynamics applied to gradient theories involving the accumulated plastic strain: the theories of Aifantis and Fleck and Hutchinson and their generalization. J. Mech. Phys. Solids **57**, 405–421 (2009)
33. Gutkin, M.Y., Aifantis, E.C.: Screw dislocation in gradient elasticity. Scripta Mater. **35**, 1353–1358 (1996)
34. Gutkin, M.Y., Aifantis, E.C.: Edge dislocation in gradient elasticity. Scripta Mater. **36**, 129–135 (1997)
35. Gutkin, M.Y., Aifantis, E.C.: Dislocations in the theory of gradient elasticity. Scripta Mater. **40**, 559–566 (1999)
36. Karlis, G.F., Tsinopoulos, S.V., Polyzos, D., Beskos, D.E.: Boundary element analysis of mode I and mixed mode (I and II) crack problems of 2-D gradient elasticity. Comput. Methods Appl. Mech. Eng. **196**, 5092–5103 (2007)
37. Lazar, M., Maugin, G.A.: Nonsingular stress and strain fields of dislocations and disclinations in first strain gradient elasticity. Int. J. Eng. Sci. **43**, 1157–1184 (2005)
38. Lazar, M., Maugin, G.A.: Dislocations in gradient elasticity revisited. Proc. R. Soc. A **462**, 3465–3480 (2006)
39. Lazar, M., Maugin, G.A., Aifantis, E.C.: On dislocations in a special class of generalized elasticity. Phys. Status Solidi B **242**, 2365–2390 (2005)

40. Maugin, G.A.: The Thermomechanics of Nonlinear Irreversible Behaviours. World Scientific, Singapore (1999)
41. Mindlin, R.D., Eshel, N.N.: On first strain-gradient theories in linear elasticity. Int. J. Solids Struct. **4**, 109–124 (1968)
42. Morgan, A.J.A.: Some properties of media defined by constitutive equations in implicit form. Int. J. Eng. Sci. **4**, 155–178 (1966)
43. Peerlings, R.H.J., de Borst, R., Brekelmans, W.A.M., de Vree, J.H.P.: Gradient-enhanced damage for quasi-brittle materials. Int. J. Numer. Methods Eng. **39**, 3391–3403 (1996)
44. Polizzotto, C.: Gradient elasticity and nonstandard boundary conditions. Int. J. Solids Struct. **40**, 7399–7423 (2003)
45. Polizzotto, C.: Interfacial energy effects within the framework of strain gradient plasticity. Int. J. Solids Struct. **46**, 1685–1694 (2009)
46. Pontes, J., Walgraef, D., Aifantis, E.C.: On dislocation patterning: multiple slip effects in the rate equation approach. Int. J. Plast. **25**, 1486–1505 (2006)
47. Rajagopal, K.R., Srinivasa, A.R.: On a class of non-dissipative materials that are not hyperelastic. Proc. R. Soc. A **465**, 493–500 (2009)
48. Ru, C.Q., Aifantis, E.C.: A simple approach to solve boundary value problems in gradient elasticity. Acta Mech. **101**, 59–68 (1993)
49. Ru, C.Q., Aifantis, E.C.: Internal Report, Michigan Technological University (1993)
50. Shi, M.X., Huang, Y., Hwang, K.C.: Fracture in a higher-order elastic continuum. J. Mech. Phys. Solids **48**, 2513–2538 (2000)
51. Stamoulis, K., Giannakopoulos, A.E.: Size effects on strength, toughness and fatigue crack growth of gradient elastic solids. Int. J. Solids Struct. **45**, 4921–4935 (2008)
52. Valanis, K.C.: A gradient theory of internal variables. Acta Mech. **116**, 1–14 (1996)
53. Vardoulakis, I., Aifantis, E.C.: Gradient dependent dilatancy and its implications in shear banding and liquefaction. Ing.-Arch. **59**, 197–208 (1989)
54. Voyiadjis, G.Z., Deliktas, B.: Formulation of strain gradient plasticity with interface energy in a consistent thermodynamic framework. Int. J. Plast. **25**, 1997–2024 (2009)
55. Walgraef, D., Aifantis, E.C.: On the gradient theory of elasticity and dislocation dynamics. In: El-Azab, A. (ed.) Proc. 4th Int. Conference on Multiscale Materials Modeling/MMM2008, Dept. of Scientific Computing, Florida State University, pp. 720–725 (2008)
56. Walgraef, D., Aifantis, E.C.: On dislocation patterning: revisiting the W–A model—Part I: The role of gradient terms in dislocation dynamics. J. Mech. Behav. Mater. **19**, 49–66 (2009)
57. Walgraef, D., Aifantis, E.C.: On dislocation patterning: revisiting the W–A model—Part II: The role of stochastic terms in dislocation dynamics. J. Mech. Behav. Mater. **19**, 67–82 (2009)

Chapter 21
Review and Critique of the Stress Gradient Elasticity Theories of Eringen and Aifantis

Harm Askes and Inna M. Gitman

Abstract This chapter discusses the stress gradient theories of Eringen and Aifantis in terms of their original formulations, their differing dispersion properties in dynamics, and their finite-element implementation. A proposed combination of the two provides a dynamically consistent gradient enrichment while avoiding implementation difficulties.

21.1 Introduction

Classical elasticity, in which stresses are related to strains without the incorporation of higher-order spatial or temporal derivatives, is an appropriate model if the scale of observation is significantly larger than the characteristic dimensions of the underlying material microstructure. However, when processes are modeled on a scale similar to the size of the microstructure, unrealistic results may be found when classical elasticity is used. For instance, classical elasticity predicts singular stresses and strains at the tips of sharp cracks and at dislocation cores, non-dispersive propagation of waves, and size-independent structural behavior. These anomalies can be rectified if the field equations are extended with appropriate higher-order gradients of the state variables—thus, so-called *gradient theories* are obtained.

Gradient elasticity has been used successfully to remove singularities from crack tips and dislocation cores. It is also possible to predict elastic wave dispersion and

H. Askes (✉)
Department of Civil and Structural Engineering, University of Sheffield, Mappin Street, Sheffield S1 3JD, UK
e-mail: h.askes@sheffield.ac.uk

I.M. Gitman
Department of Mechanical Engineering, University of Sheffield, Mappin Street, Sheffield S1 3JD, UK
e-mail: i.gitman@sheffield.ac.uk

G.A. Maugin, A.V. Metrikine (eds.), *Mechanics of Generalized Continua*, Advances in Mechanics and Mathematics 21,
DOI 10.1007/978-1-4419-5695-8_21, © Springer Science+Business Media, LLC 2010

size-dependent mechanical behavior with gradient elasticity. Many formats of gradient elasticity exist that are equipped with additional gradients of strain or acceleration, but in this contribution we wish to discuss two theories of gradient elasticity with stress gradients, namely the 1983 theory of Eringen [13] and the 1992 theory of Aifantis [1] with its operator split formulation due to Ru and Aifantis for displacements [21], strains [16, 17] and, in particular, *stresses* [15, 9]. The gradient enrichment of the stresses in these two theories takes the same form, that is,

$$\sigma_{ij} - \ell^2 \sigma_{ij,mm} = C_{ijkl} \varepsilon_{kl}, \tag{21.1}$$

where σ_{ij} are the stresses, $\varepsilon_{kl} = \frac{1}{2}(u_{k,l} + u_{l,k})$ are the strains, C_{ijkl} are the elastic moduli, and ℓ is the internal length scale parameter that represents the microstructure. Since (21.1) appears in the gradient elasticity theories of Eringen as well as Aifantis, one could conclude that these two theories are identical. However, the two theories are, in fact, subtly different, and in this contribution we aim to clarify these differences and their consequences.

In Sect. 21.2, we will treat the continuum equations of the two stress gradient elasticity theories, the extent to which they coincide and how they differ. Two important implications of these differences are (i) the prediction of wave dispersion and (ii) the complexity of subsequent finite element implementations, which are treated in Sects. 21.3 and 21.4, respectively. It follows that Eringen's theory is better suited to use in dynamics, whereas Aifantis' theory is to be preferred for finite element implementations. In Sect. 21.5, we discuss how the advantageous aspects of the two theories can be combined. Throughout, the emphasis of the discussion is on offering clarification of the similarities and differences of the two theories. For the sake of brevity, we will not present a historical context nor will we attempt to provide an appropriate overview of all the available literature.

21.2 Continuum Field Equations

Although boundary conditions are an important aspect of gradient theories, the discussion below focuses on the field equations. We will first revisit the most important equations of the theories of Eringen and Aifantis in isolation, after which we will compare them in a unified manner.

Eringen's Theory

The majority of Eringen's work on nonlocal elasticity is concerned with integral formulations, in which volume averages of relevant state variables are computed. For example, the nonlocal stress tensor σ^g_{ij} is computed from its local counterpart σ^c_{ij} as

$$\sigma^g_{ij}(\mathbf{x}) = \int_V \alpha(\mathbf{s}) \sigma^c_{ij}(\mathbf{x}+\mathbf{s})\, \mathrm{d}V, \tag{21.2}$$

where $\alpha(\mathbf{s})$ is a kernel function that is non-negative and normally adopts decreasing values for increasing values of $\mathbf{s}$.

In 1983, Eringen formulated an equivalent theory of nonlocal elasticity in which the spatial integrals are replaced by differential operators. For certain kernel functions, the transition in going from integral nonlocality to gradient nonlocality occurs, in fact, without approximation error, thus highlighting the equivalence between the two types of nonlocality. In particular, nonlocal stresses can be computed from local stresses via

$$\sigma^g_{ij} - \ell^2 \sigma^g_{ij,kk} = \sigma^c_{ij}, \tag{21.3}$$

where the local stresses are related to the local strains ε_{ij} as usual.

Aifantis' Theory

In 1992, Aifantis suggested elastic constitutive relations extended with strain gradients as

$$\sigma_{ij} = C_{ijkl}\big(\varepsilon_{kl} - \ell^2 \varepsilon_{kl,mm}\big) \tag{21.4}$$

that lead to equilibrium equations as

$$C_{ijkl}\big(u^g_{k,jl} - \ell^2 u^g_{k,jlmm}\big) + b_i = 0, \tag{21.5}$$

where b_i are the body force components and u_i are the displacement components which carry a superscript g to indicate that they are affected by the nonlocality.

In a subsequent study, Ru and Aifantis argued that the fourth-order partial differential equations (21.5) can be solved as a sequence of two sets of second-order partial differential equations, namely

$$C_{ijkl}u^c_{k,jl} + b_i = 0 \tag{21.6}$$

followed by

$$u^g_k - \ell^2 u^g_{k,mm} = u^c_k. \tag{21.7}$$

Equation (21.6) represents the equilibrium equations of classical elasticity, hence a superscript c is used for the associated displacements u^c_i. Once u^c_i are known, they are to be used as a source term in (21.7) by which the gradient-enriched displacements u^g_i can be computed.

By differentiation, (21.7) can also be expressed in terms of strains [16, 17, 9] or, after multiplication with the elastic moduli C_{ijkl}, indeed in terms of stresses [15, 9]:

$$\sigma^g_{ij} - \ell^2 \sigma^g_{ij,kk} = \sigma^c_{ij}, \tag{21.8}$$

where $\sigma^c_{ij} = C_{ijkl}u^c_{k,l}$. The use of (21.8) instead of the original (21.7) has significant advantages: the variationally consistent boundary conditions of (21.7) are insufficient to remove all singularities from crack tip fields, whereas the boundary conditions consistent with (21.8) result in crack tip fields that are free of singularities [9].

Comparison

Expressions (21.3) and (21.8) are obviously identical, which raises the question to which extent the theories of Eringen and Aifantis coincide. To answer this, it is important to note the format of the momentum balance equations used in the two theories. Eringen's theory was formulated for use in dynamics, and the equations of motion contain the divergence of the *nonlocal* stress tensor σ^g_{ij}:

$$\sigma^g_{ij,j} + b_i = \rho \ddot{u}_i, \tag{21.9}$$

where ρ is the mass density. In contrast, Aifantis' theory was originally formulated to be used in statics; the equilibrium equations include the divergence of the *local* stress tensor σ^c_{ij}:

$$\sigma^c_{ij,j} + b_i = 0. \tag{21.10}$$

Equations (21.3) and (21.9) are solved together with the constitutive relation in terms of the local stress $\sigma^c_{ij} = C_{ijkl}\varepsilon_{kl}$ and the kinematic equation $\varepsilon_{k,l} = \frac{1}{2}(u_{k,l} + u_{l,k})$; the same holds for (21.8) and (21.10). Superscripts c and g have been dropped from the displacements since there is only one displacement field in either theory.

An overview of all field equations is given in Table 21.1. For each theory, the resulting equations can be expressed in a *reducible form* of two sets of second-order equations with unknowns u_i and σ^g_{ij}, or in an *irreducible form* of fourth-order equations with unknowns u_i only. For the Eringen theory, the reducible form is *coupled*, that is, u_i and σ^g_{ij} appear in both sets of equations and the two sets of equations must thus be solved simultaneously [7]. The reducible form of the Aifantis theory is *uncoupled* since σ^g_{ij} does not appear in (21.6). Thus, the two sets of equations can be solved sequentially, which is a great advantage for numerical implementations [22, 9], see also Sect. 21.4.

The irreducible form of Eringen's theory is obtained by taking the Laplacian of (21.9), multiplying with ℓ^2 and subtracting the result from (21.9). Assuming that the

Table 21.1 Stress gradient elasticity theories of Eringen and Aifantis—overview of field equations

	Eringen	Aifantis
Balance of momentum	$\sigma^g_{ij,j} + b_i = \rho\ddot{u}_i$ (dynamics)	$\sigma^c_{ij,j} + b_i = 0$ (statics)
Gradient enrichment	$\sigma^g_{ij} - \ell^2\sigma^g_{ij,mm} = \sigma^c_{ij}$	
Constitutive relation	$\sigma^c_{ij} = C_{ijkl}\varepsilon_{kl}$	
Kinematics	$\varepsilon_{kl} = \frac{1}{2}\left(u_{k,l} + u_{l,k}\right)$	
Reducible form	$\begin{cases} \sigma^g_{ij,j} + b_i = \rho\ddot{u}_i \\ \sigma^g_{ij} - \ell^2\sigma^g_{ij,mm} = C_{ijkl}u_{k,l} \end{cases}$	$\begin{cases} C_{ijkl}u_{k,jl} + b_i = 0 \\ \sigma^g_{ij} - \ell^2\sigma^g_{ij,mm} = C_{ijkl}u_{k,l} \end{cases}$
Irreducible form	$C_{ijkl}u_{k,jl} + b_i = \rho\left(\ddot{u}_i - \ell^2\ddot{u}_{i,mm}\right)$	$C_{ijkl}\left(u_{k,jl} - \ell^2 u_{k,jlmm}\right) + b_i = 0$

second derivatives of the body forces vanish, we end up with

$$C_{ijkl}u_{k,jl} = \rho\left(\ddot{u}_i - \ell^2\ddot{u}_{i,mm}\right), \tag{21.11}$$

that is, the stress gradients have been replaced by inertia gradients. It is immediately clear that such an irreducible form is only available in dynamics—in statics, the gradient-dependence would be lost. Finally, the irreducible form of Aifantis' theory is simply the original expression (21.5).

Both theories are capable of avoiding singularities in the state variables around the tips of sharp cracks and/or dislocations, see, for instance, [13, 14, 7] and [21, 2, 17, 9], respectively. However, important differences concern wave dispersion and finite element implementations, which we will discuss in the next two sections.

21.3 Dispersion Analysis

The simplest way to clarify the differences between the two theories is to carry out a one-dimensional dispersion analysis, thereby excluding body forces. We use the irreducible forms of Table 21.1, adding a standard inertia term to the Aifantis model, and substitute a general harmonic solution $u = A\exp(ik(x - ct))$, with amplitude A, wave number k and phase velocity c. For Eringen's theory, this results in

$$\frac{c^2}{c_e^2} = \frac{1}{1 + k^2\ell^2}, \tag{21.12}$$

and for Aifantis' theory we obtain

$$\frac{c^2}{c_e^2} = 1 + k^2\ell^2, \tag{21.13}$$

where $c_e = \sqrt{E/\rho}$. It is easily seen that in both theories the phase velocity c depends on the wave number k, hence both theories are *dispersive*.

The two expressions are plotted in Fig. 21.1 together with the response of a discrete lattice (see [13] for details on the relation between gradient elasticity and the discrete lattice response). It is clear that Eringen's model provides a better approximation of the discrete model. In Aifantis' model, the phase velocities are unbounded for increasing wave numbers.

It is sometimes claimed that the Aifantis model is also capable of providing a good approximation of the dynamic behavior of the discrete model [3, 2], but this would require to change the sign of the gradient term in (21.4) from negative to positive, which would make the model unstable [10, 18]—thus, we do not recommend such a modification. However, it must be emphasized that the straightforward extension of (21.5) to dynamics was not the original motivation to include gradients, and it would be unfair to judge the Aifantis theory solely on its performance in dynamics.

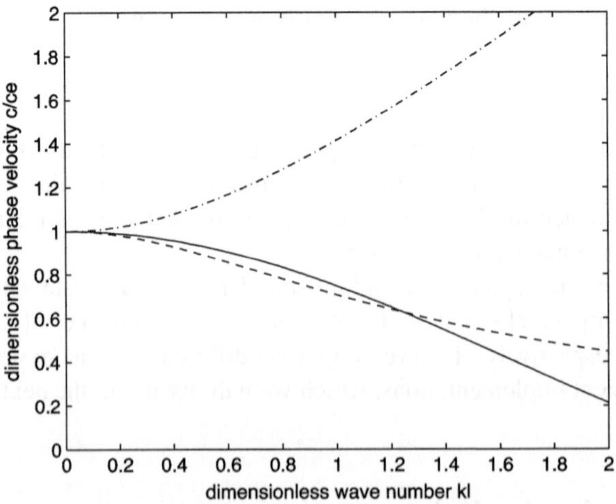

Fig. 21.1 Normalized phase velocity c/c_e versus normalized wave number $k\ell$ for discrete lattice (*solid*), Eringen's model (*dashed*) and Aifantis' model (*dot-dashed*)

21.4 Finite Element Implementation

In Sect. 21.2, it was shown that the reducible forms of the two theories are a set of coupled equations (Eringen theory) and uncoupled equations (Aifantis theory). The implications for numerical implementations, such as with the finite element method, are significant.

The coupled equations of the Eringen model are of the so-called *mixed type*, whereby the nonlocal stresses are the primary variables and the displacements are the Lagrange multipliers. The requirements put on the finite element implementation of equivalent classical elasticity models has been studied for decades, and turns out to be nontrivial matter. To avoid numerical instabilities, the stresses need to be interpolated with a polynomial order larger than that of the displacements, but for optimal convergence rates additional requirements based on the symmetry of the nonlocal stresses σ_{ij}^{g} need to be imposed. This has resulted in many attempts at efficient interpolations, see, for instance, the earlier discussion in [4] and the more recent efforts in [5, 12], but the resulting implementations are quite complicated with a mixture of nodal degrees of freedom, edge degrees of freedom and element degrees of freedom. Unfortunately, no simple effective finite element implementation seems to be available as yet. The fact that the two fields need to be interpolated with different polynomial orders remains a significant drawback.

On the other hand, the reducible form of the Aifantis model is uncoupled, which implies that the two interpolations can be chosen independently. The most straightforward option is to employ the same polynomial order and the same finite element discretization to the displacements and the nonlocal stresses [9]. This facilitates the implementation of this theory in existing finite element software enormously.

21.5 Closure and Outlook: Dynamically Consistent Gradient Elasticity

Gradient elasticity theories have been motivated for a variety of reasons, among others to dispense with singularities, to describe dispersive wave propagation and to predict size-dependent response of structures. In addition to these physical requirements, we also wish to emphasize that simplicity of the formulation and of subsequent finite element implementations are highly desirable.

In this chapter, we have discussed two theories with stress gradients, namely the ones due to Eringen and Aifantis. Both of these are simple in that they contain only one length scale parameter, which facilitates interpretation compared to theories with a multitude of length scales. Furthermore, both theories are able to remove singularities from the stresses around crack tips and dislocations. The advantage of Eringen's model is its better performance in dynamics, although it is emphasized that Aifantis' model was not originally intended to be used in dynamics. On the other hand, the advantage of the Aifantis model is that finite element implementations are straightforward and simple.

The irreducible forms presented in Table 21.1 give an indication how to combine the best aspects of the theories of Eringen and Aifantis, namely a simultaneous inclusion of inertia gradients and strain gradients. This idea was already present in Mindlin's 1964 theory of elasticity with microstructure [20]. Such a model was also derived from a discrete lattice [18, 19] and the simultaneous appearance of strain gradients and inertia gradients was denoted as *dynamically consistent* gradient enrichment. In a simple version of such a model the equations of motion can be written as

$$C_{ijkl}\big(u_{k,jl} - \ell_s^2 u_{k,jlmm}\big) + b_i = \rho\big(\ddot{u}_i - \ell_d^2 \ddot{u}_{i,mm}\big), \tag{21.14}$$

where two length scales ℓ_s and ℓ_d accompany the strain gradients and inertia gradients, respectively. The simultaneous presence of the two length scales ℓ_s and ℓ_d allows controlling the dispersive behavior of the model [18, 8, 19].

Recently, we have devised an operator split that allows rewriting the fourth-order equations (21.14) as a coupled set of symmetric second-order equations [6], that is,

$$C_{ijkl}u^m_{k,jl} + b_i = \rho\left(\tfrac{\ell_d^2}{\ell_s^2}\ddot{u}^m_i - \tfrac{\ell_d^2-\ell_s^2}{\ell_s^2}\ddot{u}^M_i\right), \tag{21.15}$$

$$0 = \rho\left(-\tfrac{\ell_d^2-\ell_s^2}{\ell_s^2}\ddot{u}^m_i + \tfrac{\ell_d^2-\ell_s^2}{\ell_s^2}\ddot{u}^M_i - \big(\ell_d^2 - \ell_s^2\big)\ddot{u}^M_{i,mm}\right). \tag{21.16}$$

This set of equations is coupled but *not* of the mixed type, thus avoiding the implementational difficulties of mixed formulations [6]. The original unknowns u_i from (21.14) are identified as the macroscopic displacements u^M_i in (21.15)–(21.16), accompanied by additional unknowns which turn out to be the microscopic displacements u^m_i, see [11] for details. Thus, (21.15)–(21.16) represent a multiscale formulation of gradient elasticity, in which the microscale and macroscale are fully coupled, and in which the gradient effects of the theories of Eringen and Aifantis are unified.

References

1. Aifantis, E.C.: On the role of gradients in the localization of deformation and fracture. Int. J. Eng. Sci. **30**, 1279–1299 (1992)
2. Altan, B.S., Aifantis, E.C.: On some aspects in the special theory of gradient elasticity. J. Mech. Behav. Mater. **8**, 231–282 (1997)
3. Altan, S.B., Aifantis, E.C.: On the structure of the mode III crack-tip in gradient elasticity. Scripta Metall. Mater. **26**, 319–324 (1992)
4. Arnold, D.N., Douglas Jr., J., Gupta, C.P.: A family of higher-order mixed finite element methods for plane elasticity. Numer. Math. **45**, 1–22 (1984)
5. Arnold, D.N., Winther, R.: Mixed finite elements for elasticity. Numer. Math. **92**, 401–419 (2002)
6. Askes, H., Bennett, T., Aifantis, E.C.: A new formulation and C_0-implementation of dynamically consistent gradient elasticity. Int. J. Numer. Methods Eng. **72**, 111–126 (2007)
7. Askes, H., Gutiérrez, M.A.: Implicit gradient elasticity. Int. J. Numer. Methods Eng. **67**, 400–416 (2006)
8. Askes, H., Metrikine, A.V.: One-dimensional dynamically consistent gradient elasticity models derived from a discrete microstructure. Part 2: Static and dynamic response. Eur. J. Mech. A, Solids **21**, 573–588 (2002)
9. Askes, H., Morata, I., Aifantis, E.C.: Finite element analysis with staggered gradient elasticity. Comput. Struct. **86**, 1266–1279 (2008)
10. Askes, H., Suiker, A.S.J., Sluys, L.J.: A classification of higher-order strain gradient models—linear analysis. Arch. Appl. Mech. **72**, 171–188 (2002)
11. Bennett, T., Gitman, I.M., Askes, H.: Elasticity theories with higher-order gradients of inertia and stiffness for the modelling of wave dispersion in laminates. Int. J. Fract. **148**, 185–193 (2007)
12. Carstensen, C., Günther, D., Reininghaus, J., Thiele, J.: The Arnold–Winther mixed FEM in linear elasticity. Part I: Implementation and numerical verification. Comput. Methods Appl. Mech. Eng. **197**, 3014–3023 (2008)
13. Eringen, A.C.: On differential equations of nonlocal elasticity and solutions of screw dislocation and surface waves. J. Appl. Phys. **54**, 4703–4710 (1983)
14. Eringen, A.C.: Nonlocal Continuum Field Theories. Springer, Berlin (2002)
15. Gutkin, M.Y.: Nanoscopics of dislocations and disclinations in gradient elasticity. Rev. Adv. Mater. Sci. **1**, 27–60 (2000)
16. Gutkin, M.Y., Aifantis, E.C.: Edge dislocation in gradient elasticity. Scripta Mater. **36**, 129–135 (1997)
17. Gutkin, M.Y., Aifantis, E.C.: Dislocations in the theory of gradient elasticity. Scripta Mater. **40**, 559–566 (1999)
18. Metrikine, A.V., Askes, H.: One-dimensional dynamically consistent gradient elasticity models derived from a discrete microstructure. Part 1: Generic formulation. Eur. J. Mech. A, Solids **21**, 555–572 (2002)
19. Metrikine, A.V., Askes, H.: An isotropic dynamically consistent gradient elasticity model derived from a 2D lattice. Philos. Mag. **86**, 3259–3286 (2006)
20. Mindlin, R.D.: Micro-structure in linear elasticity. Arch. Ration. Mech. Anal. **16**, 52–78 (1964)
21. Ru, C.Q., Aifantis, E.C.: A simple approach to solve boundary-value problems in gradient elasticity. Acta Mech. **101**, 59–68 (1993)
22. Tenek, L.T., Aifantis, E.C.: A two-dimensional finite element implementation of a special form of gradient elasticity. Comput. Mod. Eng. Sci. **3**, 731–741 (2002)

Chapter 22
On Natural Boundary Conditions in Linear 2nd-Grade Elasticity

Francesco Froiio, A. Zervos, and Ioannis Vardoulakis

Abstract This work aims at drawing the attention of mechanicians interested in the development of extended continuum theories on the unresolved issue of the physical interpretation of the additional boundary conditions introduced by 2nd-grade models. We discuss this issue in the context of the linearized theory of elasticity as an appropriate platform for discussion. Apart from lineal densities of edge-forces, 2nd-grade models allow for the prescription of force-like quantities energy-conjugated to the gradient of the velocity field on the boundary. Previous works proposed reductionistic interpretations, treating 2nd-grade models as particular cases of continua with affine microstructure; from the latter one can deduce field equations reminiscent of 2nd-grade models, either in the "low-frequency, medium wavelength" limit or by constraining the microstructural degrees of freedom to the gradient of the velocity field. The interpretation we propose here is based on the concept of *ortho-fiber*, and has the merit of achieving a simple physical interpretation of the boundary conditions without recourse to extensive algebra or the need to invoke microstructure.

F. Froiio (✉)
Ecole Centrale de Lyon, Ecully, France
e-mail: francesco.froiio@ec-lyon.fr

A. Zervos
University of Southampton, Southampton, UK
e-mail: az@soton.ac.uk

I. Vardoulakis (March 22nd, 1949—September 19th, 2009), formerly at the National Technical University of Athens, Greece.

G.A. Maugin, A.V. Metrikine (eds.), *Mechanics of Generalized Continua*, Advances in Mechanics and Mathematics 21,
DOI 10.1007/978-1-4419-5695-8_22, © Springer Science+Business Media, LLC 2010

22.1 Introduction

22.1.1 Notation

Boldface lowercase Latin characters will denote Euclidean vectors, i.e., tensors of order one, while boldface lowercase Greek characters will be reserved for tensors of higher order, e.g., $\boldsymbol{\tau} \in \mathrm{Lin}$ or $\boldsymbol{\xi} \in \mathrm{LIN}$ where Lin and LIN represent the linear space of 2nd-order tensors and of 3rd-order tensors, respectively. We will later mention the linear space of symmetric and of antisymmetric 2nd-order tensors, Sym and Asym, along with the linear space $\mathrm{SYM} = \{\boldsymbol{\alpha} \in \mathrm{LIN} \,|\, \alpha_{ijk} = \alpha_{ikj}\}$ of 3rd-order tensors with symmetry in the last two orders (subscripts denote Cartesian components). Scalar products in the respective linear spaces will be denoted by interposed single dots, while the column symbol will refer to the mixed product between 2nd-order tensors and 3rd-order tensors: for $\boldsymbol{\alpha} \in \mathrm{LIN}$ and $\boldsymbol{\beta} \in \mathrm{Lin}$ we define $\boldsymbol{\alpha} : \boldsymbol{\beta} = \alpha_{ijk}\beta_{jk}\mathbf{e}_i$ and $\boldsymbol{\beta} : \boldsymbol{\alpha} = \alpha_{jki}\beta_{jk}\mathbf{e}_i$, where $\{\mathbf{e}_i\}_{i=1,2,3}$ is the canonical basis of the Euclidean vector space and in which further use of Einstein summation convention is anticipated. A cross will denote the inner product in the Euclidean vector space.

22.1.2 Problem Statement

By 2nd-grade elasticity we refer to the class of elasticity models in which the stored energy W is defined as a quadratic functional of the first two gradients of the displacement $\nabla\mathbf{u}$ and $\nabla\nabla\mathbf{u}$:

$$W[\mathbf{u}] = \int_V w(\nabla\mathbf{u}, \nabla\nabla\mathbf{u})\,\mathrm{d}V,$$

where w is the stored energy density and V is the Euclidean region occupied by the body. We recall here briefly the context established by previous works (e.g., [15, 9–11, 1, 5–7]), in particular the papers of Mindlin and coworkers among them.

Toupin [15] pointed out the question of the proper extension of classical elasticity models when postulating internal stress fields energy-coupled to local rotations (couple-stress fields). This theory was not of the Cosserat type since no microstructure was introduced and the considered rotation rates were intended as rotors of the linear velocity field. The choice expressed in Toupin's paper led to some important indeterminacies in the formulation and opened a number of questions that have been debated since.

A first important issue concerns the analytical form and physical interpretation of the boundary conditions for 2nd-grade models. For ease of discussion, we focus in the following on the linearized formulation and require the displacement field and its first two gradients to be small. We compute the first variation of the stored energy functional in the reference, undeformed configuration:

$$\delta W = \int_V (\boldsymbol{\tau} \cdot \delta \nabla \mathbf{u} + \boldsymbol{\xi} \cdot \delta \nabla\nabla \mathbf{u})\, \mathrm{d}V \tag{22.1}$$

in which

$$\boldsymbol{\tau} := \frac{\partial w}{\partial \nabla \mathbf{u}} \in \mathrm{Sym} \quad \text{and} \quad \boldsymbol{\xi} := \frac{\partial w}{\partial \nabla\nabla \mathbf{u}} \in \mathrm{SYM}$$

are generalized forces in the sense of analytical mechanics. Integration by parts of the right hand side of (22.1) gives the classical expression obtained in [9]:

$$\delta W = \int_V (-\mathrm{div}\, \boldsymbol{\tau} + \mathrm{div}\,(\mathrm{div}\, \boldsymbol{\xi})) \cdot \delta \mathbf{u}\, \mathrm{d}V \tag{22.2}$$

$$+ \int_S \big((\boldsymbol{\tau} - \mathrm{div}\, \boldsymbol{\xi})\mathbf{n} - \mathrm{div}^s(\boldsymbol{\xi}\mathbf{n}) + (\mathrm{div}^s\mathbf{n})(\boldsymbol{\xi} : \mathbf{n} \otimes \mathbf{n})\big) \cdot \delta \mathbf{u}\, \mathrm{d}S \tag{22.3}$$

$$+ \int_S (\boldsymbol{\xi} : \mathbf{n} \otimes \mathbf{n}) \cdot \delta \mathrm{D}\mathbf{u}\, \mathrm{d}S \tag{22.4}$$

$$+ \int_E (\boldsymbol{\xi} : [\![\mathbf{m} \otimes \mathbf{n}]\!]) \cdot \delta \mathbf{u}\, \mathrm{d}E, \tag{22.5}$$

in which div^s denotes the surface divergence operator, $\mathbf{n}$ is the unit-normal vector field for the boundary S of V and the *edge tensor* $[\![\mathbf{m} \otimes \mathbf{n}]\!] \in \mathrm{Sym}$ is a geometrical descriptor of the edge E formed by discontinuities along S.

Comparing the above expression to the analogous one for 1st-grade materials one notices, apart from more complex forms of terms (22.2) and (22.3), the presence of two additional integrals (22.4) and (22.5). We assume, as a starting point, a potential of the external loads P of the same form as for 1st-grade materials:

$$P[\mathbf{u}] = \int_V \mathbf{F} \cdot \mathbf{u}\, \mathrm{d}V + \int_S \mathbf{t} \cdot \mathbf{u}\, \mathrm{d}S. \tag{22.6}$$

In this case, the principle of virtual work, i.e., the requirement that $\delta P - \delta W = 0$ for arbitrary variations $\delta \mathbf{u}$, results in the following Euler's equations:

$$-\mathrm{div}\, \boldsymbol{\tau} + \mathrm{div}\,(\mathrm{div}\, \boldsymbol{\xi}) = \mathbf{F} \quad \text{in } V, \tag{22.7}$$

$$(\boldsymbol{\tau} - \mathrm{div}\, \boldsymbol{\xi})\mathbf{n} - \mathrm{div}^s(\boldsymbol{\xi}\mathbf{n}) + (\mathrm{div}^s\mathbf{n})(\boldsymbol{\xi} : \mathbf{n} \otimes \mathbf{n}) = \mathbf{t} \quad \text{in } S, \tag{22.8}$$

$$\boldsymbol{\xi} : \mathbf{n} \otimes \mathbf{n} = 0 \quad \text{in } S, \tag{22.9}$$

$$\boldsymbol{\xi} : [\![\mathbf{m} \otimes \mathbf{n}]\!] = 0 \quad \text{in } E. \tag{22.10}$$

It is evident that the question of the physical interpretation of the quantities in integrals (22.4) and (22.5) has to be addressed in order to explore the full range of applications for 2nd-grade models. It was clear since [9] and [10] that (22.5) is linked to the application of lineal force densities along the edge (though some aspects of this interpretation still deserve a rigorous systematization, cf. [2]). It can be asserted in this sense that most of the controversy concerns integral (22.4) and in particular the interpretation of the normal derivative $\mathrm{D}\mathbf{u}$ as a degree of freedom of the system; we focus on this specific question in the following. As a starting point of the controversy one usually refers to [9], where Toupin's theory is framed by Mindlin as the limit of a "low-frequency, long wavelength approximation" of a continuum theory with affine microstructure. In the most general form of Mindlin's theory, each material point was assumed to undergo homogeneous "micro-strains" and rigid "micro-rotations", as well as displacements. In this form, Mindlin's theory is more general than Toupin's, as the latter invokes only the rotor of the displacement field. However, [9] did not clearly answer the question of the interpretation of the normal derivative $D\mathbf{u}$, neither did [11] where further development was presented.

In the following, we try to give a convincing—and possibly simple—physical interpretation of the terms in the integral (22.4) by introducing the concept of *ortho-fiber*. We propose two examples in the context of linear 2nd-grade isotropic elasticity. Our derivations are presented succinctly and we refer the interested reader to the forthcoming publication [4] for a more complete presentation.

22.2 Ortho-Fiber Kinematics

By the term *ortho-fiber* we refer to a thin, notional fiber of a material, starting at the surface and extending inward along the direction of $\mathbf{n}$. We assume the length l of the ortho-fiber to be small compared to the macroscopic characteristic length of the problem. In that limit, i.e., for $l \to 0$, the kinematics of the fiber is fully determined by a scalar measure of stretch Σ and a measure of rotation $\boldsymbol{\Omega} \in \mathrm{Orth}$. According to the above definition, for the measure of stretch one must use

$$\Sigma = \sqrt{((\nabla\chi)\mathbf{n}) \cdot ((\nabla\chi)\mathbf{n})} - 1,$$

where χ is the deformation. The inherent measure of rotation should express the change of orientation of the ortho-fiber under the action of χ. The structure of $\boldsymbol{\Omega}$ can be given in the terms of Beatty's formula, i.e.,

$$\boldsymbol{\Omega} := \mathbf{I} - (\sin\theta)\boldsymbol{\varepsilon}\mathbf{r} + (1 - \cos\theta)(\mathbf{r} \otimes \mathbf{r} - \mathbf{I}),$$

where

$$\theta = \arcsin\left|\mathbf{n} \times \frac{\nabla\chi\mathbf{n}}{|\nabla\chi\mathbf{n}|}\right|$$

is the rotation angle, $\boldsymbol{\varepsilon}$ is the permutation tensor and

$$\mathbf{r} = \frac{\mathbf{n} \times (\nabla\chi\mathbf{n})}{|\mathbf{n} \times (\nabla\chi\mathbf{n})|} \tag{22.11}$$

is the axial vector. Notice that, by virtue of (22.11), $\mathbf{r}$ is orthogonal to $\mathbf{n}$, excluding rotations of the ortho-fiber about its own axis. In the limit $\|\nabla\mathbf{u}\| \to 0$, one can derive the respective infinitesimal expressions

$$\Sigma = \varsigma + O(\|\nabla\mathbf{u}\|), \quad \varsigma := D\mathbf{u} \cdot \mathbf{n},$$

as well as

$$\boldsymbol{\Omega} = (\mathbf{I} + \boldsymbol{\omega}) + O(\|\nabla \mathbf{u}\|),$$
$$\boldsymbol{\omega} := (\mathrm{D}\mathbf{u}) \otimes \mathbf{n} - \mathbf{n} \otimes (\mathrm{D}\mathbf{u}) \in \mathrm{Asym}.$$

In the following, we use the infinitesimal formulation and call ς the *infinitesimal ortho-fiber stretch* and $\boldsymbol{\omega}$ the *infinitesimal ortho-fiber rotation*. These two quantities are shown schematically in Fig. 22.1. The above definitions allow us to split the normal gradient of the displacement and its variation as

$$\mathrm{D}\mathbf{u} = \varsigma\mathbf{n} + \boldsymbol{\omega}\mathbf{n}, \qquad \delta\mathrm{D}\mathbf{u} = (\delta\varsigma)\mathbf{n} + (\delta\boldsymbol{\omega})\mathbf{n}. \tag{22.12}$$

The importance of this decomposition can be appreciated by introducing (22.12) in (22.4). Algebraic manipulation then gives

$$\int_S (\boldsymbol{\xi} : \mathbf{n} \otimes \mathbf{n}) \cdot \delta\mathrm{D}\mathbf{u}\, \mathrm{d}S$$
$$= \int_S (\boldsymbol{\xi} \cdot \mathbf{n} \otimes \mathbf{n} \otimes \mathbf{n})\delta\varsigma\, \mathrm{d}S + \int_S \mathrm{asym}\big((\boldsymbol{\xi} : \mathbf{n} \otimes \mathbf{n}) \otimes \mathbf{n}\big) \cdot \delta\boldsymbol{\omega}\, \mathrm{d}S.$$

The above decomposition identifies the additional degrees of freedom evoked by the term (22.4) with the infinitesimal ortho-fiber stretch and rotation, addressing the issue of the physical interpretation of the boundary conditions.

The term (22.5), on the other hand, accounts for the special status that forces applied on slope discontinuities of the boundary have, compared to forces applied on smooth parts of the boundary. The *edge tensor* $[\![\mathbf{m} \otimes \mathbf{n}]\!]$ quantifies such discontinuities; however, a more extensive discussion has to be postponed till [4].

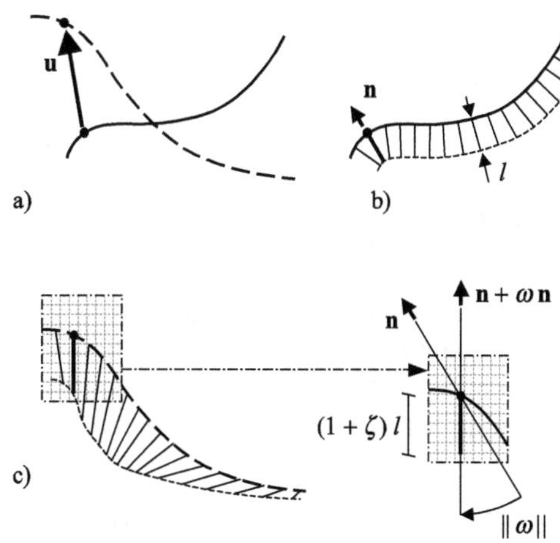

Fig. 22.1 (**a**) Material surface under the displacement field $\mathbf{u}$. (**b**) Embedded ortho-fiber field. (**c**) Effect of the ortho-fiber stretch ς and of the ortho-fiber rotation $\boldsymbol{\omega}$

22.3 Principle of Virtual Work and Balance Laws

We now extend the potential of the external loads by introducing additional terms energy-conjugated to the variations of the degrees of freedom identified in Sect. 22.2. We substitute (22.6) by

$$P[\mathbf{u}] = \int_V \mathbf{F}\cdot\mathbf{u}\,\mathrm{d}V + \int_S (\mathbf{t}\cdot\mathbf{u} + f\varsigma + \boldsymbol{\mu}\cdot\boldsymbol{\omega})\,\mathrm{d}S + \int_E \mathbf{s}\cdot\mathbf{u}\,\mathrm{d}E \qquad (22.13)$$

and call f the *ortho-fiber tension*, $\boldsymbol{\mu}$ *the ortho fiber couple* and $\mathbf{s}$ the *edge-* or *line traction*. Notice that the ortho-fiber couple $\boldsymbol{\mu}$ and rotation $\boldsymbol{\omega}$, being energy-conjugated, vary in the same linear space, that is,

$$\{\boldsymbol{\alpha}\in \mathrm{Asym} \mid \forall \mathbf{b} : (\mathbf{I}-\mathbf{n}\otimes\mathbf{n})\boldsymbol{\alpha}\mathbf{b} = 0\}.$$

In other words, neither ortho-fiber rotation about the axis $\mathbf{n}$ is allowed, nor couples can be prescribed that have axial vectors with direction $\mathbf{n}$ (i.e., "torques" about $\mathbf{n}$). In this context, the principle of virtual work leads again to Euler's equations (22.7) and (22.8). On the other hand, (22.9) and (22.10) are now substituted by, respectively,

$$\begin{aligned} \boldsymbol{\xi}\cdot\mathbf{n}\otimes\mathbf{n}\otimes\mathbf{n} &= f, \\ \mathrm{asym}\big((\boldsymbol{\xi} : \mathbf{n}\otimes\mathbf{n})\otimes\mathbf{n}\big) &= \boldsymbol{\mu} \quad \text{in } S, \end{aligned} \qquad (22.14)$$

$$\boldsymbol{\xi} : [\![\mathbf{m}\otimes\mathbf{n}]\!] = \mathbf{s} \quad \text{in } E. \qquad (22.15)$$

A robust way to introduce balance laws is to deduce them from objectivity requirements posed on energetic quantities [12, 8]. When dealing with extended continuum theories, to which our intuition is less accustomed, this practice is particularly useful. Requiring Galilean invariance of the potential of the external loads (22.13) gives in particular the force balance law for the external actions:

$$\int_V \mathbf{F}\,\mathrm{d}V + \int_S \mathbf{t}\,\mathrm{d}S + \int_E \mathbf{s}\,\mathrm{d}E = 0.$$

Following loosely the terminology adopted in [13], we refer to Leibnitz invariance as the property of a scalar quantity to be measured identically by two observers in relatively constant angular motion. Requiring the potential P in (22.13) to have such invariance gives a momentum balance law in the form:

$$\int_V \mathbf{p}\times\mathbf{F}\,\mathrm{d}V + \int_S (\mathbf{p}\times\mathbf{t} + \boldsymbol{\mu})\,\mathrm{d}S + \int_E \mathbf{p}\times\mathbf{s}\,\mathrm{d}E = 0$$

(cf. [3]), where $\mathbf{p}$ denotes the position vector with respect to a fixed pole.

22.4 Stored Energy Density

To achieve a suitable format for the expression of the stored energy density, we first seek an efficient decomposition of the native tensor spaces of its two variables. The problem being trivial for $\nabla\mathbf{u}$, we only focus on $\nabla\nabla\mathbf{u} \in \mathrm{SYM}$ and adopt a procedure inspired by Spencer's representation formula [14]. The projection operators $(\cdot)^S$ and $(\cdot)^A$, with

$$(\alpha_{ijk})^S = \frac{1}{3}(\alpha_{ijk} + \alpha_{jki} + \alpha_{kij}), \qquad (\alpha_{ijk})^A = \alpha_{ijk} - \alpha^S_{ijk}, \quad \boldsymbol{\alpha} \in \mathrm{SYM}$$

define by their images the linear spaces $\mathrm{SYM^S}$ of totally symmetric 3rd-order tensors and $\mathrm{SYM^A}$ of partially antisymmetric 3rd-order tensors. Notice that $\mathrm{SYM^S}$ and $\mathrm{SYM^A}$ provide a direct-sum decomposition of SYM and a finer one can be achieved by introducing further projection operators $(\cdot)^{SS}$, $(\cdot)^{SD}$, $(\cdot)^{AS}$ and $(\cdot)^{AD}$, where the additional superscripts evoke the terms "spherical" and "deviatoric". We define in particular

$$\begin{aligned}(\alpha_{ijk})^{SS} &= \frac{1}{5}(\alpha_{ill}\delta_{jk} + \alpha_{ljl}\delta_{ki} + \alpha_{llk}\delta_{ij}),\\ (\alpha_{ijk})^{SD} &= \alpha_{ijk} - \alpha^{SS}_{ijk}, \quad \boldsymbol{\alpha} \in \mathrm{SYM^S}\end{aligned}$$

and

$$\begin{aligned}(\alpha_{ijk})^{AS} &= \frac{1}{2}(\alpha_{ill}\delta_{jk} + \alpha_{ljl}\delta_{ki} + \alpha_{llk}\delta_{ij}),\\ (\alpha_{ijk})^{AD} &= \alpha_{ijk} - \alpha^{AS}_{ijk}, \quad \boldsymbol{\alpha} \in \mathrm{SYM^A}\end{aligned}$$

which lead to the direct-sum decomposition

$$\mathrm{SYM} = \underbrace{\mathrm{SYM^{SS}} \oplus \mathrm{SYM^{SD}}}_{\mathrm{SYM^S}} \oplus \underbrace{\mathrm{SYM^{AS}} \oplus \mathrm{SYM^{AD}}}_{\mathrm{SYM^A}}, \tag{22.16}$$

where $\mathrm{SYM^{SS}}$, $\mathrm{SYM^{SD}}$, $\mathrm{SYM^{AS}}$, and $\mathrm{SYM^{AD}}$ are the images of $\mathrm{SYM^S}$ and $\mathrm{SYM^A}$ under the action of corresponding projection operators.

The above results are exploited in the expression

$$\begin{aligned} w = {} & \frac{3\lambda + 2\mu}{2}(\mathrm{sph}\nabla\mathbf{u}) \cdot (\mathrm{sph}\nabla\mathbf{u}) + \mu(\mathrm{dev}\nabla\mathbf{u}) \cdot (\mathrm{dev}\nabla\mathbf{u})\\ & + \frac{G^{(1)}}{2}(\nabla\nabla\mathbf{u})^{SS} \cdot (\nabla\nabla\mathbf{u})^{SS} + \frac{G^{(2)}}{2}(\nabla\nabla\mathbf{u})^{SD} \cdot (\nabla\nabla\mathbf{u})^{SD}\\ & + \frac{G^{(3)}}{2}(\nabla\nabla\mathbf{u})^{AS} \cdot (\nabla\nabla\mathbf{u})^{AS} + \frac{G^{(4)}}{2}(\nabla\nabla\mathbf{u})^{AD} \cdot (\nabla\nabla\mathbf{u})^{AD}\\ & + G^{(5)}\big((\nabla\nabla\mathbf{u})^{SS} : \mathbf{I}\big) \cdot \big((\nabla\nabla\mathbf{u})^{AS} : \mathbf{I}\big), \end{aligned}$$

where the stored energy density w is expressed in terms of mutually orthogonal components of $\nabla\mathbf{u}$ and $\nabla\nabla\mathbf{u}$. Along with Lame's μ and λ one counts five additional constitutive parameters, from $G^{(1)}$ to $G^{(5)}$, that complete the set of material constants. By virtue of the direct-sum decomposition (22.16), one proves that the inequalities

$$G^{(1)} > 0, \qquad G^{(2)} > 0, \qquad G^{(3)} > 0, \qquad G^{(4)} > 0 \tag{22.17}$$

and

$$-\sqrt{\frac{1}{5}G^{(1)}G^{(3)}} < \frac{2}{3}G^{(5)} < \sqrt{\frac{1}{5}G^{(1)}G^{(3)}} \tag{22.18}$$

complete the set of positive-definiteness conditions for the stored energy density w along with the classical inequalities for Lame's constants. The latter, along with (22.17) and (22.18), are assumed to hold in the following sections.

22.5 Bolted-Layers

The system we consider here is given by an infinite layer of thickness H. We treat the latter as a periodic assembly of rectangular prisms of dimensions L_1, L_2 and H along the directions $\mathbf{e}_1$, $\mathbf{e}_2$ and $\mathbf{e}_3$, respectively. We therefore restrict our attention to the base element of the paving, the prism centered at the origin of our reference. We call $S^{[-1]}$, $S^{[+1]}$, $S^{[-2]}$, $S^{[+2]}$, $S^{[-3]}$, and $S^{[+3]}$ the six rectangular portions of the boundary of the base element, where the superscripts refer to the corresponding outward-oriented unit normals: $\mathbf{e}_1$, $-\mathbf{e}_1$, $\mathbf{e}_2$, $-\mathbf{e}_2$, $\mathbf{e}_3$, and $-\mathbf{e}_3$, respectively.

In the problem considered here, the displacements are inhibited on the "lower boundary" $S^{[-3]}$ while negative (compressive) ortho-fiber tension is applied on both $S^{[-3]}$ and on the "upper boundary" $S^{[+3]}$. We imagine that this condition is realized physically by regular arrays of bolts implanted along the two boundaries, anchored at a depth d which small compared to the thickness H, and properly pretensioned by screwing the externally accessible nuts. Representing the assigned pretension by the surface density $\hat{F}$, one can implement these conditions in the form

$$\mathbf{u} = 0 \quad \text{in } S^{[-3]}, \qquad f = -\hat{F}d \quad \text{in } S^{[-3]}, \qquad f = -\hat{F}d \quad \text{in } S^{[+3]}. \tag{22.19}$$

To have the considered base element representative of the behavior of the whole layer we assign periodic boundary conditions in the form

$$\mathbf{u}|_{S^{[-1]}}\left(\frac{-L_1}{2},\cdot,\cdot\right) = \mathbf{u}|_{S^{[+1]}}\left(\frac{L_1}{2},\cdot,\cdot\right),$$
$$\mathbf{u}|_{S^{[-2]}}\left(\cdot,\frac{-L_2}{2},\cdot\right) = \mathbf{u}|_{S^{[+2]}}\left(\cdot,\frac{L_2}{2},\cdot\right),$$
$$\varsigma|_{S^{[-1]}}\left(\frac{-L_1}{2},\cdot,\cdot\right) = \varsigma|_{S^{[+1]}}\left(\frac{L_1}{2},\cdot,\cdot\right),$$

$$\varsigma|_{S^{[-2]}}\left(\cdot,\frac{-L_2}{2},\cdot\right)=\varsigma|_{S^{[+2]}}\left(\cdot,\frac{L_2}{2},\cdot\right),$$
$$\boldsymbol{\omega}|_{S^{[-1]}}\left(\frac{-L_1}{2},\cdot,\cdot\right)=\boldsymbol{\omega}|_{S^{[+1]}}\left(\frac{L_1}{2},\cdot,\cdot\right),$$
$$\boldsymbol{\omega}|_{S^{[-2]}}\left(\cdot,\frac{-L_2}{2},\cdot\right)=\boldsymbol{\omega}|_{S^{[+2]}}\left(\cdot,\frac{L_2}{2},\cdot\right).$$

Attempting to solve Euler's equations (22.7), (22.8), (22.14), and (22.15) demands computing $\boldsymbol{\tau}$ and $\boldsymbol{\xi}$ according to the expression of the elastic energy density given in Sect. 22.4. Using in addition the boundary condition introduced in this section will give a solution of the form

$$u_3=-\frac{\hat{F}d}{2(\lambda+2\mu)}\left(1+\frac{\sinh(x_3/\hat{l})}{\sinh(H/(2\hat{l}))}\right),\qquad \hat{l}:=\sqrt{\frac{3G^{(1)}+2G^{(2)}}{5(\lambda+2\mu)}},\tag{22.20}$$

where $\hat{l}$ plays the role of internal length. Notice that the displacement field in (22.20) is characterized by a "vertical" strain $\partial u_3/\partial x_3$ with exponential decay with respect to the distance from the upper and lower boundaries. We also remark that the solution in (22.20) is unique, as it can be shown in view of the boundary conditions and of the assumed positive-definiteness of the stored energy density.

22.6 Eccentric Boundaries

Under the same geometry as in Sect. 22.5, we consider now a different type of boundary conditions apt to implement "rugosity" or other physical situations assimilable to an eccentricity e of the surface tractions of magnitude comparable to internal lengths of the model. We focus on the shear layer problem. We assign

$$\begin{aligned}\mathbf{t}&=-\tilde{F}\mathbf{e}_1\quad\text{in } S^{[-3]},\\ \mathbf{t}&=\tilde{F}\mathbf{e}_1\quad\text{in } S^{[+3]},\end{aligned}\tag{22.21}$$

and implement the eccentricity e of the boundary in the form of ortho-fiber couples:

$$\begin{aligned}\boldsymbol{\mu}&=\operatorname{asym}\big(\mathbf{t}\otimes(-e\mathbf{e}_3)\big)\quad\text{in } S^{[-3]},\\ \boldsymbol{\mu}&=\operatorname{asym}\big(\mathbf{t}\otimes(e\mathbf{e}_3)\big)\quad\text{in } S^{[+3]}.\end{aligned}\tag{22.22}$$

We additionally require antisymmetry of the displacements on the upper and lower boundaries in order to exclude uncontrolled rigid motions:

$$\mathbf{u}|_{S^{[-3]}}\left(\cdot,\cdot,\frac{-H_1}{2}\right)=\mathbf{u}|_{S^{[+3]}}\left(\cdot,\cdot,\frac{H_1}{2}\right).\tag{22.23}$$

Following the same procedure addressed in Sect. 22.5 but using (22.21)–(22.23) instead of (22.19), one obtains the solution

$$u_1 = \frac{\tilde{F}}{\mu}\left(x_3 + e\frac{\sinh(x_3/\tilde{l})}{\sinh(H/(2\tilde{l}))}\right),$$
$$\tilde{l} = \sqrt{\frac{1}{3\mu}\left[\frac{G^{(1)}}{5} + \frac{4}{5}G^{(2)} + G^{(3)} + G^{(4)} + \frac{4}{3}G^{(5)}\right]}, \tag{22.24}$$

where the internal length $\tilde{l}$ is given evidence. Once more the solution is unique. We remark, moreover, that (22.24) differs from the classical (1st-grade) solution of the shear-layer problem for an additional amount of shear strain, computed as $(\partial u_3/\partial x_1 + \partial u_1/\partial x_3)/2$, decaying exponentially with respect to the distance from the upper, "eccentric" boundary.

22.7 Conclusions

In this work we proposed an interpretation of the additional boundary conditions allowed by 2nd-grade models on the normal-to-the-boundary derivative of the displacement. This is based on the concept of a material ortho-fiber that can independently extend or compress and rotate. Example boundary value problems were used to demonstrate the effect of these additional boundary conditions and how they can be applied in different settings. The advantage over past work is that the proposed concept of ortho-fiber has a clear physical meaning, allowing a more confident use of higher-order boundary conditions for general boundary value problems.

References

1. Bluenstein, J.L.: A note on boundary conditions of Toupin's second gradient theory. Int. J. Solids Stuct. **3**, 1053–1057 (1967)
2. Degiovanni, M., Marzocchi, A., Musesti, A.: Edge-forces and second-order powers. Ann. Math. Pura Appl. **185**, 81–103 (2006)
3. Fried, E., Gurtin, M.: Traction, balances and boundary conditions for nonsimple materials with applications to liquid at small length-scales. Arch. Ration. Mech. Anal. **182**, 513–554 (2006)
4. Froiio, F., Zervos, A., Vardoulakis, I.: Linear 2nd-grade elasticity. J. Elast. (2009, submitted)
5. Germain, P.: La méthode de puissances virtuelles en mécanique des milieux continus. Première partie: théorie du second gradient. J. Méc. **12**, 235–274 (1973)
6. Germain, P.: La méthode de puissances virtuelles en mécanique des milieux continus. Première partie: théorie du second gradient. J. Méc. **25**, 556–575 (1973)
7. Germain, P.: The method of virtual power in continuum mechanics. Part 2: Microstructure. SIAM J. Appl. Math. **25**, 556–575 (1973)
8. Green, A.E., Rivlin, R.: On Cauchy equation of motion. Z. Angew. Math. Phys. **15**, 290–292 (1964)

9. Mindlin, R.D.: Microstructre in linear elasticity. Arch. Ration. Mech. Anal. **16**, 51–78 (1995)
10. Mindlin, R.D.: Second gradient of strain and surface tension in linear elasticity. Int. J. Solids Stuct. **1**, 417–438 (1964)
11. Mindlin, R.D., Eshel, N.N.: On first strain-gradient theory in linear elasticity. Int. J. Solids Stuct. **4**, 109–124 (1968)
12. Noll, W.: La mécanique classique, basée sur un axiome d'objectivité. In: Actes Quatrième Colloq. Internat. de Logique et Philos. Sci., Paris (1959), pp. 47–56. Gauthier-Villars, Paris (1963)
13. Podio-Guidugli, P.: Inertia and invariance. Ann. Mat. Pura Appl. **172**, 103–124 (1997)
14. Spencer, A.J.M.: Isotropic polynomial invariants and tensor functions. In: Boheler, J.P. (ed.) Application of Tensor Functions in Continuum Mechanics. CISM Courses and Lectures, vol. 292. Springer, Berlin (1987)
15. Toupin, R.A.: Elastic material with couple stress. Arch. Ration. Mech. Anal. **11**, 385–414 (1962)

Chapter 23
Gradient Theory of Media with Conserved Dislocations: Application to Microstructured Materials

Sergey Lurie, Petr Belov, and Natalia Tuchkova

Abstract In the paper, a rigorous continuous media model with conserved dislocations is developed. The important feature of the newly developed classification is a new kinematic interpretation of dislocations, which reflects the connection of dislocations with distortion, change in volume (porosity), and free forming. Our model generalizes those previously derived by Mindlin, Cosserat, Toupin, Aero–Kuvshinskii and so on, and refines some assertions of these models from the point of view of the account of adhesive interactions.

23.1 Introduction

Gradient theories have their origin in work written over a half century ago, see the review of Mindlin (1972) [17]. In most applications of gradient theories, second gradient theories have been used for describing of the size effects in the miniaturized structures and devices [2, 1, 4, 3, 5–8, 16, 17, 13–15] because of their relative simplicity. It has been shown that gradient theories are quite effective in the analyses of the media at the nano- and micro-levels. We surmise that gradient corrections to the physical equations of continuous media (the Laplace equation, the Maxwell equations, etc.) can also significantly extend the scope of their validity and make them suitable for predicting and interpreting the properties of nanostructured media, materials, and small-size devices.

S. Lurie (✉) · P. Belov
Inst. of Applied Mechanics of RAS, Russia, Moscow
e-mail: lurie@ccas.ru

P. Belov
e-mail: Petr.Belov@boing.com

N. Tuchkova
Dorodnicyn Computing Centre of RAS, Russia, Moscow
e-mail: tuchkova@ccas.ru

G.A. Maugin, A.V. Metrikine (eds.), *Mechanics of Generalized Continua*, Advances in Mechanics and Mathematics 21,
DOI 10.1007/978-1-4419-5695-8_23, © Springer Science+Business Media, LLC 2010

In the present work, the second-order unified gradient model of the continuous media with conserved dislocations is used to describe the spectrum of cohesion and the superficial phenomena. Generalization of the more common isotropic gradient Mindlin's model with a complete description of corresponding adhesive properties of the boundary surfaces for the elastic bodies is offered. Actually, the gradient continual theory of adhesive interactions is being developed. Variation formulations of the generalized applied models of the medium of Toupin's type are received. They are constructed using of the additional hypothesis which is generalization of a hypothesis of Aero–Kuvshinskii about proportionality of spins and vortices of displacements. These models have only one additional physical parameter quantifying strain-gradient effects; the models are simple enough and can be used for simulation in mechanics of filled composites because they give full enough description of the cohesion field with adhesion interactions in the contact zone of different components [9–12].

Let us assume that the kinematic relations of the media under consideration are defined by, generally speaking, asymmetric Cauchy relations $d^0_{in} = R_{i,n}$ and Papkovich non-homogeneous equations [4, 3]:

$$d_{in,m}\Theta_{nmj} = \Xi_{ij}. \tag{23.1}$$

Here $d_{in} = \gamma_{in} + (1/3)\theta\delta_{in} - \omega_k\Theta_{ink}$, R_i is a continuous displacement vector, $d^0_{ij} = R_{i,j}$ is a distortion tensor, γ_{in} is a strain deviator tensor, θ is the change in volume, ω_k is a vector of elastic spins, Θ_{ijk} is the permutation symbol, δ_{ij} is the Kronecker's delta; Ξ_{ij} is a second-rank pseudo-tensor which defines incompatibility of displacements.

The tensor Ξ_{ij} is the dislocations density tensor [4, 3, 9, 10]. This means that in the case under consideration the continuum with a field of defects, dislocations, is considered. Solution of the non-homogeneous equations (23.1) can be presented as the sum $d_{ij} = R_{i,j} + d^{\Xi}_{ij}$, where $d^0_{in,m}\Theta_{nmj} = 0$, d^{Ξ}_{ij} is a partial solution of the non-homogeneous equation ($\Xi_{ij} \neq 0$). Along with a continuous component, d^{Ξ}_{ij}, one can consider quantities γ^{Ξ}_{ij}, ω^{Ξ}_k and θ^{Ξ} as generalized displacements, $[\gamma^{\Xi}_{in} + (1/3)\theta^{\Xi}\delta_{in} - \omega^{\Xi}_{k\ ink}]_{m}\Theta_{nmj} = \Xi_{ij} \neq 0$. The general solution of the Papkovich non-homogeneous (23.1) can be written in the symmetrized form:

$$\begin{aligned}
\gamma_{ij} &= \frac{1}{2}(R_{i,j} + R_{j,i}) - \frac{1}{3}R_{k,k}\delta_{ij} + \gamma^{\Xi}_{ij},\\
\omega_k &= -\frac{1}{2}R_{i,j}\Theta_{ijk} + \omega^{\Xi}_k,\\
\theta &= R_{k,k} + \theta^{\Xi}.
\end{aligned}$$

Three types of dislocations are defined, namely related to γ^{Ξ}_{ij}, ω^{Ξ}_k and θ^{Ξ}:

$$\begin{aligned}
\Xi_{ij} &= \left[\gamma^{\Xi}_{in} + \frac{1}{3}\theta^{\Xi}\delta_{in} - \omega^{\Xi}_k\Theta_{ink}\right]_{,m}\Theta_{nmj}\\
&= \Xi^{\gamma}_{ij} + \Xi^{\theta}_{ij} + \Xi^{\omega}_{ij}.
\end{aligned}$$

Here

$$\Xi_{ij}^{\gamma} = \gamma_{in,m}^{\Xi}\Theta_{nmj}, \qquad \Xi_{ij}^{\theta} = \frac{1}{3}\theta_{,m}^{\Xi}\Theta_{imj}, \qquad \Xi_{ij}^{\omega} = -\omega_{k,m}^{\Xi}\Theta_{nki}\Theta_{nmj}.$$

The quantities Ξ_{ij}^{γ}, Ξ_{ij}^{θ}, Ξ_{ij}^{ω} are the sources of the corresponding three types of dislocations: γ-dislocations, θ-dislocations ω-dislocations. The law of conservation of dislocations holds: $\Xi_{ij,j}^{\gamma} = 0$, $\Xi_{ij,j}^{\theta} = 0$, $\Xi_{ij,j}^{\omega} = 0$.

In the recent works [2, 8–11], the minimal variant of the theory of the media with conserved dislocations was elaborated on the base of the kinematic variational principle [4, 9–11]. It was shown that for such continuous media it is generally possible to write the following equation for potential energy densities in the volume V and on the surface F:

$$U_V = U_V\big(d_{ij}^0; d_{ij}^{\Xi}; \Xi_{ij}\big), \qquad U_F = U_F\big(d_{ij}^{\Xi}, d_{ij}^0\big),$$
$$U = \iiint U_V \,\mathrm{d}V + \iiint U_F \,\mathrm{d}F.$$

For physically linear continuous media, the potential energy densities in the volume and on the surface is a quadratic form of its arguments:

$$\begin{aligned} U_V &= \frac{1}{2}\big(C_{ijnm}^{11}R_{n,m}R_{i,j} + 2C_{ijnm}^{12}R_{n,m}d_{ij}^{\Xi} \\ &\quad + C_{ijnm}^{22}d_{nm}^{\Xi}d_{ij}^{\Xi} + C_{ijnm}^{33}\Xi_{nm}\Xi_{ij}\big), \\ U_F &= \frac{1}{2}\big(A_{ijnm}^{11}d_{nm}^{0}d_{ij}^{0} + A_{ijnm}^{12}d_{nm}^{0}d_{ij}^{\Xi} + A_{ijnm}^{22}d_{nm}^{\Xi}d_{ij}^{\Xi}\big). \end{aligned} \tag{23.2}$$

Note that the kinematic variables of the model (23.2) are defined by $R_i d_{ij}^{\Xi}$, and the kinematic parameters of the model are R_i, d_{ij}^{Ξ}, gradients of displacements gradients $R_{i,j}$ and density of dislocations Ξ_{ij}. The tensors of moduli C_{ijnm}^{pq}, $A_{ijnm}^{\alpha\beta}$ in (23.3) define the physical constants of the model (constitutive equations) in the volume and on the surface. The structure of these tensors will be presented in more detail below. The mathematical formulation of the model (constitutive equations, equilibrium equations and boundary value problem as a whole) is completely defined by the potential energy. Therefore, for brevity, we will formulate only the density of potential energy in a volume and on a surface.

23.2 Variational Model of Generalized Continuous Media

Let us examine the more general Mindlin model of the medium with conserved dislocations rather than the model (23.2). In the general model, the displacement vector R_i and free distortion tensor d_{ij}^{Ξ} are also independent kinematic variables. The kinematic variables R_i, d_{ij}^{Ξ} and their gradients $R_{i,j}$, $d_{ij,k}^{\Xi}$ are the independent kinematic parameters [16]. Let us propose that the energy density U_V depends on the full curvature tensor $d^{\Xi}_{in,k}$. Then the potential energy densities in the volume and on the surface, U_V and U_F, can be written in the following form:

$$U_V = \frac{1}{2}\big[C^{11}_{ijnm} R_{i,j} R_{n,m} + 2C^{12}_{ijnm} R_{i,j} d^{\Xi}_{nm} + C^{22}_{ijnm} d^{\Xi}_{ij} d^{\Xi}_{nm} + C_{ijknml} d^{\Xi}_{ij,k} d^{\Xi}_{nm,l}\big], \tag{23.3}$$

$$U_F = \frac{1}{2}\big[A^{11}_{ijnm} R_{i,j} R_{n,m} + 2A^{12}_{ijnm} R_{i,j} d^{\Xi}_{nm} + A^{22}_{ijnm} d^{\Xi}_{ij} d^{\Xi}_{nm}\big],$$

where $C^{\alpha\beta}_{ijnm}$ is the tensor of generalized elastic constants of the forth order $(\alpha, \beta = 1, 2)$, C_{ijknml} is a Mindlin's tensor quantifying gradient effects, $A^{\alpha\beta}_{ijnm}$ is the tensor of the physical constants of the model on the surface. Note, that in the particular case (see (23.2)) the potential energy density depended only on the anti-symmetric tensor Ξ_{ij} of curvature tensors $d^{\Xi}_{in,k}$ with respect to indices n, k. By analogy with the dislocations density tensor, the disclinations density tensor is naturally entered: $T_{ijk} = d_{ija,b}\Theta_{abk}$, where d_{ija} are the curvature tensors, $d_{ijk} = (R_{i,j} + d^{\Xi}_{ij})_{,k}$. Let us consider a further gradient elastic medium where the following conditions are satisfied: $T_{ijk} = d_{ija,b}\Theta_{abk} = 0$. These integrability conditions represent the existence criterion for the tensor potential of the tensor of the curvatures $d_{ija} = (R_{i,j} + d^{\Xi}_{ij})_{,a}$. The equation $T_{ijk} = 0$ is a necessary and sufficient condition for the absence of the disclinations field in the considered medium. Thus, the generalized model (23.3) also concerns the mediums with conserved dislocations because the disclinations are absent [3]. The conservation law takes place for the dislocation tensor Ξ_{ij}, too.

For an isotropic, generally asymmetrical model of a continuum, tensors $C^{\alpha\beta}_{ijnm}$ in (23.3) have the usual structure

$$C^{\alpha\beta}_{ijnm} = \lambda^{\alpha\beta}\delta_{ij}\delta_{nm} + (\mu^{\alpha\beta} + \chi^{\alpha\beta})\delta_{in}\delta_{jm} + (\mu^{\alpha\beta} - \chi^{\alpha\beta})\delta_{im}\delta_{jn}, \quad (1, 2, 3),$$

where $\mu^{\alpha\beta}$ are analogs of the shear modulus; $2\mu^{\alpha\beta} + 3\lambda^{\alpha\beta}$ are spherical tensors; $\chi^{\alpha\beta}$ are the third Lame coefficients for an asymmetrical model. The tensor C_{ijknml} has the same structure as in Mindlin theory, with 11 independent moduli. The tensor moduli $A^{\alpha\beta}_{ijnm}$ for the surface of the body with normal vector n_i $(\eta_{mp} = \delta_{mp} - n_m n_p)$ can be represented as follows [4, 9]:

$$A^{\alpha\beta}_{ijnm} = \lambda^{\alpha\beta}_F \eta_{ij}\eta_{nm} + \delta^{\alpha\beta}_F n_i n_n \eta_{jm} + (\mu^{\alpha\beta}_F + \chi^{\alpha\beta}_F)\eta_{in}\eta_{jm} + (\mu^{\alpha\beta}_F - \chi^{\alpha\beta}_F)\eta_{im}\eta_{jn}. \tag{23.4}$$

Taking into account (23.3), we can find the constitutive equations and write the variational equation of the problem $\delta L = \delta(A - U) = 0$:

$$\delta L = \iiint \big[(C^{11}_{ijnm} R_{n,mj} + C^{12}_{ijnm} d^{\Xi}_{nm,j} + P^V_i)\delta R_i + (C_{ijknml} d^{\Xi}_{nm,lk} - C^{12}_{ijnm} R_{n,m} - C^{22}_{ijnm} d^{\Xi}_{nm})\delta d^{\Xi}_{ij}\big]\, \mathrm{dV}$$

$$
\begin{aligned}
&+ \oiint \{ [P_i^F - (C_{ijnm}^{11} R_{n,m} + C_{ijnm}^{12} d_{nm}^{\Xi}) n_j \\
&+ (A_{ijnm}^{11} R_{n,mj} + A_{ijnm}^{12} d_{nm,j}^{\Xi})] \delta R_i \\
&- [A_{ijnm}^{12} R_{n,m} + A_{ijnm}^{22} d_{nm}^{\Xi} + C_{ijknml} n_k d_{nm,l}^{\Xi}] \delta d_{ij}^{\Xi} \} \, \mathrm{d}F \\
&- \sum \oint (A_{ijnm}^{11} R_{n,m} + A_{ijnm}^{12} d_{nm}^{\Xi}) v_j \delta R_i \, \mathrm{d}s = 0,
\end{aligned}
\tag{23.5}
$$

where v_j is a normal to the bars on the surface, A is the work of external forces in the volume and on the surface, $A = \iiint P_i^V R_i \, \mathrm{d}V + \oiint P_i^F R_i \, \mathrm{d}F$; P_i^V, P_i^F are the given densities of volume and surface forces. It is important to point out that the dimension of moduli C_{ijknml} differs by a square of length from the dimension of moduli $C_{ijnm}^{\alpha\beta}$. As a result, the account of the contribution of the invariants of the dislocation tensor $C_{ijknml} d_{ij,k}^{\Xi} d_{nm,l}^{\Xi}$ in the expression for the potential energy (23.3) inevitability results in the volumetric scale effects. It should be also noted that the dimension of moduli of adhesion $A_{ijnm}^{\alpha\beta}$ in (23.3) and (23.5) also differs from the bulk moduli by the dimension of length. Thus, the account of the component of adhesion in the expression for potential energy (23.4) results in the surface scale effects.

23.3 Decomposition of Equilibrium Equations

Let us introduce the potential vector-function φ_i using the following equations

$$
d_{nm}^{\Xi} = C_{nmab}^{-22} C_{abpq}^{12} \varphi_{p,q}, \qquad R_n = e_{nkpq} \varphi_{p,qk} - \varphi_n,
\tag{23.6}
$$

where $e_{ijnm} = \lambda^* \delta_{ij} \delta_{nm} + (\mu^* + \chi^*) \delta_{in} \delta_{jm} + (\mu^* - \chi^*) \delta_{im} \delta_{jn}$ is the normalized tensor of the effective moduli for the volumetric gradient effects. For isotropic media, the components of the tensor e_{ijnm} can be submitted as linear combinations from components of the tensor $C_{ijknml} C_{nmab}^{-22} C_{abpq}^{12}$. Substituting (23.6) into equations of equilibrium for the moments in (23.5), $C_{ijknml} d_{nm,lk}^{\Xi} - C_{ijnm}^{12} R_{n,m} - C_{ijnm}^{22} d_{nm}^{\Xi} = 0$, we can see that the nine equations of equilibrium for the moments in (23.5) are automatically satisfied for an arbitrary vector, φ_n. Consider the equilibrium equations for forces in (23.5), $C_{ijnm}^{11} R_{n,mj} + C_{ijnm}^{12} d_{nm,j}^{\Xi} + P_i^{3D} = 0$. Substituting (23.6) into these equations and rearranging terms, we obtain the equations of equilibrium of isotropic strain-gradient elasticity in decomposed form:

$$
L_{ij}(H_{jk} \varphi_k) + P_i^V = 0,
\tag{23.7}
$$

where $L_{ij}(\cdot) = (\mu + \chi)(\cdot)_{,pp} \delta_{ij} + (\mu + \lambda - \chi)(\cdot)_{,ij}$ is the operator of the classical theory of elasticity; the effective moduli $(\mu + \chi)(\mu + \lambda - \chi)$ are defined by the relations

$$
C_{ijpq} = C_{ijpq}^{11} - C_{ijnm}^{12} C_{nmab}^{-22} C_{abpq}^{12} = \lambda \delta_{ij} \delta_{nm} + (\mu + \chi) \delta_{in} \delta_{jm} + (\mu - \chi) \delta_{im} \delta_{jn},
$$

$H_{ij}(\cdot)$ is a generalized Helmholtz operator,

$$H_{ij}(\cdot) = l_1^2(\cdot)_{,pp}\delta_{ij} + \left(l_2^2 - l_1^2\right)(\cdot)_{,ij} - \delta_{ij},$$

$$l_1^2 = (\mu^* + \chi^*)\frac{\mu^{11} + \chi^{11}}{\mu + \chi}, \qquad l_2^2 = (2\mu^* + \lambda^*)\frac{2\mu^{11} + \lambda^{11}}{2\mu + \lambda}.$$

Equation (23.7) states that the general integral of equations of equilibrium can be written as a sum of that of the classical equations of equilibrium and that of generalized Helmholtz equations. It is seen from the structure of the operator (23.7) that the total solution of the boundary problem can be represented in two different forms. The first form of solution allows defining the classical field U_j which satisfies the classical equations of equilibrium $H_{jk}\varphi_k = U_j$, $L_{ij}U_j + P_i^V = 0$. The second form of solution gives the definition of the local cohesive field u_j satisfying the generalized Helmholtz equations $L_{jk}\varphi_k/(\mu+\chi) = u_j$, $(\mu + \chi)H_{ij}u_j + P_i^V = 0$. Here l_1^2, l_2^2 are the only two nonclassical physical constants of the model that determine the cohesive interactions [9].

Note that we use classification of kinematics, which allows us to introduce three types of dislocations: γ-dislocations, ω-dislocations, and θ-dislocations. The offered classification allows prediction of certain particular cases of dislocations and examination of some interesting exact special cases of the theory when only one or two types of dislocations are dominating in the medium. In special models, it is possible to reduce the number of degrees of freedom, which essentially simplifies the research of separate properties of the mediums with conserved dislocations.

Let us consider the Cosserat media model as the theory of ω-dislocations. It is a special case of the general model of medium with the conserved defects in which the tensor of free distortion $d_{ij}^{\Xi} = \gamma_{ij}^{\Xi} + (1/3)\theta^{\Xi}\delta_{ij} - \omega_k^{\Xi} i_{ijk}$ is determined only by free rotations $d_{ij}^{\Xi} = -\omega_k^{\Xi}\Theta_{ijk}$, free deformations γ_{ij}^{Ξ} and θ^{Ξ} are equal to zero, $\gamma_{ij}^{\Xi} = 0$, $\theta^{\Xi} = 0$. Then the potential energy of the generalized Cosserat model with adhesive properties of a surface is defined by the energy densities ($C_{ijnm}^{33} = C_{abjcdm}\Theta_{abi}\Theta_{cdn}$):

$$\begin{aligned} U_V &= \big(C_{ijnm}^{11} R_{i,j} R_{n,m} - 2C_{ijnm}^{12}\Theta_{nmq} R_{i,j}\omega_q^{\Xi} \\ &\quad + C_{ijnm}^{22}\Theta_{ijp}\Theta_{nmq}\omega_p^{\Xi}\omega_q^{\Xi} + C_{ijnm}^{33}\omega_{i,j}^{\Xi}\omega_{n,m}^{\Xi}\big)/2, \\ U_F &= \big(A_{ijnm}^{11} R_{i,j} R_{n,m} - 2A_{ijnm}^{12}\Theta_{nmq} R_{i,j}\omega_q^{\Xi} + A_{ijnm}^{22}\Theta_{ijp}\Theta_{nmq}\omega_p^{\Xi}\omega_q^{\Xi}\big)/2. \end{aligned}$$

Similarly, the porous media model as the theory of θ-dislocations can be considered. For this model, in the kinematic model only scalar porosity characteristic prevails $d_{ij}^{\Xi} = (1/3)\theta^{\Xi}\delta_{ij}$, $\theta^{\Xi} \neq 0$, $\gamma_{ij}^{\Xi} = 0$, $\omega_k^{\Xi} = 0$. As a consequence, the media with the free forming (γ-dislocations) can be considered as such a medium in which γ-dislocations prevail, so that $\gamma_{ij}^{\Xi} \neq 0$, $\theta^{\Xi} = 0$ and $\omega_k^{\Xi} = 0$. As a result, it is easy to write the density of potential energies in the volume and on the surfaces for the generalized porous model and for the model of the media with the free forming.

The "minimal" model of the media with conserved dislocations (23.2) can be obtained also as a particular case of the common model (23.3). Assume that the tensor C_{ijknml} has the following specific structure: $C_{ijknml} = C^{33}_{ianb}\Theta_{jka}\Theta_{mlb}$. Then (23.3) gives the strain energy densities of the "minimal" model (23.2).

In this model, the volume density depends on the quadratic form of the dislocation density Ξ_{ij}. Thus the model (23.2) is a model of the media with conserved dislocations of the minimal complexity. Nevertheless, it allows studying interaction of various types of dislocations contained in the potential energy. Moreover, it remains as full as the generalized Mindlin's model [16] because both characteristic lengths are present in this theory.

23.4 Generalized Toupin Model as Applied Model of Mindlin Theory

To describe the generalized Toupin model of media [16], let us introduce a generalization of the hypothesis about the proportionality of spins and vortices for the continuous media with the conserved dislocations (generalized Aero–Kuvshinskii hypothesis [1]):

$$\begin{aligned} d^{\Xi}_{ij} &= a_{ijnm}R_{n,m}, \\ a_{ijnm} &= a\delta_{ij}\delta_{nm} + (b+c)\delta_{in}\delta_{jm} + (b-c)\delta_{im}\delta_{jn}, \end{aligned} \tag{23.8}$$

where a, b, c are some constants. The densities of energy for the generalized Toupin model are obtained as a special case of the general theory by taking into account (23.5) and (23.8):

$$\begin{gathered} U_V = \frac{1}{2}(E_{ijnm}R_{i,j}R_{n,m} + E_{ijknml}R_{i,jk}R_{n,ml}), \\ U_F = \frac{1}{2}A_{ijnm}R_{i,j}R_{n,m}, \\ \begin{aligned} E_{ijnm} &= C^{11}_{ijnm} + \left(C^{12}_{abnm}a_{abij} + C^{12}_{abij}a_{abnm}\right) + C^{22}_{abcd}a_{abij}a_{cdnm} \\ &= E_{nmij}, \end{aligned} \\ E_{ijknml} = C_{abkcdl}a_{abij}a_{cdnm} = E_{nmlijk}, \\ \begin{aligned} A_{ijnm} &= A^{11}_{ijnm} + \left(A^{12}_{abnm}a_{abij} + A^{12}_{abij}a_{abnm}\right) + A^{22}_{abcd}a_{abij}a_{cdnm} \\ &= A_{nmij}. \end{aligned} \end{gathered} \tag{23.9}$$

Let us notice that the structure of the tensor A_{ijnm} here differs from the structure tensor of the adhesive moduli A^{pq}_{ijnm} in (23.4). If the tensor A^{pq}_{ijnm} satisfies $A^{pq}_{ijnm}n_j = 0$ and $A^{pq}_{ijnm}n_m = 0$, the tensor A_{ijnm} does not possess such properties any more. The tensor of the adhesive properties here possesses properties of the general tensor of the fourth rank, which is the transversal-isotropic tensor corresponding to the normal to a surface:

$$\begin{aligned}A_{ijnm} = {} & \big[\lambda^F(\delta_{ij} - n_i n_j)(\delta_{nm} - n_n n_m) + \delta^F n_i n_n(\delta_{jm} - n_j n_m) \\ & + (\mu^F + \chi^F)(\delta_{in} - n_i n_n)(\delta_{jm} - n_j n_m) \\ & + (\mu^F - \chi^F)(\delta_{im} - n_i n_m)(\delta_{jn} - n_j n_n)\big] \\ & + \alpha\big(n_i n_m(\delta_{jn} - n_j n_n) + n_n n_j(\delta_{im} - n_i n_m)\big) \\ & + \beta\big(n_i n_j(\delta_{nm} - n_n n_m) + n_n n_m(\delta_{ij} - n_i n_j)\big) \\ & + A n_i n_j n_n n_m + B(\delta_{in} - n_i n_n) n_j n_m. \end{aligned} \tag{23.10}$$

In the square brackets, the part of adhesion tensor coinciding in structure with the tensor A^{pq}_{ijnm} in (23.4) is allocated.

Let us examine a special case of the general Toupin model (23.9) which was used for construction of the interphase layer model and for prediction of the micro- and nanocomposites' properties [10–12]. Assume that $E_{ijknml} = E_{ijrk}E_{nmrl}/C$. As a result, in accordance with (23.9), we obtain the following energy densities:

$$\begin{aligned} U_V &= \frac{1}{2}\left[E_{ijnm}R_{i,j}R_{n,m} + \frac{1}{C}(E_{ijrk}R_{i,jk})(E_{nmrl}R_{n,mk})\right], \\ U_F &= \frac{1}{2}A_{ijnm}R_{i,j}R_{n,m}. \end{aligned} \tag{23.11}$$

Equilibrium equations for the forces for the model (23.11) can be written:

$$L_{ij}H_{jk}R_k + P_i^V = 0, \qquad H_{jk} = (\cdot)\delta_{jk} - L_{jk}(\cdot)/C. \tag{23.12}$$

It is seen from the structure of operator (23.12) that the total solution of the boundary problem can be represented as the sum of two components: the classical field $U_j = R_j - L_{jk}R_k/C$ satisfying the classical equations of equilibrium of elasticity theory $L_{ij}U_j + P_i^V = 0$ and the local cohesive field $u_j = -L_{jk}R_k/C$ satisfying the generalized Helmholtz equations $L_{ij}u_j - Cu_i + P_i^V = 0$. Here C is an additional physical constant of the model (the nonclassical module) that determines the cohesive interactions (see [9]). It was shown that the physical constant C can be defined through specific surface energy and other parameters of fracture mechanics: the length of the Barenblatt's zone, stress intensity factor, and so on. Gradient cohesion solution u_j allows receiving the strong theoretical justification for the famous hypothesis of Barenblatt in fracture mechanics [9]. Let us notice that the mathematical formulation of the applied model of the interphase layer (23.12) can be obtained from the more general model (23.5)–(23.7) as a special case if we assume that the following relations take place: $\varphi_n = -R_n$ instead of more common relations $R_n = e_{nkpq}\varphi_{p,qk} - \varphi_n$. Nevertheless, the advantage of the interphase layer model is its relative simplicity and that the accurate physical objects, namely classical and cohesion displacement fields, are used in it.

23.5 Conclusions

The generalized model of the Mindlin's media model with corresponding spectrum of adhesive interactions is constructed. The energetically consistent continual model of adhesive interactions is offered. For comparison we note that the variant of the continual model of adhesive interactions of Young–Laplace [6] is not consistent in the framework of the classical theory of elasticity. This model demands definition of the second derivative of displacements on a boundary surface. Such smoothness of the solution cannot be provided by the classical Lame operator. The suggested gradient model of adhesion is completely consistent and is full enough. In the works [4, 9] it is shown that this model describes all known adhesion effects (superficial tension, capillarity, a meniscus, and so forth). This model also models the adhesion damage effects on the interfaces with the aid of parameters $An_in_jn_nn_m + B(\delta_{in} - n_in_n)n_jn_m$ in (23.10). It is expected that using the adhesion model with parameters $An_in_jn_nn_m+B(\delta_{in}-n_in_n)n_jn_m$ will allow carrying out the real optimization of technological processes for the micro-structured materials.

It was shown that in the generalized Mindlin model all gradient effects in a volume are described only by two scale parameters, $l_1 \neq 0$ and $l_2 \neq 0$, which correspond to scale effects for a potential and vortical component of the displacements. All other gradient parameters (there are 9 of them) are included into boundary conditions. In this regard, "adhesive generalization" of the gradient models is rather relevant. Adhesive components in boundary conditions define the main effects as adhesive moduli differ from the classical moduli by a length; at the same time, the gradient moduli defined in a volume differ from the classical by a square of length.

The "minimal" gradient model (23.2) and its applied model of an interphase layer (23.11)–(23.12), in which parameters l_1, l_2 are proportional, also describe all scale effects characteristic for the general Mindlin model.

In spite of the fact that in these models there is only one additional gradient parameter in comparison with the theory of elasticity, they describe a full spectrum of the scale effects in a volume and a full spectrum of the adhesive interactions, characteristic for the general Mindlin model.

As a result, the applied gradient model of the interfacial layer is very effective in modeling of the properties of micro-and nano-structured materials [9–12].

References

1. Aero, L., Kuvshinskii, E.V.: The fundamental equations of the theory of the elasticity of a medium of particles with rotary interactions. Fiz. Tverd. Tela **2**, 1399–1409 (1960)
2. Aifantis, E.C.: The physics of plastic-deformation. Int. J. Plast. **3**, 211–247 (1987)
3. Belov, P., Lurie, A.: Continuous model of micro-heterogeneous media. Physical Mesomech. **10**(6), 49–61 (2007)
4. Belov, P., Lurie, A.: To a general geometrical theory of defect-containing media. Appl. Math. Mech. **73**(5), 833–848 (2009)

5. Cosserat, E., Cosserat, F.: Théorie des Corps Déformables. Hermann, Paris (1909)
6. Duan, H.L., Wang, J., Huang, Z.P., Karihaloo, B.L.: Size-dependent effective elastic constants of solids containing nano-inhomogeneities with interface stress. J. Mech. Phys. Solids **53**, 1574–1596 (2005)
7. Evans, A.G., Hutchinson, J.W.: A critical assessment of theories of strain gradient plasticity. Acta Mater. **57**(5), 1675–1688 (2009)
8. Fleck, N.A., Hutchinson, J.W.: Strain gradient plasticity. Adv. Appl. Mech. **33**, 295–361 (1997)
9. Lurie, S., Belov, P.: Cohesion field: Barenblatt's hypothesis as formal corollary of theory of continuous media with conserved dislocations. Int. J. Fract. **150**(1–2), 181–194 (2008)
10. Lurie, S., Belov, P., Tuchkova, N.: The application of the multiscale models for description of the dispersed composites. Int. J. Comput. Mater. Sci. A **36**(2), 145–152 (2005)
11. Lurie, S., Belov, P., Volkov-Bogorodsky, D., Tuchkova, N.: Interphase layer theory and application in the mechanics of composite materials. J. Mater. Sci. **41**(20), 6693–6707 (2006)
12. Lurie, S., Belov, P., Volkov-Bogorodsky, D., Zubov, V., Tuchkova, N.: Advanced theoretical and numerical multiscale modeling of cohesion/adhesion interactions in continuum mechanics and its applications for filled nanocomposites. Comput. Mater. Sci. **45**(3), 709–714 (2009)
13. Maugin, G.A.: Material forces: concepts and applications. Appl. Mech. Rev. **48**, 213–245 (1995)
14. Maugin, G.A.: On the structure of the theory of polar elasticity. Philos. Trans. R. Soc. Lond. A **356**, 1367–1395 (1998)
15. Maugin, G.A., Alshits, V.I., Kirchner, H.O.K.: Elasticity of multilayers: properties of the propagator matrix and some applications. Math. Mech. Solids **6**(5), 481–502 (2001)
16. Mindlin, R.D.: Micro-structure in linear elasticity. Arch. Ration. Mech. Anal. **16**(1), 51–78 (1964)
17. Mindlin, R.D.: Elasticity, piezoelectricity and crystal lattice dynamic. J. Elast. **2**(4), 217–282 (1972)

Part VI
Complex Structured Media (Often with Application to Dislocations)

Chapter 24
Dislocations in Generalized Continuum Mechanics

Markus Lazar

Abstract An overview on dislocations in the framework of different types of generalized continua is given. We consider Cosserat elasticity, gradient Cosserat elasticity, strain gradient elasticity, and dislocation gauge theory. We review similarities and differences between these generalized continuum theories. In fact, we discuss the constitutive relations for the linear and isotropic case. Moreover, we demonstrate how the characteristic length scales are given in terms of the material parameters. Also, we discuss the mathematical solutions of the elastic fields of a screw dislocation.

24.1 Introduction

In this paper, we want to give an overview of theories of generalized elasticity. The theory of Cosserat elasticity [3] is maybe the simplest generalization of 'classical' elasticity. It describes a continuum with orientation (rotational microstructure). Also, it possesses asymmetric force stresses and moment stresses. Characteristic length scale parameters appear in Cosserat elasticity.

Other generalized continuum theories are micromorphic elasticity [18, 6] for more complicated microstructures and gradient elasticity [18, 1, 2] in order to model microstructures and to succeed regularization of stress and strain fields. Nonlocal theories [12, 7, 13] contain nonlocal constitutive relation to describe the discreteness coming from the lattice. Dislocation gauge theory [10, 4, 17] turns out to be a straightforward theory of dislocations.

In this context, the following question comes up: What is the proper framework of a dislocation theory? First of all, we have to mention that a dislocation field the-

M. Lazar (✉)
Emmy Noether Research Group, Department of Physics, Darmstadt University of Technology, Hochschulstr. 6, 64289 Darmstadt, Germany
e-mail: lazar@fkp.tu-darmstadt.de

G.A. Maugin, A.V. Metrikine (eds.), *Mechanics of Generalized Continua*, Advances in Mechanics and Mathematics 21,
DOI 10.1007/978-1-4419-5695-8_24, © Springer Science+Business Media, LLC 2010

ory is not a Cosserat theory in contrast to what is often claimed in the literature. First, Günther [8] formulated the dislocation density tensor as the incompatibility condition in the framework of Cosserat elasticity. Schaefer [20] pointed out that constitutive equations for the dislocation density tensor are missing in conventional dislocation theories. Later, Eringen and Claus [5] showed that a dislocation theory is contained in micromorphic elasticity as a special case. Recently, Lazar and Anastassiadis [17] have proven that the (static) translation gauge theory of dislocations can be derived from micromorphic elasticity. Therefore, a static dislocation field theory is a special version of micromorphic elasticity.

24.2 Cosserat Elasticity

Cosserat elasticity [3, 19] can be used to describe materials with three translational degrees of freedom and three micro-rotational degrees of freedom. In Cosserat elasticity, the total deformation tensors (relative deformation tensor γ^{T}_{ij} and wryness tensor κ^{T}_{ij}) are given in terms of the displacement vector u_i and the microrotation axial vector ϕ_i, and may be decomposed into the elastic $(\gamma_{ij}, \kappa_{ij})$ and plastic parts $(\gamma^{\mathrm{P}}_{ij}, \kappa^{\mathrm{P}}_{ij})$

$$\gamma^{\mathrm{T}}_{ij} = \partial_j u_i + \varepsilon_{ijk}\phi_k = \gamma_{ij} + \gamma^{\mathrm{P}}_{ij}, \tag{24.1}$$

$$\kappa^{\mathrm{T}}_{ij} = \partial_j \phi_i = \kappa_{ij} + \kappa^{\mathrm{P}}_{ij}. \tag{24.2}$$

The stored energy reads

$$W = \frac{1}{2}\sigma_{ij}\gamma_{ij} + \frac{1}{2}\mu_{ij}\kappa_{ij}. \tag{24.3}$$

The force and moment stress tensors are defined by

$$\sigma_{ij} = \frac{\partial W}{\partial \gamma_{ij}}, \qquad \mu_{ij} = \frac{\partial W}{\partial \kappa_{ij}}. \tag{24.4}$$

The equilibrium conditions turn out to be

$$\partial_j \sigma_{ij} = 0 \qquad \text{(force equilibrium)}, \tag{24.5}$$

$$\partial_j \mu_{ij} - \varepsilon_{ijk}\sigma_{jk} = 0 \quad \text{(moment equilibrium)}. \tag{24.6}$$

The local, linear, isotropic constitutive relations with six material parameters are:

$$\sigma_{ij} = \lambda\delta_{ij}\gamma_{kk} + (\mu + \eta)\gamma_{ij} + (\mu - \eta)\gamma_{ji}, \tag{24.7}$$

$$\mu_{ij} = \alpha\delta_{ij}\kappa_{kk} + (\beta + \gamma)\kappa_{ij} + (\beta - \gamma)\kappa_{ji}. \tag{24.8}$$

One can define two characteristic lengths l and h in terms of the six material parameters according to:

$$l^2 = \frac{(\mu+\eta)(\beta+\gamma)}{4\mu\eta}, \qquad h^2 = \frac{\alpha+2\beta}{4\eta}. \tag{24.9}$$

The solution of a screw dislocation in Cosserat elasticity was given by Kessel [11] and Nowacki [19]. For a straight screw dislocation with the dislocation density tensor, $\alpha_{zz} = b\delta(x)\delta(y)$, the force stress reads ($r = \sqrt{x^2+y^2}$)

$$\sigma_{z\phi} = \frac{b}{2\pi}\frac{1}{r}\left[\mu + \eta\frac{r}{h}K_1\left(\frac{r}{h}\right)\right], \tag{24.10}$$

$$\sigma_{\phi z} = \frac{b}{2\pi}\frac{1}{r}\left[\mu - \eta\frac{r}{h}K_1\left(\frac{r}{h}\right)\right] \tag{24.11}$$

and has $1/r$-singularities. On the other hand, the moment stress is given by

$$\mu_{rr} = -\frac{b}{2\pi}\left[\frac{\beta}{r^2}\left(1-\frac{r}{h}K_1\left(\frac{r}{h}\right)\right) + \frac{\alpha+2\beta}{2}K_0\left(\frac{r}{h}\right)\right], \tag{24.12}$$

$$\mu_{\phi\phi} = \frac{b}{2\pi}\left[\frac{\beta}{r^2}\left(1-\frac{r}{h}K_1\left(\frac{r}{h}\right)\right) - \frac{\alpha}{2}K_0\left(\frac{r}{h}\right)\right], \tag{24.13}$$

$$\mu_{zz} = \frac{b\alpha}{4\pi h^2}K_0\left(\frac{r}{h}\right), \tag{24.14}$$

and it possesses $1/r^2$- and $\ln r$-singularities.

24.3 Gradient Cosserat Elasticity

Due to the appearance of singularities in the stress fields of dislocations, Lazar et al. [14, 16] have introduced a simple version of gradient Cosserat elasticity in order to regularize the force and moment stress fields of dislocations in Cosserat elasticity. The stored energy reads

$$W = W(\gamma_{ij}, \kappa_{ij}, \partial_k\gamma_{ij}, \partial_k\kappa_{ij}) \tag{24.15}$$

and

$$W = \frac{1}{2}\sigma_{ij}\gamma_{ij} + \frac{1}{2}\mu_{ij}\kappa_{ij} + \frac{1}{2}\varepsilon^2\partial_k\sigma_{ij}\partial_k\gamma_{ij} + \frac{1}{2}\varepsilon^2\partial_k\mu_{ij}\partial_k\kappa_{ij}, \quad \varepsilon \geq 0. \tag{24.16}$$

Such a theory possesses three internal length scales h, l and ε. In addition to the constitutive relations (24.7) and (24.8), we have additional constitutive relations for the double force and double moment stress tensors with the appearance of the gradient parameter ε:

$$\tau_{ijk} = \frac{\partial W}{\partial(\partial_k\gamma_{ij})} = \varepsilon^2\partial_k\sigma_{ij}, \qquad \lambda_{ijk} = \frac{\partial W}{\partial(\partial_k\kappa_{ij})} = \varepsilon^2\partial_k\mu_{ij}. \tag{24.17}$$

Thus, such a gradient Cosserat elasticity has 6 + 1 material parameters. The field equations of gradient Cosserat elasticity reduce to:

$$\partial_j\big(\sigma_{ij}-\partial_k\tau_{ijk}\big)=0 \qquad \text{(force equilibrium)}, \tag{24.18}$$

$$\partial_j\big(\mu_{ij}-\partial_k\lambda_{ijk}\big)-\varepsilon_{ijk}\big(\sigma_{jk}-\partial_l\tau_{jkl}\big)=0 \quad \text{(moment equilibrium)}. \tag{24.19}$$

The dislocation density of a screw dislocation in gradient Cosserat elasticity reads: $\alpha_{zz}=[b/(2\pi\varepsilon^2)]K_0(r/\varepsilon)$. The force stress of a screw dislocation is calculated as

$$\sigma_{z\phi}=\frac{b}{2\pi}\frac{1}{r}\bigg\{\mu\bigg[1-\frac{r}{\varepsilon}K_1\Big(\frac{r}{\varepsilon}\Big)\bigg]+\frac{\eta h^2}{h^2-\varepsilon^2}\bigg[\frac{r}{h}K_1\Big(\frac{r}{h}\Big)-\frac{r}{\varepsilon}K_1\Big(\frac{r}{\varepsilon}\Big)\bigg]\bigg\}, \tag{24.20}$$

$$\sigma_{\phi z}=\frac{b}{2\pi}\frac{1}{r}\bigg\{\mu\bigg[1-\frac{r}{\varepsilon}K_1\Big(\frac{r}{\varepsilon}\Big)\bigg]-\frac{\eta h^2}{h^2-\varepsilon^2}\bigg[\frac{r}{h}K_1\Big(\frac{r}{h}\Big)-\frac{r}{\varepsilon}K_1\Big(\frac{r}{\varepsilon}\Big)\bigg]\bigg\}, \tag{24.21}$$

and the moment stress reads

$$\begin{aligned}\mu_{rr}=&-\frac{b}{2\pi}\frac{1}{r^2}\bigg\{\beta-\frac{1}{h^2-\varepsilon^2}\bigg[\beta\Big(hrK_1\Big(\frac{r}{h}\Big)-\varepsilon rK_1\Big(\frac{r}{\varepsilon}\Big)\Big)\\&+\frac{\alpha+2\beta}{2}r^2\Big(K_0\Big(\frac{r}{h}\Big)-K_0\Big(\frac{r}{\varepsilon}\Big)\Big)\bigg]\bigg\},\end{aligned} \tag{24.22}$$

$$\begin{aligned}\mu_{\phi\phi}=&\frac{b}{2\pi}\frac{1}{r^2}\bigg\{\beta-\frac{1}{h^2-\varepsilon^2}\bigg[\beta\Big(hrK_1\Big(\frac{r}{h}\Big)-\varepsilon rK_1\Big(\frac{r}{\varepsilon}\Big)\Big)\\&-\frac{\alpha r^2}{2}\Big(K_0\Big(\frac{r}{h}\Big)-K_0\Big(\frac{r}{\varepsilon}\Big)\Big)\bigg]\bigg\},\end{aligned} \tag{24.23}$$

$$\mu_{zz}=\frac{b\alpha}{4\pi}\frac{1}{h^2-\varepsilon^2}\bigg[K_0\Big(\frac{r}{h}\Big)-K_0\Big(\frac{r}{\varepsilon}\Big)\bigg]. \tag{24.24}$$

In can be seen that the stress fields do not possess singularities. They are zero or finite at the dislocation line (see Fig. 24.1).

24.4 Strain Gradient Elasticity

As a special case of the simple version of gradient Cosserat elasticity, we obtain the version of strain gradient elasticity derived by Aifantis [1, 2] and Lazar & Maugin [15]. With $\phi_i=0$ and $e_{ij}=\beta_{(ij)}$, the gradient Cosserat elasticity reduces to the strain gradient elasticity with the following strain energy density

$$W=W(e_{ij},\partial_k e_{ij})=\frac{1}{2}\sigma_{ij}e_{ij}+\frac{1}{2}\varepsilon^2\partial_k\sigma_{ij}\partial_k e_{ij},\quad \varepsilon\geq 0 \tag{24.25}$$

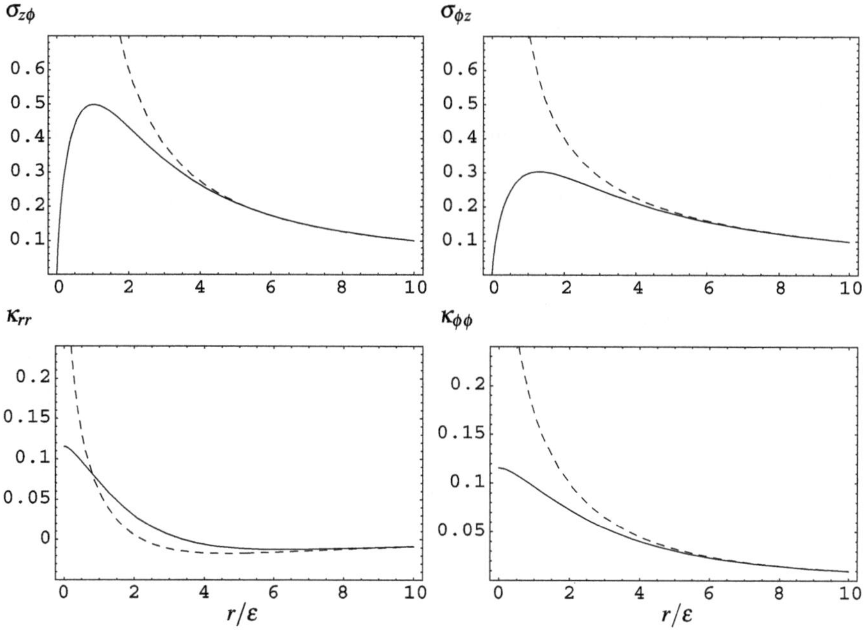

Fig. 24.1 Characteristic fields of a screw dislocation in gradient Cosserat elasticity. Choice of the parameters: $h = 2\varepsilon$, $\mu = 3\eta$

with only one internal length scale ε. It is obvious that it is a particular case of Mindlin's gradient elasticity [18] with two length parameters.

The linear, isotropic constitutive relations with $2 + 1$ material parameters are given by

$$\sigma_{ij} = \lambda\delta_{ij}e_{kk} + 2\mu e_{ij}, \quad \sigma_{ij} = \sigma_{ji}, \tag{24.26}$$

$$\tau_{ijk} = \varepsilon^2\partial_k\sigma_{ij}, \quad \tau_{ijk} = \tau_{jik}, \tag{24.27}$$

and the field equations reduce to

$$2\partial_j\big(\sigma_{ij} - \partial_k\tau_{ijk}\big) = 0 \quad \text{(force equilibrium)}. \tag{24.28}$$

For a screw dislocation with the dislocation density tensor, $\alpha_{zz} = [b/(2\pi\varepsilon^2)] \times K_0(r/\varepsilon)$, the force stress reads [9, 15]

$$\sigma_{z\phi} = \frac{\mu b}{2\pi}\frac{1}{r}\left\{1 - \frac{r}{\varepsilon}K_1\left(\frac{r}{\varepsilon}\right)\right\}. \tag{24.29}$$

It does not have a singularity.

24.5 Dislocation Field Theory

In this section, we want to demonstrate how one can derive a dislocation field theory from a compatible micromorphic elasticity using the symmetry of the translation group. In the framework of micromorphic elasticity, each material point is endowed with three translational degrees of freedom u_i and a microdeformation tensor ϕ_{ij} with nine independent components. In the compatible micromorphic elasticity [18, 6], we have three deformation tensors:

$$\text{(classical strain tensor)} \qquad e_{ij} = \partial_{(i}u_{j)}, \tag{24.30}$$
$$\text{(relative deformation tensor)} \qquad \gamma_{ij} = \partial_j u_i + \phi_{ij}, \tag{24.31}$$
$$\text{(microdeformation gradient)} \qquad \kappa_{ijk} = \partial_k \phi_{ij}. \tag{24.32}$$

Using the following statements, we can derive the translation gauge theory of dislocations as a particular version of micromorphic elasticity:

- Postulate a local $T(3)$-transformation for the displacement vector u_i:

$$u_i^* = u_i + \tau_i(x), \quad \tau_i(x)\text{-local (or space-dependent) translations.} \tag{24.33}$$

- Identify the microdeformation tensor ϕ_{ij} with the translational gauge potential (negative plastic deformation) which possesses an 'inhomogeneous' transformation under the translation group:

$$\phi_{ij}^* = \phi_{ij} - \partial_j \tau_i(x). \tag{24.34}$$

- Require translation-gauge invariance of the physical state quantities.

Using the argument of translational invariance, the state quantities of the dislocation theory are derived from the micromorphic deformation tensors (24.30)–(24.32) as follows:

$$\text{(incompatible elastic deformation tensor)} \quad \gamma_{ij} \to \beta_{ij} := \partial_j u_i + \phi_{ij}, \tag{24.35}$$
$$\text{(dislocation density tensor)} \quad \kappa_{ijk} \to T_{ijk} := -2\kappa_{i[jk]} = \partial_j \phi_{ik} - \partial_k \phi_{ij} \tag{24.36}$$

and $\alpha_{ij} = \frac{1}{2}\varepsilon_{jkl}T_{ikl}$. The classical strain tensor e_{ij} is not a state quantity because it is not gauge invariant. Thus, the relative deformation tensor reduces to the incompatible elastic distortion tensor and the microdeformation gradient to the dislocation density tensor. The stored energy of the dislocation field theory is a quadratic function of the state variables β_{ij}, T_{ijk}:

$$W = W(\beta_{ij}, T_{ijk}). \tag{24.37}$$

The corresponding stress tensors are the force stress tensor and the pseudomoment stress tensor given by

$$\sigma_{ij} = \frac{\partial W}{\partial \beta_{ij}}, \qquad H_{i[jk]} = 2\frac{\partial W}{\partial T_{i[jk]}}. \tag{24.38}$$

Moreover, σ_{ij} is a dislocation stress. The linear, isotropic constitutive relations possess six material parameters:

$$\sigma_{ij} = \lambda\delta_{ij}\beta_{kk} + (\mu+\gamma)\beta_{ij} + (\mu-\gamma)\beta_{ji}, \tag{24.39}$$
$$H_{ijk} = c_1 T_{ijk} + c_2(T_{jki} + T_{kij}) + c_3(\delta_{ij}T_{llk} + \delta_{ik}T_{ljl}). \tag{24.40}$$

Boundary conditions can be considered by means of a so-called null Lagrangian:

$$W_{\text{bg}} = \partial_j(\sigma^0_{ij}u_i) = (\partial_j\sigma^0_{ij})u_i + \sigma^0_{ij}\partial_j u_i \longrightarrow \sigma^0_{ij}\beta_{ij}, \tag{24.41}$$

where σ^0_{ij} is the classical stress of dislocations which plays the role of a background stress or applied stress. The classical stress acts as a source term in the gauge theory of dislocations. Using (24.37) and (24.41), the Euler–Lagrange equations are:

$$\frac{\delta W}{\delta\beta_{ij}} \equiv \frac{\partial W}{\partial\beta_{ij}} - \partial_k\frac{\partial W}{\partial\beta_{ij,k}} = 0. \tag{24.42}$$

The equation of motion in terms of the force stress and pseudomoment stress tensors has the form

$$H_{ijk,k} + \sigma_{ij} = \sigma^0_{ij} \quad \text{(pseudomoment equilibrium)}. \tag{24.43}$$

Thus, the source stress of the pseudomoment stress is the sum of the dislocation stress and the background stress. The force equilibrium is a continuity equation and is a trivial consequence of (24.43) since the pseudomoment stress tensor is skew-symmetric in the last two indices

$$\sigma_{ij,j} = 0 \quad \text{(force equilibrium)}. \tag{24.44}$$

It is obvious that the translation gauge theory of dislocations is not a Cosserat theory. Moreover, the pseudomoment equilibrium may be written in terms of the elastic distortion as

$$\begin{aligned} &c_1(\beta_{ik,jk} - \beta_{ij,kk}) + c_2(\beta_{ji,kk} - \beta_{jk,ki} + \beta_{kj,ki} - \beta_{ki,kj}) \\ &\quad + c_3\big[\delta_{ij}(\beta_{lk,kl} - \beta_{ll,kk}) + \beta_{ll,ji} - \beta_{lj,li}\big] + \sigma_{ij} = \sigma^0_{ij}, \end{aligned} \tag{24.45}$$

and in terms of the force stress tensor as

$$\begin{aligned} &\left[(c_1 - c_2 + 2c_3)\frac{2\gamma\nu}{1+\nu} - 2c_3\gamma\right](\delta_{ij}\sigma_{ll,kk} - \sigma_{ll,ij}) \\ &- \big[c_1(\gamma+\mu) - c_2(\gamma-\mu)\big]\sigma_{ij,kk} + \big[c_1(\gamma-\mu) - c_2(\gamma+\mu)\big](\sigma_{ki,kj} - \sigma_{ji,kk}) \\ &+ \big[2c_2\mu - c_3(\gamma+\mu)\big]\sigma_{kj,ki} + 4\mu\gamma\sigma_{ij} = 4\mu\gamma\sigma^0_{ij}. \end{aligned} \tag{24.46}$$

Equation (24.46) is a system of coupled partial differential equations with six material parameters.

At this point the following question comes up: How many length scales can we define in terms of six material moduli? To answer to this question, we calculate the

Green tensor of the field equation (24.46)

$$\sigma_{ij} = G_{ijkl} * \sigma^0_{kl}, \qquad \sigma^0_{ij} = L_{ij}\delta(x)\delta(y)\delta(z). \tag{24.47}$$

The three-dimensional Green tensor is found as

$$\begin{aligned} G_{ijkl} = \frac{1}{8\pi}\bigg\{ &(\delta_{ik}\delta_{jl} + \delta_{il}\delta_{jk})\frac{\mathrm{e}^{-r/\ell_1}}{\ell_1^2 r} + (\delta_{ik}\delta_{jl} - \delta_{il}\delta_{jk})\frac{\mathrm{e}^{-r/\ell_4}}{\ell_4^2 r} \\ &- (\delta_{ij}\Delta - \partial_i\partial_j)\delta_{kl}\left[\frac{1}{r}\left(\mathrm{e}^{-r/\ell_1} - \mathrm{e}^{-r/\ell_2}\right)\right] \\ &- (\delta_{jl}\partial_i + \delta_{il}\partial_j)\partial_k\left[\frac{1}{r}\left(\mathrm{e}^{-r/\ell_1} - \mathrm{e}^{-r/\ell_3}\right)\right] \\ &- (\delta_{jl}\partial_i - \delta_{il}\partial_j)\partial_k\left[\frac{1}{r}\left(\mathrm{e}^{-r/\ell_4} - \mathrm{e}^{-r/\ell_3}\right)\right]\bigg\}, \end{aligned} \tag{24.48}$$

where the characteristic length scales are:

$$\ell_1^2 = \frac{c_1 - c_2}{2\mu}, \qquad \ell_2^2 = \frac{(1-\nu)(c_1 - c_2 + 2c_3)}{2\mu(1+\nu)}, \tag{24.49}$$

$$\ell_3^2 = \frac{(\mu+\gamma)(c_1 - c_2 + c_3)}{4\mu\gamma}, \qquad \ell_4^2 = \frac{c_1 + c_2}{2\gamma}. \tag{24.50}$$

Thus, the dislocation field theory possesses four length scales. Here, ℓ_1 is the length scale for shear effects, ℓ_2 is the length scale for dilatation effects and ℓ_3, ℓ_4 depend on γ and they are Cosserat-like length scales (compare with ℓ and h in (24.9)).

For a screw dislocation in the framework of dislocation gauge theory the field equation (24.45) simplifies to

$$\left[1 - \frac{c_1}{\mu+\gamma}\Delta\right]\beta_{z\phi} = \beta^0_{z\phi}, \qquad \left[1 + \frac{c_2}{\mu-\gamma}\Delta\right]\beta_{z\phi} = \beta^0_{z\phi} \tag{24.51}$$

with the elastic distortion of a Volterra screw dislocation as source term: $\beta^0_{z\phi} = b/(2\pi r)$. From (24.51) we find the relation

$$c_2 = -\frac{\mu-\gamma}{\mu+\gamma}c_1. \tag{24.52}$$

The length scale of anti-plane strain reduces to

$$\ell_1^2 = \ell_4^2 = \frac{c_1}{\mu+\gamma}. \tag{24.53}$$

Finally, the incompatible elastic distortion reads

$$\beta_{z\phi} = \frac{b}{2\pi}\frac{1}{r}\left[1 - \frac{r}{\ell_1}K_1\left(\frac{r}{\ell_1}\right)\right], \tag{24.54}$$

and the asymmetric force stress is given by

$$\sigma_{z\phi} = \frac{(\mu+\gamma)b}{2\pi}\frac{1}{r}\left[1 - \frac{r}{\ell_1}K_1\left(\frac{r}{\ell_1}\right)\right], \tag{24.55}$$

$$\sigma_{\phi z} = \frac{(\mu-\gamma)b}{2\pi}\frac{1}{r}\left[1 - \frac{r}{\ell_1}K_1\left(\frac{r}{\ell_1}\right)\right]. \tag{24.56}$$

The dislocation density tensor reads

$$\alpha_{zz} = T_{zxy} = \frac{b}{2\pi\ell_1^2}K_0\left(\frac{r}{\ell_1}\right), \tag{24.57}$$

and the pseudomoment stress is

$$\begin{aligned} H_{xyz} = H_{yzx} &= -\frac{(\mu-\gamma)b}{2\pi}K_0\left(\frac{r}{\ell_1}\right), \\ H_{zxy} &= \frac{(\mu+\gamma)b}{2\pi}K_0\left(\frac{r}{\ell_1}\right). \end{aligned} \tag{24.58}$$

24.6 Conclusions

We have reviewed the following generalized continuum theories of linear elasticity:

- Cosserat elasticity: six material moduli, two length scales, stresses and strains of dislocations are singular.
- Gradient Cosserat elasticity: $6+1$ material moduli, $2+1$ length scales, stresses and strains of dislocations are nonsingular.
- Strain gradient elasticity: $2+1$ material moduli, one length scale, force stresses and strains of dislocations are nonsingular.
- T(3)-gauge theory of dislocations: six material moduli, four length scales, force stresses and strains of dislocations are nonsingular.

The dislocation field theory, Cosserat elasticity and strain gradient elasticity are contained in the theory of micromorphic elasticity as shown in Fig. 24.2.

(a) Micromorphic elasticity: $H_{ijk,k} + \sigma_{ij} = 0$	(hyperstress) $H_{ijk} \sim \kappa_{ijk} = \phi_{ij,k}$
(b) Dislocation field theory: $H_{i[jk],k} + \sigma_{ij} = 0$	(pseudomoment stress) $H_{i[jk]} \sim T_{i[jk]} = \phi_{ik,j} - \phi_{ij,k}$
(c) Cosserat elasticity: $H_{[ij]k,k} + \sigma_{[ij]} = 0$	(moment stress) $H_{[ij]k} \sim \kappa_{[ij]k} = \phi_{[ij],k}$
(d) Strain gradient elasticity: $H_{(ij)k,k} + \sigma_{(ij)} = 0$	(double stress) $H_{(ij)k} \sim \kappa_{(ij)k} = \phi_{(ij),k} \equiv e_{(ij),k}$

Fig. 24.2 Flow chart for hyperstresses in generalized elasticity contained in micromorphic elasticity

Acknowledgements The author has been supported by an Emmy Noether grant of the Deutsche Forschungsgemeinschaft (Grant No. La1974/1-2).

References

1. Aifantis, E.C.: On the role of gradients in the localization of deformation and fracture. Int. J. Eng. Sci. **30**, 1279–1299 (1992)
2. Altan, B.C., Aifantis, E.C.: On some aspects in the special theory of gradient elasticity. J. Mech. Behav. Mater. **8**, 231–282 (1997)
3. Cosserat, E., Cosserat, F.: Théorie des corps déformables. Hermann, Paris (1909)
4. Edelen, D.G.B., Lagoudas, D.C.: Gauge Theory and Defects in Solids. North-Holland, Amsterdam (1988)
5. Eringen, A.C., Claus Jr., D.W.: A micromorphic approach to dislocation theory and its relation to several existing theories. In: Simmons, J.A., deWit, R., Bullough, R. (eds.) Fundamental Aspects of Dislocation Theory, vol. 1. Natl. Bur. Stand. (U.S.), Spec. Publ., vol. 317, p. 1023. Natl. Bur. of Standford, Washington (1970)
6. Eringen, A.C.: Microcontinuum Field Theories I: Foundations and Solids. Springer, New York (1999)
7. Eringen, A.C.: Nonlocal Continuum Field Theories. Springer, New York (2002)
8. Günther, W.: Zur Statik und Kinematik des Cosseratschen Kontinuums. Abh. Braunschweig. Wiss. Ges. **10**, 195–213 (1958)
9. Gutkin, M.Y., Aifantis, E.C.: Screw dislocation in gradient elasticity. Scripta Mater. **35**, 1353–1358 (1996)
10. Kadić, A., Edelen, D.G.B.: A Gauge Theory of Dislocations and Disclinations. Lecture Notes in Physics, vol. 174. Springer, Berlin (1983)
11. Kessel, S.: Spannungsfelder einer Schraubenversetzung und einer Stufenversetzung im Cosserat Kontinuum. Z. Angew. Math. Mech. **50**, 547–553 (1970)
12. Kröner, E., Datta, B.K.: Nichtlokale Elastostatik: Ableitung aus der Gittertheorie. Z. Phys. **196**, 203–211 (1966)
13. Kunin, I.A.: Elastic Media with Microstructure II: Three-Dimensional Models. Springer, Berlin (1983)
14. Lazar, M., Maugin, G.A.: Defects in gradient micropolar elasticity: I. Screw dislocation. J. Mech. Phys. Solids **52**, 2263–2284 (2004)
15. Lazar, M., Maugin, G.A.: Nonsingular stress and strain fields of dislocations and disclinations in first strain gradient elasticity. Int. J. Eng. Sci. **43**, 1157–1184 (2005)
16. Lazar, M., Maugin, G.A., Aifantis, E.C.: On dislocations in a special class of generalized elasticity. Phys. Status Solidi (b) **242**, 2365–2390 (2005)
17. Lazar, M., Anastassiadis, C.: The gauge theory of dislocations: static solutions of screw and edge dislocations. Philos. Mag. **89**, 199–231 (2009)
18. Mindlin, R.D.: Micro-structure in linear elasticity. Arch. Ration. Mech. Anal. **16**, 51–78 (1964)
19. Nowacki, W.: Theory of Asymmetric Elasticity. Pergamon, Oxford (1986)
20. Schaefer, H.: Eine Feldtheorie der Versetzungen im Cosserat-Kontinuum. J. Appl. Math. Phys. (ZAMP) **20**, 891–899 (1969)

Chapter 25
Higher-Order Mesoscopic Theories of Plasticity Based on Discrete Dislocation Interactions

Robert H.J. Peerlings, Y. Kasyanyuk, A. Roy, and M.G.D. Geers

Abstract In this chapter, a rigorous analysis is given as a reference in which dislocations are treated as discrete entities. Then the transition towards a crystalline continuum is made in a number of steps by subsequent averaging along and perpendicular to the slip direction. This procedure eliminates the short-range dislocation interactions. Based on the considered idealized configuration, a back-stress term is derived which allows the conventional theory to predict finite-size pile-ups and turns out to be virtually identical to the result obtained by means of statistical arguments.

25.1 Introduction

Conventional crystal plasticity theories fail to correctly predict the pile-up of dislocations against, e.g., hard particles in a plastically deforming matrix. Instead of a gradually increasing slip gradient (and therefore dislocation density) towards the particle, they predict a jump in slip (i.e., a continuous array of super-dislocations) at the particle-matrix interface. As a consequence, the interaction of the stress fields within the particle and matrix is incorrectly described and size effects, such as the Hall–Petch effect, cannot be captured. A number of higher-order crystal plasticity theories have been proposed in recent years which repair this shortcoming, e.g., [1–3]. Most of these theories are phenomenological, and the relationship between the introduced higher-order terms and the underlying dislocation interactions is not always very clear.

In this contribution, we aim to pinpoint the precise reasons for the limitations of the classical theory by studying an idealized pile-up configuration of infinite edge dislocation walls. A rigorous analysis, in which the dislocations are treated as dis-

R.H.J. Peerlings (✉) · Y. Kasyanyuk · A. Roy · M.G.D. Geers
Department of Mechanical Engineering, Eindhoven University of Technology, PO Box 513, 5600 MB Eindhoven, The Netherlands
e-mail: R.H.J.Peerlings@tue.nl

G.A. Maugin, A.V. Metrikine (eds.), *Mechanics of Generalized Continua*, Advances in Mechanics and Mathematics 21,
DOI 10.1007/978-1-4419-5695-8_25, © Springer Science+Business Media, LLC 2010

crete entities, serves as a reference. A transition towards a (crystalline) continuum is then made in a number of steps by subsequent averaging along and perpendicular to the slip direction. It is shown that the latter, i.e., averaging out the internal structure of an individual wall, eliminates the short-range dislocation interactions which govern the pile-up response. Based on the idealized configuration considered, a back-stress term is derived which, when inserted in the conventional theory, allows it to predict finite-size pile-ups and which turns out to be virtually identical to that proposed by Groma et al. [2] based on statistical arguments.

25.2 Discrete Analysis

The discrete dislocation configuration on which our analysis is based is shown in Fig. 25.1. It consists of a series of dislocation walls in an infinite linear elastic medium. A single slip system is assumed, with discrete slip planes perpendicular to the y coordinate at a constant spacing h. The dislocation lines are straight, infinitely long and perpendicular to the xy-plane. They are organized in planar walls which are perpendicular to the x-direction. The wall at $x = 0$ (indicated in red in Fig. 25.1) is immobilized and the other walls pile-up against it under the influence of an externally applied shear stress σ. Here we limit ourselves to the case of edge dislocations with Burgers vector in the positive x-direction; see [6] for the screw dislocation case, as well as for details of the present analysis.

The relevant stress component for dislocation motion is the shear stress σ_{xy}. As the dislocations sit on the discrete slip planes, this stress component should be

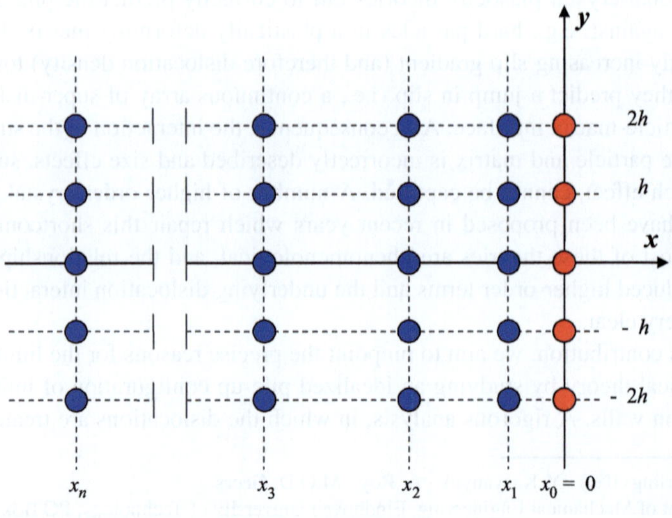

Fig. 25.1 Discrete dislocation configuration considered. The (*red*) dots at $x = 0$ represent the immobile dislocations which form the barrier against which the (*blue*) dislocations pile-up

evaluated on such a plane. Given the periodicity of the configuration, we can freely choose $y = 0$. The shear stress acting on this slip plane due to a single dislocation wall at $x = x_j$ is given by [4]:

$$\tau(x, x_j) = \frac{G}{2(1-\nu)} \frac{b}{h} \frac{\bar{x}}{\sinh^2 \bar{x}}, \tag{25.1}$$

where G, ν are the elastic constants, b is the length of the Burgers vector and $\bar{x} = \pi(x - x_j)/h$. This stress field decays rapidly (exponentially) with x and is of a short-range nature. The infinite walls considered here can indeed be shown not to generate a long-range stress field as they do not introduce a net incompatibility.

An equilibrium state is reached if for each of the n mobile walls in the pile-up the applied stress σ is balanced by the sum of the interaction stresses τ exerted by all other dislocation walls. This condition results in a set of n nonlinear equations in terms of the n unknowns x_i:

$$\sigma + \sum_{\substack{j=1 \\ j \neq i}}^{n} \tau(x_i, x_j) = 0 \quad (i = 1, \ldots, n). \tag{25.2}$$

These equations have been solved numerically [6]. The numerical solution for $n = 50$ and an external stress of $\sigma = 50Gb/(2h(1-\nu))$ is shown in the diagram of Fig. 25.1 by the circular markers. Along the vertical axis of the diagram the local dislocation wall density f has been plotted, which is defined at each wall position x_i $(0 < i < n)$ as $f(x_i) = 2/(x_{i-1} - x_{i+1})$. Both axes have been made dimensionless using the slip plane spacing h. The diagram clearly shows the typical expected pile-up response, with an increasing dislocation density (decreasing wall spacing) while approaching the barrier at $x = 0$.

25.3 Averaging Towards Conventional Crystal Plasticity

We now examine the consequences of the transition from the above, fully discrete analysis to a continuum description in which the precise positions of the individual dislocations are no longer traced. First, the discreteness of the individual walls within the pile-up is gradually removed by increasing the number of walls while reducing the Burgers vector of the dislocations inversely proportionally. Results of this exercise are shown in Fig. 25.2. They show that "smearing out" of the walls to—in the limit—a continuous distribution of walls leaves the shape of the pile-up essentially unchanged and is therefore allowed.

If, however, instead of the (horizontal) discreteness of the walls, we attempt to average out the (vertical) internal discreteness of the walls, a different conclusion is reached. Continuous walls are obtained by taking the limit $b \to 0$, $h \to 0$ while keeping constant the ratio b/h. In this limit, the interaction stress τ as defined above vanishes for all $\bar{x} > 0$. This implies that individual walls can no longer "sense" each

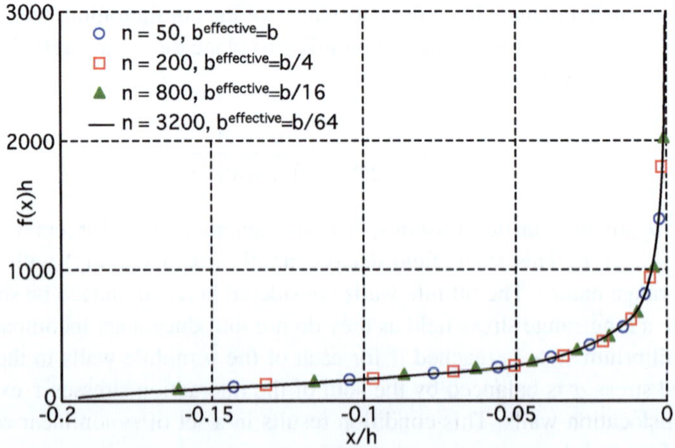

Fig. 25.2 Discrete pile-up solution ($n = 50$) and distributions obtained as the discreteness of the walls is gradually removed by repeatedly doubling reducing the number of dislocations n while reducing $b^{\text{effective}}$ so that the dislocation content nb remains constant

other and the discrete balance equations can no longer be satisfied for finite wall spacings. As a consequence, the external stress σ drives all walls into the barrier at $x = 0$, thus creating a super-dislocation wall there. Removing the discreteness of the individual slip systems thus results in a non-physical response because it removes the short-range stress field associated with the wall.

Conventional crystal plasticity theories are fully continuous and thus combine both limits as discussed above. As a result of their failure to account for the short-range stresses associated with individual dislocations, they also predict a pile-up of vanishing width.

25.4 Averaging Towards Higher-Order Crystal Plasticity

The shortcoming of conventional crystal plasticity theories as discussed above can be partially repaired by incorporating the effect of short-range interactions via a back-stress. To this end, focus on one of the walls in the discrete dislocation configuration of Fig. 25.2 and the interaction with its nearest neighbors only. The resultant of the interaction stresses τ exerted on the central wall at x_i by the nearest neighbors at x_{i+1} and x_{i-1} can be approximated by

$$\tau(x_i, x_{i+1}) + \tau(x_i, x_{i-1}) \approx -\frac{Gb}{2\pi(1-\nu)}\left[\frac{1}{x_{i-1} - x_i} - \frac{1}{x_i - x_{i+1}}\right], \quad (25.3)$$

where a first-order approximation has been introduced for the hyperbolic sine in (25.1) and the definition of $\bar{x}$ has been employed.

The discrete distances $x_{i-1} - x_i$ and $x_i - x_{i+1}$ may be rewritten in terms of overall (mesoscopic) quantities by defining the dislocation density ρ at x_i as

$$\rho = \frac{2}{h(x_{i-1} - x_{i+1})}. \tag{25.4}$$

A signed, net dislocation density ρ_{GND}, also termed Geometrically Necessary Dislocation (GND) density, may be introduced analogically. In the configuration considered here (Fig. 25.1), where all dislocations are positive, we have $\rho_{\mathrm{GND}} = \rho$, but for negative dislocations we would have $\rho_{\mathrm{GND}} = -\rho$ where $\rho \geq 0$; in the more general case of a mixed distribution, the absolute values of the two quantities no longer coincide, and we can only say that $\rho \geq |\rho_{\mathrm{GND}}|$. The—again signed—spatial derivative of the GND density is defined consistently with the above as

$$\frac{\partial \rho_{\mathrm{GND}}}{\partial x} = \frac{2}{x_{i-1} - x_{i+1}} \left[\frac{1}{h(x_{i-1} - x_i)} - \frac{1}{h(x_i - x_{i+1})} \right]. \tag{25.5}$$

Combining Expressions (25.4) and (25.5) allows us to eliminate the factor between brackets in (25.3) to finally obtain the following estimate for the net effect of short-range interactions with the neighboring walls:

$$\tau_{\mathrm{b}} = -\frac{Gb}{2\pi(1-\nu)} \frac{1}{\rho} \frac{\partial \rho_{\mathrm{GND}}}{\partial x}. \tag{25.6}$$

This influence of short-range internal stress should be added to the mesoscopic stress, i.e., the stress field which is the combined effect of external loads and internal (long-range) stresses due to mesoscopic incompatibilities (or geometrically necessary dislocations). It therefore enters the crystal plasticity framework like a back-stress—hence the symbol τ_{b} used for it.

Interestingly, Expression (25.6) virtually coincides with that derived by Groma et al. [2] based on statistical arguments. Inserting it in a crystal plasticity formulation results in pile-ups of finite width, as in the discrete analysis [2, 6, 5]. For the idealized pile-up configuration considered here, this can be seen as follows. No long-range internal stresses exist in this configuration and the stress responsible for dislocation (wall) motion thus equals the sum of the externally applied stress and the back stress. Rather than by the discrete balance equations (25.2), the equilibrium state reached at constant σ is therefore now governed by the continuous balance equation $\sigma + \tau_{\mathrm{b}} = 0$. Substituting the back-stress according to (25.6) and using $\rho_{\mathrm{GND}} = \rho$ this equation can be rewritten as

$$\frac{Gb}{2\pi(1-\nu)} \frac{\mathrm{d}\rho}{\mathrm{d}x} - \sigma\rho = 0. \tag{25.7}$$

The solution of this equation clearly is of the form $\rho(x) = A\exp(x/\lambda)$, with $\lambda = Gb/(2\pi(1-\nu)\sigma)$ and A a constant which remains to be determined. Thus, rather than the collapse into a wall of super-dislocations predicted by the conventional theory, an exponential increase of the dislocation density towards the barrier is now predicted, which appears to qualitatively agree with the discrete solutions of Fig. 25.2. Note also that the length of the pile-up is predicted to be inversely proportional to the applied stress by the above analysis.

25.5 Discussion and Concluding Remarks

A more detailed comparison with numerically generated solutions of the discrete pile-up problem has revealed that the exponential pile-up predicted by the higher-order theory does not fit very well to some of the discrete results [6]. Recent discrete simulations (not shown here) show that an exponential pile-up is obtained if only nearest-neighbor interactions are taken into account in the discrete balance equations (25.2). This suggests that the assumption that nearest neighbors dominate the short-range stress fields may be too crude in pile-up situations.

It can be concluded that, whereas higher-order crystal plasticity theories based on the back-stress (25.6) do repair an essential shortcoming of the conventional theories, further work needs to be done to make their predictions to be in agreement with the underlying discrete response.

References

1. Evers, L.P., Brekelmans, W.A.M., Geers, M.G.D.: Non-local crystal plasticity model with intrinsic SSD and GND effects. J. Mech. Phys. Solids **52**, 2379–2401 (2004)
2. Groma, I., Csikor, F.F., Zaiser, M.: Spatial correlations and higher-order gradient terms in a continuum description of dislocation dynamics. Acta Mater. **51**, 1271–1281 (2003)
3. Gurtin, M.E.: A gradient theory of single-crystal viscoplasticity that accounts for geometrically necessary dislocations. J. Mech. Phys. Solids **50**, 5–32 (2002)
4. Hirth, J.P., Lothe, J.: Theory of Dislocations, 2nd edn. Krieger, Melbourne (1992)
5. Kasyanyuk, Y., Peerlings, R.H.J., Geers, M.G.D.: In preparation
6. Roy, A., Peerlings, R.H.J., Geers, M.G.D., Kasyanyuk, Y.: Continuum modelling of dislocation interactions: Why discreteness matters? Mater. Sci. Eng. A **486**, 653–661 (2008)

Part VII
Numerical Problems

Chapter 26
An Approach Based on Integral Equations for Crack Problems in Standard Couple-Stress Elasticity

H.G. Georgiadis and P.A. Gourgiotis

Abstract The distributed dislocation technique proved to be in the past an effective approach in studying crack problems within classical elasticity. The present work aims at extending this technique in studying crack problems within standard couple-stress elasticity (or Cosserat elasticity with constrained rotations), i.e., within a theory accounting for effects of microstructure. This extension is not an obvious one since rotations and couple-stresses are involved in the theory employed to analyze the crack problems. Here, the technique is introduced to study the case of a Mode I crack. Due to the nature of the boundary conditions that arise in couple-stress elasticity, the crack is modeled by a continuous distribution of climb dislocations and wedge disclinations that create both standard stresses and couple stresses in the body. In particular, it is shown that the Mode I case is governed by a system of coupled singular integral equations with both Cauchy and logarithmic kernels. The numerical solution of this system shows that a cracked solid governed by couple-stress elasticity behaves in a more rigid way (having increased stiffness) as compared to a solid governed by classical elasticity.

26.1 Introduction

The present work is concerned with the problem of a Mode I crack in a material with microstructure. We assume that the response of the material is governed by standard couple-stress elasticity (Cosserat elasticity with constrained rotations). This theory falls into the category of generalized continuum theories and is a particular case of the general approach of Mindlin [13]. One of the earlier works that advanced ideas underlying couple-stress elasticity was the treatise by the Cosserat

H.G. Georgiadis (✉)
Department of Mechanics, National Technical University of Athens, Zographou Campus, Zographou 15773, Greece
e-mail: georgiad@central.ntua.gr

G.A. Maugin, A.V. Metrikine (eds.), *Mechanics of Generalized Continua*, Advances in Mechanics and Mathematics 21,
DOI 10.1007/978-1-4419-5695-8_26, © Springer Science+Business Media, LLC 2010

brothers [3]. Other important contributions include the work by Toupin [17] and Mindlin and Tiersten [14].

Earlier applications of couple-stress elasticity, mainly on stress-concentration problems, met with some success providing solutions physically more adequate than solutions based on classical elasticity [18]. Work employing couple-stress theories on elasticity and plasticity problems is also continued in recent years [5, 11, 7]. Nevertheless, there is only a limited number of studies concerning the effects of couple-stresses in crack problems. One of the earlier works on this subject is that of Sternberg and Muki [15], who considered the Mode I finite-length crack by employing the method of dual integral equations. They provided only asymptotic results and showed that both stress and couple-stress fields exhibit a square-root singularity while the rotation field is bounded at the crack-tip. More recently, Huang et al. [9] also provided near-tip asymptotic fields for Mode I and Mode II crack problems by using the method of eigenfunction expansion. In addition, Huang et al. [10] obtained full-field solutions for a *semi-infinite* crack by using the Wiener–Hopf method.

Here, we aim at providing a full-field solution to the Mode I *finite-length* crack problem within couple-stress elasticity by introducing an approach based on distributed dislocations and disclinations. Since the pioneering work of Bilby and Eshelby [2], the *distributed dislocation* technique has been employed to analyze various crack problems in classical elasticity (see, e.g., Hills et al. [8]). The strength of this analytical/numerical technique lies in the fact that it gives very detailed information about crack problems at the expense of relatively little computational demands as compared to the methods of Finite Elements and Boundary Elements. Although the technique proved to be very successful in studying crack problems within classical elasticity, it appears that there is no work at all in modeling cracks with distribution of dislocations and/or disclinations in materials with microstructure. Therefore, the present work aims at extending the technique in solving crack problems within couple-stress elasticity. This extension is not an obvious one since rotations and couple-stresses are involved in the theory employed. In another recent work by the present authors [6], cracks under Mode II and Mode III conditions were also considered within the same framework. A comparison between the Mode I case studied here and the Mode II case [6] leads to the conclusion that the opening mode is mathematically more involved than the shear mode. This is in some contrast with situations of classical elasticity where the two plane-strain crack modes involve equivalent mathematical effort.

As in analogous situations of classical elasticity, a superposition scheme will be followed. Thus, the solution to the basic problem (body with a traction-free crack under a remote constant tension) will be obtained by the superposition of the stress and couple-stress fields arising in an un-cracked body (of the same geometry) to the 'corrective' stresses induced in the cracked body with a loading only along the crack faces. Due to the nature of the boundary conditions, it will be shown that in order to obtain the corrective solution, we need to distribute not only climb dislocations but also constant discontinuities of the rotation along the crack faces. We name the latter discontinuities *constrained* wedge disclinations. The term 'constrained' refers to the requirement of zero normal displacement along the disclination plane. Notice that in the classical sense of a wedge disclination [1], the normal displacement along the disclination plane increases linearly with distance from the core and is unbounded at infinity. The Green's functions of our problem (i.e., the stress fields of a discrete climb dislocation and a discrete constrained wedge disclination) are obtained by the use of Fourier transforms. Finally, it is shown that the continuous distribution of the discontinuities along the crack faces results in a system of coupled singular integral equations with both Cauchy and logarithmic kernels. The numerical solution of this system shows that a cracked solid governed by couple-stress elasticity behaves in a more rigid way (having increased stiffness) as compared to a solid governed by classical elasticity. Also, the stress level at the crack-tip region is appreciably higher than the one predicted by classical elasticity.

26.2 Basic Equations of Plane Strain in Couple-Stress Elasticity

For a body that occupies a domain in the (x, y)-plane and is under conditions of plane strain the displacement components are $u_x \equiv u_x(x,y) \neq 0$, $u_y \equiv u_y(x,y) \neq 0$, and $u_z = 0$. We generally assume the absence of inertial effects and body-force and body-couple fields. The constitutive equations for the non-vanishing components of the *asymmetric* force-stress and couple-stress tensors are written as [15, 12]

$$\sigma_{xx} = (\lambda + 2\mu)\frac{\partial u_x}{\partial x} + \lambda\frac{\partial u_y}{\partial y}, \qquad \sigma_{yy} = (\lambda + 2\mu)\frac{\partial u_y}{\partial y} + \lambda\frac{\partial u_x}{\partial x}, \tag{26.1}$$

$$\begin{aligned} \sigma_{yx} &= \mu\left(\frac{\partial u_x}{\partial y} + \frac{\partial u_y}{\partial x}\right) + \mu\ell^2\left(\frac{\partial^3 u_y}{\partial x^3} - \frac{\partial^3 u_x}{\partial x^2 \partial y} + \frac{\partial^3 u_y}{\partial x \partial y^2} - \frac{\partial^3 u_x}{\partial y^3}\right), \\ \sigma_{xy} &= \mu\left(\frac{\partial u_x}{\partial y} + \frac{\partial u_y}{\partial x}\right) - \mu\ell^2\left(\frac{\partial^3 u_y}{\partial x^3} - \frac{\partial^3 u_x}{\partial x^2 \partial y} + \frac{\partial^3 u_y}{\partial x \partial y^2} - \frac{\partial^3 u_x}{\partial y^3}\right), \end{aligned} \tag{26.2}$$

$$m_{xz} = 2\mu\ell^2\left(\frac{\partial^2 u_y}{\partial x^2} - \frac{\partial^2 u_x}{\partial x \partial y}\right), \qquad m_{yz} = 2\mu\ell^2\left(\frac{\partial^2 u_y}{\partial x \partial y} - \frac{\partial^2 u_x}{\partial x^2}\right), \tag{26.3}$$

where the moduli (λ, μ) have the same meaning with the Lamé constants of classical elasticity theory, and $\ell \equiv \sqrt{\eta/\mu}$ is a characteristic material length with η being a modulus accounting for couple-stress effects.

Then, the equations of equilibrium written in terms of displacements are expressed as follows

$$\begin{aligned} &\frac{1}{1-2\nu}\frac{\partial}{\partial x}\left[2(1-\nu)\frac{\partial u_x}{\partial x} + \frac{\partial u_y}{\partial y}\right] + \frac{\partial^2 u_x}{\partial y^2} \\ &+ \ell^2\left[\frac{\partial^4 u_y}{\partial x^3 \partial y} - \frac{\partial^4 u_x}{\partial x^2 \partial y^2} + \frac{\partial^4 u_y}{\partial x \partial y^3} - \frac{\partial^4 u_x}{\partial y^4}\right] = 0, \end{aligned} \tag{26.4}$$

$$\frac{1}{1-2\nu}\frac{\partial}{\partial y}\left[2(1-\nu)\frac{\partial u_y}{\partial y}+\frac{\partial u_x}{\partial x}\right]+\frac{\partial^2 u_y}{\partial x^2}$$
$$+\ell^2\left[\frac{\partial^4 u_x}{\partial x^3\partial y}-\frac{\partial^4 u_y}{\partial x^2\partial y^2}+\frac{\partial^4 u_x}{\partial x\partial y^3}-\frac{\partial^4 u_y}{\partial y^4}\right]=0. \quad (26.5)$$

26.3 Formulation of Crack Problem

Consider a straight finite-length crack of length $2a$ embedded in the (x, y)-plane of infinite extent in a field of uniform uni-axial tension (Fig. 26.1). The crack faces are traction free and the body is considered to be in plane-strain conditions. The boundary conditions along the faces of the crack are written as

$$\sigma_{yx}(x,0)=0, \qquad \sigma_{yy}(x,0)=0, \qquad m_{yz}(x,0)=0, \quad \text{for } |x|<a, \quad (26.6)$$

while the regularity conditions at infinity are given as

$$\sigma_{yx}^{\infty},\sigma_{xy}^{\infty},\sigma_{xx}^{\infty}\to 0, \qquad \sigma_{yy}^{\infty}\to\sigma_0, \qquad m_{xz}^{\infty},m_{yz}^{\infty}\to 0 \quad \text{as } r\to\infty, \quad (26.7)$$

where $r\equiv\sqrt{x^2+y^2}$.

Then, the crack problem is decomposed into the following two auxiliary problems.

The Un-Cracked Body The un-cracked body is in a state of uniform tension σ_0. It can readily be verified that the appropriate Mindlin's stress functions [12] are

$$\Phi=\frac{\sigma_0}{2}x^2, \qquad \Psi=0. \quad (26.8)$$

There are *no* couple-stresses induced in the un-cracked body under uniform tension.

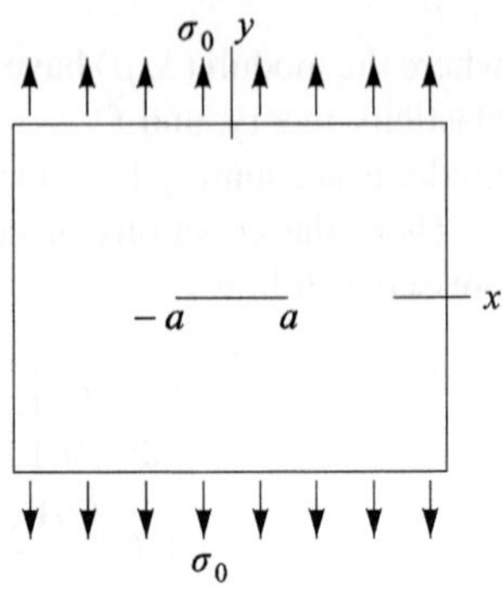

Fig. 26.1 Cracked body loaded by a remote constant tension field σ_0

The Corrective Solution Consider a body geometrically identical to the initial cracked body but with no remote loading now. The only loading applied is along the crack faces. This consists of equal and opposite tractions to those generated in the un-cracked body. The boundary conditions along the faces of the crack are written as

$$\sigma_{yy}(x,0) = -\sigma_0, \qquad m_{yz}(x,0) = 0, \qquad \sigma_{yx}(x,0) = 0 \quad \text{for } |x| < a. \tag{26.9}$$

Notice that in classical elasticity it would suffice to continuously distribute climb dislocations with Burger's vector $\mathbf{b} = (0, b, 0)$ in order to produce the desired normal stress in $(26.9)_1$. However, this is not the case in couple-stress elasticity because a discrete climb dislocation produces both normal stresses σ_{yy} and couple-stresses m_{yz} along the crack line $y = 0$. Therefore, it is not possible to satisfy both $(26.9)_1$ and $(26.9)_2$ just by a continuous distribution of climb dislocations. On the other hand, within the framework of couple-stress elasticity, we know that the work conjugates of the reduced force traction $P_y \equiv \sigma_{yy} n_y$ and the tangential couple traction $R_z \equiv m_{yz} n_y$ are the normal displacement u_y and the rotation ω respectively. In light of the above, we are led to the conclusion that in order to satisfy the boundary conditions in (26.9) we should distribute both discontinuities in the displacement (climb dislocations) and in the rotation (constrained wedge disclinations) along the crack faces.

Next, our aim is determining the stress and couple-stress fields, along the crack-line $y = 0$, induced by a discrete climb dislocation and a discrete constrained wedge disclination both located at the origin of the (x, y)-plane. The above stress fields will serve as the Green's functions of our problem.

26.4 Green's Functions

We impose at the origin of the (x, y)-plane a discrete climb dislocation of strength b and a discrete constrained wedge disclination of strength Ω. In the upper half-plane, the climb dislocation gives rise to the boundary value problem

$$u_y(x, 0^+) = -\frac{b}{2} H(x), \qquad \omega(x, 0^+) = 0, \qquad \sigma_{yx}(x, 0^+) = 0, \tag{26.10}$$

whereas the constrained wedge disclination generates the boundary value problem

$$u_y(x, 0^+) = 0, \qquad \omega(x, 0^+) = \frac{\Omega}{2} H(x), \qquad \sigma_{yx}(x, 0^+) = 0, \tag{26.11}$$

where $H(x)$ is the Heaviside step-function. We emphasize that the term 'constrained wedge disclination' is justified from the fact that the discontinuity in the rotation $(26.11)_2$ does not affect the normal displacement in $(26.11)_1$. This concept departs from the one of the classical 'wedge disclination', which generates a field where the jump in rotation implies a discontinuity in the normal displacement too (see, e.g., Anthony [1]).

Both boundary value problems are attacked with Fourier transforms. The inversion procedure involves use of some distribution theory. By a superposition of the solutions of the above boundary value problems, we obtain the expressions for the normal stress and the couple-stress

$$\begin{aligned}\sigma_{yx}(x,y=0^+) &= \frac{\mu b}{2\pi(1-\nu)x} + 2\frac{\mu b}{\pi x}\left[2\frac{\ell^2}{x^2} - K_2(|x|/\ell)\right]\\ &\quad - 2\frac{\mu\Omega}{\pi}\left[2\frac{\ell^2}{x^2} - K_2(|x|/\ell)\right] + 2\frac{\mu\Omega}{\pi}K_0(|x|/\ell), \end{aligned}\tag{26.12}$$

$$\begin{aligned}m_{yx}(x,y=0^+) &= -\frac{\mu b}{\pi}\left[2\frac{\ell^2}{x^2} - K_2(|x|/\ell)\right] - \frac{\mu b}{\pi}K_0(|x|/\ell)\\ &\quad - \mu\ell\Omega + \frac{\mu\ell\Omega}{2\pi}\operatorname{sgn}(x) G^{2,1}_{1,3}\left(\frac{x^2}{4\ell^2}\middle|\begin{matrix}1\\ -1/2, 1/2, 0\end{matrix}\right), \end{aligned}\tag{26.13}$$

where $K_i(x/\ell)$ is the ith order modified Bessel function of the second kind, and $G^{a,b}_{c,d}[x^2/(4\ell^2)|[\;],[\;]]$ is the Meijer G-function [4]. Equations (26.12) and (26.13) are the Green's functions of the crack problem.

26.5 Reduction of the Problem to a System of Singular Integral Equations

In order to satisfy the boundary conditions of the corrective solution (26.9), we continuously distribute climb dislocations of strength b and constrained wedge disclinations of strength Ω along the crack-faces ($|x| < a$). The normal stresses σ_{yy} and the couple-stresses m_{yz} induced by the continuous distribution of dislocations and disclinations are obtained as integrals of (26.12) and (26.13). We also notice that $(26.9)_3$ is automatically satisfied since neither the discrete dislocation nor the constrained wedge disclination produce shear stresses σ_{yx} along the crack-line. Then, satisfaction of the boundary conditions $(26.9)_1$ and $(26.9)_2$ leads to a system of coupled singular integral equations ($|x| < a$)

$$\begin{aligned}\sigma_{yy} = -\sigma_0 &= \frac{\mu(3-2\nu)}{2\pi(1-\nu)}\int_{-a}^{+a}\frac{B(\xi)}{x-\xi}\,\mathrm{d}\xi - \frac{2\mu}{\pi a}\int_{-a}^{+a} W(\xi)\ln|x-\xi|\,\mathrm{d}\xi\\ &\quad + \frac{2\mu}{\pi a}\int_{-a}^{+a} B(\xi)K_{a1}(x,\xi)\,\mathrm{d}\xi\\ &\quad + \frac{2\mu}{\pi a}\int_{-a}^{+a} W(\xi)K_{a2}(x,\xi)\,\mathrm{d}\xi, \end{aligned}\tag{26.14}$$

$$m_{yz} = 0 = -\frac{2\mu}{\pi}\int_{-a}^{+a}\frac{W(\xi)}{x-\xi}\,\mathrm{d}\xi + \frac{\mu a}{\pi\ell^2}\int_{-a}^{+a} B(\xi)\ln|x-\xi|\,\mathrm{d}\xi$$
$$-\frac{\mu}{\pi a}\int_{-a}^{+a} B(\xi)K_{a3}(x,\xi)\,\mathrm{d}\xi$$
$$+\frac{\mu}{2\pi a}\int_{-a}^{+a} W(\xi)K_{a4}(x,\xi)\,\mathrm{d}\xi, \qquad (26.15)$$

where

$$K_{a1}(x,\xi) = \frac{a}{x-\xi}\left[\frac{2\ell^2}{(x-\xi)^2} - K_2\big(|x-\xi|/\ell\big) - \frac{1}{2}\right],$$
$$K_{a3}(x,\xi) = \frac{a^2}{\ell^2}\left[\frac{2\ell^2}{(x-\xi)^2} - K_2\big(|x-\xi|/\ell\big)\right] + \frac{a^2}{\ell^2}K_{a2}(x,\xi), \qquad (26.16)$$
$$K_{a2}(x,\xi) = K_0\big(|x-\xi|/\ell\big) + \ln|x-\xi|,$$

$$K_{a4}(x,\xi) = \frac{a}{\ell}\,\mathrm{sgn}(x-\xi)G_{1,3}^{2,1}\left(\frac{(x-\xi)^2}{4\ell^2}\,\middle|\,\begin{matrix}1\\-1/2,1/2,0\end{matrix}\right) + \frac{4a}{x-\xi}.$$

It can be readily proved using asymptotic analysis that the above kernels are regular as $x \to \xi$.

The unknown functions $B(\xi)$ and $W(\xi)$ are the dislocation and the disclination densities at a point ξ. These are defined as

$$B(\xi) = \frac{\mathrm{d}b(\xi)}{\mathrm{d}\xi} = -\frac{\mathrm{d}\Delta u_y(\xi)}{\mathrm{d}\xi}, \qquad W(\xi) = a\frac{\mathrm{d}\Omega(\xi)}{\mathrm{d}\xi} = a\frac{\mathrm{d}\Delta\omega(\xi)}{\mathrm{d}\xi}, \qquad (26.17)$$

where $\Delta u_y(x)$ represents the relative opening displacement, and $\Delta\omega(x)$ represents the relative rotation between the upper and lower crack faces. Physically, the dislocation density can be interpreted as the negative of the slope whereas the disclination density as the curvature at any point between the crack faces. Both dislocation and disclination densities can be written as a product of a regular-bounded function and a fundamental solution that takes into account the asymptotical behavior of the latter quantities. A previous asymptotic analysis [9] for the Mode I case showed that both the displacement u_y and the rotation ω behave as $\sim x^{1/2}$ near the crack tips. Consequently, the dislocation and the disclination densities have the following endpoint behavior, respectively, $B(x) \sim x^{-1/2}$ and $W(x) \sim x^{-1/2}$. Finally, in order to ensure uniqueness in the values of displacement and rotation for a closed loop around the crack, we impose pertinent closure conditions.

26.6 Numerical Results

For the numerical solution of the coupled system of singular integral equations (26.14) and (26.15) the Gauss–Chebyshev quadrature is employed with a modifi-

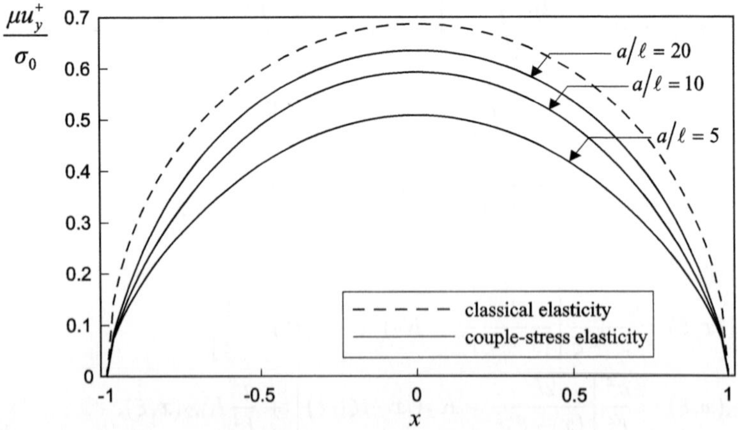

Fig. 26.2 Normalized upper-half crack-displacement profile ($\nu = 0.3$)

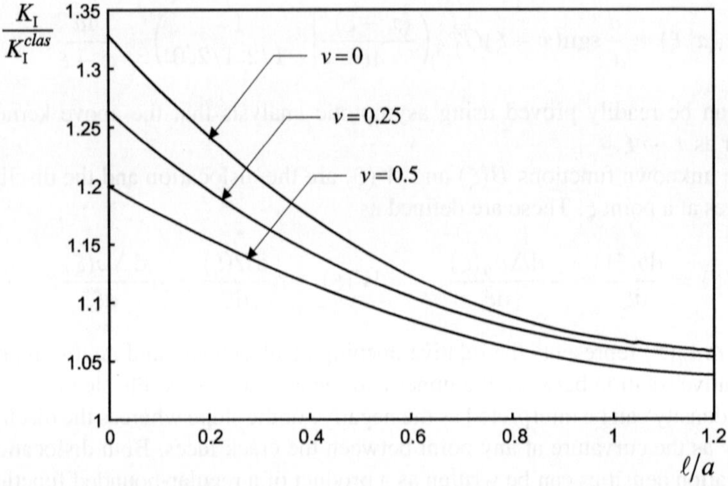

Fig. 26.3 Variation of the ratio of the stress intensity factor K_I in couple-stress elasticity to the respective one in classical elasticity

cation that takes into account the existence of logarithmic kernels [16]. Figure 26.2 depicts the dependence of the normal crack-face displacement upon the ratio a/ℓ. It is noteworthy that as the crack length becomes comparable to the characteristic length ($a < 10\ell$), the material exhibits a more *stiff* behavior, i.e., the crack-face displacement becomes smaller in magnitude as compared to the one in the classical elasticity solution.

Figure 26.3 now shows the ratio of the stress intensity factors given by couple-stress elasticity and classical elasticity. Notice that both theories predict stresses with the same order of singularity (inverse square root) near the crack tips. Further, it is observed that for $\ell/a \to 0$ and Poisson's ratio $\nu = 0.5$ there is a 20% increase in K_I

when couple-stress effects are taken into account, while for $\nu = 0.25$ and $\nu = 0$ the increase is 27% and 32%, respectively. It should also be noted that when $\ell/a = 0$ (no couple-stress effects) the above ratio becomes evidently $K_I/K_I^{\mathrm{clas}} = 1$. Therefore, the ratio plotted in Fig. 26.3 exhibits a finite jump discontinuity at $\ell/a = 0$; the ratio being increased abruptly as ℓ/a departs from zero. The same discontinuity was observed by Sternberg and Muki [15], who attributed this behavior to the severe boundary-layer effects predicted by couple-stress elasticity in singular stress-concentration problems.

26.7 Concluding Remarks

In this paper, the technique of distributed dislocations was extended in couple-stress elasticity for the solution of the Mode I crack problem. The results of our analysis indicate that when the microstructure of the material is taken into account a more rigid behavior is exhibited. In particular, in the Mode I crack problem the crack-face displacements become significantly smaller than their counterparts in classical elasticity, when the length of the crack is comparable to the characteristic length ℓ of the material (about 25% decrease for $a/\ell = 5$). On the other hand, it is observed that the stress intensity factor K_I is higher than the one predicted by classical elasticity, while the couple-stress effects are dominant within a zone of 3ℓ.

Acknowledgements Financial support under the "*ΠΕΒΕ* 2008, # 65/1695" programme of the NTU Athens is acknowledged with thanks.

References

1. Anthony, K.H.: Die Theorie der Dislokationen. Arch. Ration. Mech. Anal. **39**, 43–88 (1970)
2. Bilby, B.A., Eshelby, J.D.: Dislocations and the theory of fracture. In: Liebowitz, H. (ed.) Fracture, vol. I. Academic Press, New York (1968)
3. Cosserat, E., Cosserat, F.: Théorie des Corps Déformables. Hermann et Fils, Paris (1909)
4. Erdelyi, A.: Higher Transcendental Functions. McGraw-Hill Book, New York (1953)
5. Georgiadis, H.G., Velgaki, E.G.: High-frequency Rayleigh waves in materials with microstructure and couple-stress effects. Int. J. Solids Struct. **40**, 2501–2520 (2003)
6. Gourgiotis, P.A., Georgiadis, H.G.: Distributed dislocation approach for cracks in couple-stress elasticity: Shear Modes. Int. J. Fract. **147**, 83–102 (2007)
7. Grentzelou, C.G., Georgiadis, H.G.: Uniqueness for plane crack problems in dipolar gradient elasticity and in couple-stress elasticity. Int. J. Solids Struct. **42**, 6226–6244 (2005)
8. Hills, D.A., Kelly, P.A., Dai, D.N., Korsunsky, A.M.: Solution of Crack Problems: The Distributed Dislocation Technique. Kluwer Academic, Dordrecht (1996)
9. Huang, Y., Zhang, L., Guo, T.F., Hwang, K.: Mixed mode near tip fields for cracks in materials with strain-gradient effects. J. Mech. Phys. Solids **45**, 439–465 (1997)
10. Huang, Y., Chen, J.Y., Guo, T.F., Zhang, L., Hwang, K.-C.: Analytic and numerical studies on mode I and mode II fracture in elastic-plastic materials with strain gradient effects. Int. J. Fract. **100**, 1–27 (1999)

11. Lubarda, V.A.: The effects of couple stresses on dislocation strain energy. Int. J. Solids Struct. **40**, 3807–3826 (2003)
12. Mindlin, R.D.: Influence of couple-stresses on stress concentrations. Exp. Mech. **3**, 1–7 (1963)
13. Mindlin, R.D.: Micro-structure in linear elasticity. Arch. Ration. Mech. Anal. **16**, 51–78 (1964)
14. Mindlin, R.D., Tiersten, H.F.: Effects of couple-stresses in linear elasticity. Arch. Ration. Mech. Anal. **11**, 415–448 (1962)
15. Sternberg, E., Muki, R.: The effect of couple-stresses on the stress concentration around a crack. Int. J. Solids Struct. **3**, 69–95 (1967)
16. Theocaris, P.S., Ioakimidis, N., Chrysakis, A.C.: On the application of numerical integration rules to the solution of some singular integral equations. Comput. Methods Appl. Mech. Eng. **94**, 1–11 (1980)
17. Toupin, R.A.: Perfectly elastic materials with couple stresses. Arch. Ration. Mech. Anal. **11**, 385–414 (1962)
18. Weitsman, Y.: Strain-Gradient effects around cylindrical inclusions and cavities in a field of cylindrically symmetric tension. ASME J. Appl. Mech. **33**, 57–67 (1966)

Chapter 27
A Cosserat Point Element (CPE) for the Numerical Solution of Problems in Finite Elasticity

Mahmood Jabareen and Miles B. Rubin

Abstract The theory of a Cosserat Point is a special continuum theory that characterizes the motion of a small material region which can be modeled as a point with finite volume. This theory has been used to develop a 3-D eight-noded brick Cosserat Point Element (CPE) to formulate the numerical solution of dynamical problems in finite elasticity. The kinematics of the CPE are characterized by eight director vectors which are functions of time only. Also, the kinetics of the CPE are characterized by balance laws which include: conservation of mass, balances of linear and angular momentum, as well as balances of director momentum. The main difference between the standard Bubnov–Galerkin and the Cosserat approaches is the way that they each develop constitutive equations. In the direct Cosserat approach, the kinetic quantities are given by derivatives of a strain energy function that models the CPE as a structure and that characterizes resistance to all models of deformation. A generalized strain energy function has been developed which yields a CPE that is truly a robust user friendly element for nonlinear elasticity that can be used with confidence for 3-D problems as well as for problems of thin shells and rods.

27.1 Introduction

In 1909, the Cosserat brothers introduced the notion of a deformable body with generalized kinematics [6]. Specifically, the usual vector locating the position of

M. Jabareen (✉)
Faculty of Civil and Environmental Engineering, Technion-Israel Institute of Technology, Haifa 32000, Israel
e-mail: cvjmah@tx.technion.ac.il

M.B. Rubin
Faculty of Mechanical Engineering, Technion-Israel Institute of Technology, Haifa 32000, Israel
e-mail: mbrubin@tx.technion.ac.il

G.A. Maugin, A.V. Metrikine (eds.), *Mechanics of Generalized Continua*, Advances in Mechanics and Mathematics 21,
DOI 10.1007/978-1-4419-5695-8_27, © Springer Science+Business Media, LLC 2010

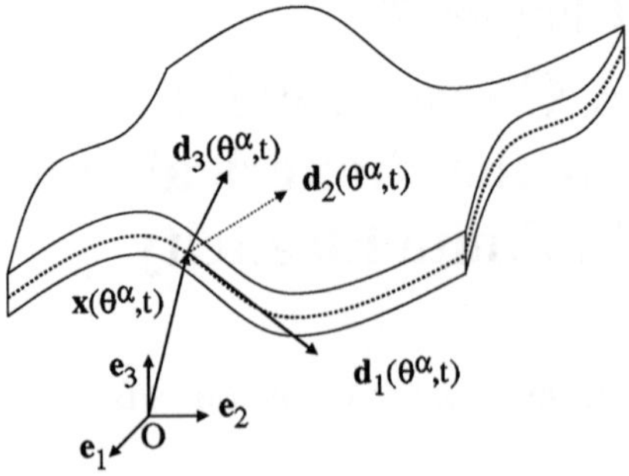

Fig. 27.1 Sketch of a Cosserat surface that is "thin" in one dimension

a material point in the body was supplemented by director vectors that characterize additional localized degrees of freedom. This concept has been used to discuss Cosserat theories that model "thin" structures which introduce director vectors to characterize material line elements in the dimensions that are reduced due to "thinness". For example, Naghdi [15] discusses the notion of a Cosserat surface (see Fig. 27.1) which is a shell-like structure that is "thin" in one dimension. The position vector $\mathbf{x}^*$ locating material points in this structure is represented in terms of three convected coordinates θ^i $(i = 1, 2, 3)$ and time t, and it is approximated by the kinematic assumption

$$\mathbf{x}^*(\theta^i, t) = \mathbf{x}(\theta^\alpha, t) + \theta^3 \mathbf{d}_3(\theta^\alpha, t). \tag{27.1}$$

In (27.1), $\mathbf{x}$ locates points on a reference surface of the shell that is characterized by two convected coordinates θ^α $(\alpha = 1, 2)$, and $\mathbf{d}_3$ is a director vector that models deformations of a material line element through the shell's thickness.

A Cosserat curve (see Fig. 27.2) is a rod-like structure that is "thin" in two dimensions. Antman [1, 2] models the cross-section of the rod as rigid using a triad of orthonormal directors. Green et al. [7, 8] model the cross-section of the rod as deformable. In this later theory, the vector $\mathbf{x}^*$ is approximated by

$$\mathbf{x}^*(\theta^i, t) = \mathbf{x}(\theta^3, t) + \sum_{\alpha=1}^{2} \theta^\alpha \mathbf{d}_\alpha(\theta^3, t), \tag{27.2}$$

where $\mathbf{x}$ locates points on a reference curve of the rod that is characterized by one convected coordinate θ^3, and $\mathbf{d}_\alpha$ are two deformable director vectors that characterize deformations of the rod's cross-section.

Following this line of thought, it is possible to introduce a model for a point-like structure that is "thin" in all three-dimensions, like a finite element. In its earliest form [16, 17], attention was limited to one-dimensional deformations and it was shown that the theory can be used to formulate the numerical solution of problems of finite elasticity. Later [18], the theory was generalized for the numerical solution of two- and three-dimensional thermomechanical problems. For example, Fig. 27.3 shows a sketch of a homogeneously deformed tetrahedral element for

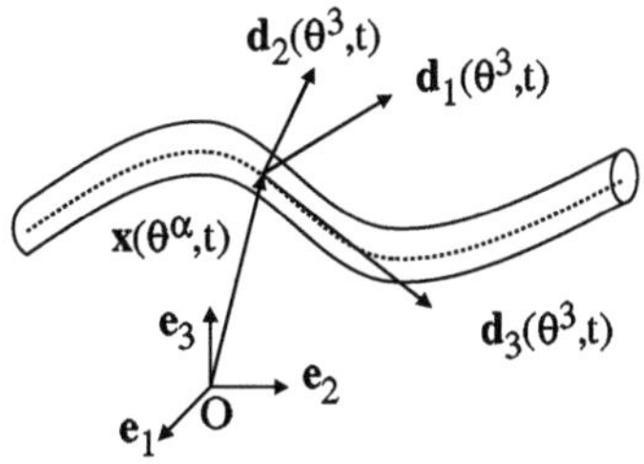

Fig. 27.2 Sketch of a Cosserat curve that is "thin" in two dimensions

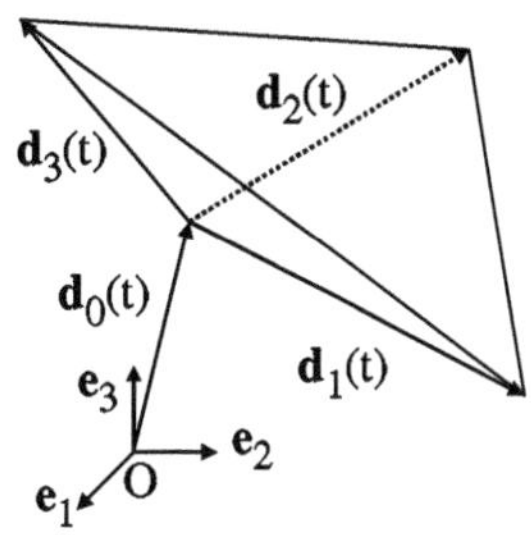

Fig. 27.3 Sketch of a Cosserat point that is "thin" in three dimensions

three-dimensional problems. The position vector $\mathbf{x}^*$ for this element is approximated by

$$\mathbf{x}^*\left(\theta^i, t\right) = \mathbf{x}(t) + \sum_{i=1}^{3} \theta^i \mathbf{d}_i(t), \tag{27.3}$$

where $\mathbf{x}$ locates one of the nodes of the element and $\mathbf{d}_i$ are three deformable director vectors that characterize deformations of the tetrahedral element. Other models for a homogeneously deformable body have also been considered in [3–5, 20–22, 24–26, 13]. A more complete discussion of these Cosserat theories of structures can be found in [19].

27.2 Basic Equations of a 3-D brick Cosserat Point Element (CPE)

Figure 27.4 shows a sketch of a 3-D eight-noded brick Cosserat Point Element (CPE) which can experience homogeneous deformations and inhomogeneous deformations that include bending, torsion and higher-order hourglassing. Within the context of the direct approach, the kinematics of the CPE are characterized by

$$\{\mathbf{D}_i, \mathbf{d}_i, \mathbf{w}_i = \dot{\mathbf{d}}_i\} \quad (i = 0, 1, \ldots, 7), \tag{27.4}$$

where the reference element director vectors $\mathbf{D}_i$ are constants, their present values $\mathbf{d}_i$ and the director velocities $\mathbf{w}_i$ are functions of time only, and a superposed $(\dot{\cdot})$ denotes time differentiation. These directors are restricted so that

$$D^{1/2} = \mathbf{D}_1 \times \mathbf{D}_2 \bullet \mathbf{D}_3, \qquad \mathbf{D}^i \bullet \mathbf{D}_j = \delta^i_j,$$
$$d^{1/2} = \mathbf{d}_1 \times \mathbf{d}_2 \bullet \mathbf{d}_3, \qquad \mathbf{d}^i \bullet \mathbf{d}_j = \delta^i_j. \tag{27.5}$$

Moreover, the inertia properties of the CPE are characterized by its constant mass m and director inertia coefficients y_{ij}

$$\dot{m} = 0, \qquad \dot{y}^{ij} = 0, \qquad y^{ji} = y^{ij} \quad (i,j = 0,1,\ldots,7). \tag{27.6}$$

The balances of director momentum are given by

$$\frac{\mathrm{d}}{\mathrm{d}t}\sum_{j=0}^{7} my^{ij}\mathbf{w}_j = m\mathbf{b}^i - \mathbf{m}^i - \mathbf{t}^i \quad (i = 0,1,\ldots,7), \qquad \mathbf{t}^0 = \mathbf{0}, \tag{27.7}$$

where $\{\mathbf{b}^i, \mathbf{m}^i\}$ are external director couples due body forces and surface tractions, respectively, and $\mathbf{t}^i$ are intrinsic director couples which require constitutive equations. Also, the reduced form of the balance of angular momentum requires the tensor $\mathbf{T}$ to be symmetric

$$\mathbf{T}^T = \mathbf{T} = d^{-1/2}\sum_{i=1}^{7} \mathbf{t}^i \otimes \mathbf{d}_i. \tag{27.8}$$

Next, introducing the additional kinematic variables

$$\mathbf{F} = \sum_{i=1}^{3} \mathbf{d}_i \otimes \mathbf{D}^i, \qquad \boldsymbol{\beta}_i = \mathbf{F}^{-1}\mathbf{d}_{i+3} - \mathbf{D}_{i+3} \quad (i = 1,2,3,4),$$
$$\mathbf{L} = \dot{\mathbf{F}}\mathbf{F}^{-1} = \sum_{i=1}^{3} \mathbf{w}_i \otimes \mathbf{d}^i, \qquad \mathbf{D} = \frac{1}{2}(\mathbf{L} + \mathbf{L}^T), \tag{27.9}$$

it can be shown that $\boldsymbol{\beta}_i$ are pure measures of inhomogeneous deformations. For an elastic CPE, the rate of material dissipation vanishes

$$\mathsf{D} = d^{1/2}\mathbf{T} + \sum_{i=1}^{4} \mathbf{F}^T\mathbf{t}^{i+3} \bullet \dot{\boldsymbol{\beta}}_i - m\dot{\Sigma} = 0 \tag{27.10}$$

for all deformations, and the strain energy function takes the form

$$\Sigma = \Sigma(\mathbf{C}, \boldsymbol{\beta}_i), \quad \mathbf{C} = \mathbf{F}^T\mathbf{F}. \tag{27.11}$$

Then, using standard arguments, the constitutive equations for an elastic CPE take the forms

$$d^{1/2}\mathbf{T} = m\mathbf{F}\frac{\partial \Sigma}{\partial \mathbf{C}}\mathbf{F}^T, \qquad \mathbf{t}^{(i+3)} = m\mathbf{F}^{-T}\frac{\partial \Sigma}{\partial \boldsymbol{\beta}_i} \quad (i = 1,2,3,4),$$
$$\mathbf{t}^i = \left(d^{1/2}\mathbf{T} - \sum_{j=4}^{7} \mathbf{t}^j \otimes \mathbf{d}_j\right)\mathbf{d}^i \quad (i = 1,2,3,4). \tag{27.12}$$

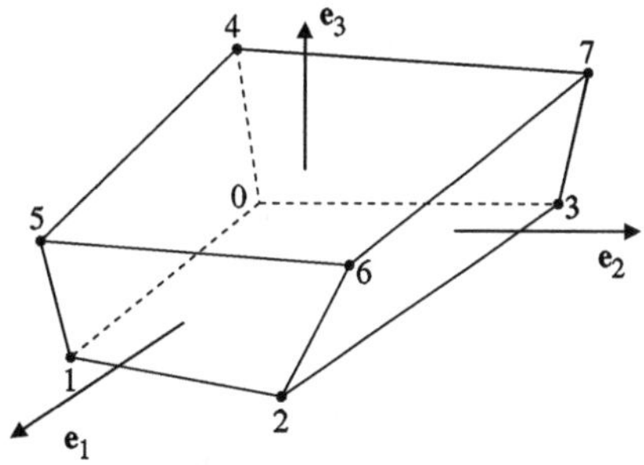

Fig. 27.4 Sketch of a general brick CPE showing the numbering of the nodes

27.3 Numerical Solutions of Problems of Finite Elasticity

An important difference between standard Bubnov–Galerkin methods and the CPE approach is that the strain energy function Σ for the CPE characterizes the response of the CPE as a structure. In particular, it was shown in [14] that can be restricted to satisfy a nonlinear patch test which ensures that the CPE predicts exact solutions for all homogeneous deformations of a uniform homogeneous elastic material and for all reference element shapes. Also, it was shown in [14] that the CPE can be formulated in terms of nodal quantities which are compatible with kinematic and kinetic conditions at nodes that are standard in finite element codes.

Loehnert et al. [12] implemented the CPE into the computer code FEAP [23] and showed that the CPE produces robust and accurate results for many problems which typically exhibit unphysical hourglassing or exhibit locking due to poor element aspect ratios or near incompressible material response. Loehnert et al. [12] also showed that the accuracy of the original CPE decreases with increased element irregularity. Determining a functional form for Σ which produces accurate results for element irregularity has proven to be challenging [9–11]. However, a generalized form for Σ was developed in [10, 11] and example problems were considered to show that the generalized CPE is truly a robust user friendly element that is free of locking and hourglass instabilities and can be used with confidence for the numerical solution of challenging problems in finite elasticity that include 3-D elements and thin shells and rods.

Acknowledgements This research was partially supported by MB Rubin's Gerard Swope Chair in Mechanics and by the fund for the promotion of research at the Technion.

References

1. Antman, S.S.: The Theory of Rods. Handbuch der Physik, vol. VIa/2. Springer, Berlin (1972)
2. Antman, S.S.: Nonlinear Problems of Elasticity. Springer, New York (1995)
3. Cohen, H.: Pseudo-rigid bodies. Utilitas Math. **20**, 221–247 (1981)
4. Cohen, H., Muncaster, R.G.: The dynamics of pseudo-rigid bodies: general structure and exact solutions. J. Elast. **14**, 127–154 (1984)

5. Cohen, H., Muncaster, R.G.: Theory of Pseudo-Rigid Bodies. Springer, Berlin (1984)
6. Cosserat, E., Cosserat, F.: Théorie des corps déformables. Hermann et Fils, Paris (1909). vi+226pp. = Appendix, pp. 953–1173 of Chwolson's Traite de Physique, 2nd edn., Paris. Also in translated form: "Theory of deformable bodies", NASA TT F-11, 561 (Washington, DC). Clearinghouse for US Federal Scientific and Technical Information, Springfield, Virginia
7. Green, A.E., Naghdi, P.M., Wenner, M.L.: On the theory of rods I: Derivations from the three-dimensional equations. Proc. R. Soc. Lond. A **337**, 451–483 (1974)
8. Green, A.E., Naghdi, P.M., Wenner, M.L.: On the theory of rods II: Developments by direct approach. Proc. R. Soc. Lond. A **337**, 485–507 (1974)
9. Jabareen, M., Rubin, M.B.: An improved 3-D Cosserat brick element for irregular shaped elements. Comput. Mech. **40**, 979–1004 (2007)
10. Jabareen, M., Rubin, M.B.: A generalized Cosserat point element (CPE) for isotropic nonlinear elastic materials including irregular 3-D brick and thin structures. United States Patent and Trademark Office, Serial No. 61/038 (2008)
11. Jabareen, M., Rubin, M.B.: A generalized Cosserat point element (CPE) for isotropic nonlinear elastic materials including irregular 3-D brick and thin structures. J. Mech. Mater. Struct. **3**, 1465–1498 (2008)
12. Loehnert, S., Boerner, E.F.I., Rubin, M.B., Wriggers, P.: Response of a nonlinear elastic general Cosserat brick element in simulations typically exhibiting locking and hourglassing. Comput. Mech. **36**, 255–265 (2005)
13. Muncaster, R.G.: Invariant manifolds in mechanics I: Zero-dimensional elastic bodies with directors. Arch. Ration. Mech. Anal. **84**, 353–373 (1984)
14. Nadler, B., Rubin, M.B.: A new 3-D finite element for nonlinear elasticity using the theory of a Cosserat point. Int. J. Solids Struct. **40**, 4585–4614 (2003)
15. Naghdi, P.M.: The theory of shells and plates. In: Truesdell, C. (ed.) S. Flugge's Handbuch der Physik, vol. VIa/2, pp. 425–640. Springer, Berlin (1972)
16. Rubin, M.B.: On the theory of a Cosserat point and its application to the numerical solution of continuum problems. J. Appl. Mech. **52**, 368–372 (1985)
17. Rubin, M.B.: On the numerical solution of one-dimensional continuum problems using the theory of a Cosserat point. J. Appl. Mech. **52**, 373–378 (1985)
18. Rubin, M.B.: Numerical solution of two- and three-dimensional thermomechanical problems using the theory of a Cosserat point. J. Math. Phys. (ZAMP) **46**, S308–S334 (1995). In Theoretical, Experimental, and Numerical Contributions to the Mechanics of Fluids and Solids, edited by J. Casey and M.J. Crochet, Brikhäuser, Basel, 1995
19. Rubin, M.B.: Cosserat Theories: Shells, Rods and Points. Solid Mechanics and its Applications, vol. 79. Kluwer Academic, Dordrecht (2000)
20. Slawianowski, J.J.: Analytical mechanics of finite homogeneous strains. Arch. Mech. **26**, 569–587 (1974)
21. Slawianowski, J.J.: Newtonian dynamics of homogeneous strains. Arch. Mech. **26**, 569–587 (1975)
22. Slawianowski, J.J.: The mechanics of the homogeneously-deformable body. Dynamical models with high symmetries. Z. Angew. Math. Mech. **62**, 229–240 (1982)
23. Taylor, R.L.: FEAP—A Finite Element Analysis Program, Version 7.5. University of California, Berkeley (2005)
24. Wozniak, C.Z.: Basic concepts of the theory of discrete elasticity. Bull. Acad. Pol. Sci. **19**, 753–758 (1971)
25. Wozniak, C.Z.: Equations of motion and laws of conservation in the discrete elasticity. Arch. Mech. **25**, 155–163 (1973)
26. Wozniak, C.Z.: Discrete elastic Cosserat media. Arch. Mech. **25**, 119–136 (1973)

Chapter 28
Discretization of Gradient Elasticity Problems Using C^1 Finite Elements

Stefanos-Aldo Papanicolopulos, A. Zervos, and Ioannis Vardoulakis

Abstract Strain-gradient theories have been used to model a variety of problems (such as elastic deformation, fracture behavior and plasticity) where size effect is of importance. Their use with the finite element method, however, has the drawback that specially designed elements are needed to obtain correct results.

This work presents an overview of the use of elements with C^1 continuous interpolation for strain-gradient models, using gradient elasticity as an example. After showing how the C^1 requirement arises and giving details concerning the implementation of specific elements, a theoretical comparison is made between elements based on this approach and elements resulting from the use of some alternative formulations.

28.1 Introduction

Strain-gradient theories have seen increased use since the seminal papers of Toupin [14], Mindlin [8] and Mindlin and Eshel [9] on gradient elasticity. Indicatively, we can mention some examples of the use of strain-gradient models in problems of elastic deformation [13, 3, 18, 15, 11], fracture behavior [1, 4, 10] and plasticity [7, 16, 17].

S.-A. Papanicolopulos (✉)
Department of Mechanics, National Technical University of Athens, Iroon Polytechneiou 5, 15773 Zografou, Greece
e-mail: stefanos@mechan.ntua.gr

A. Zervos
School of Civil Engineering and the Environment, University of Southampton, Southampton SO17 1BJ, UK
e-mail: az@soton.ac.uk

I. Vardoulakis (March 22nd, 1949 – September 19th, 2009), Formerly at the National Technical University of Athens, Greece.

G.A. Maugin, A.V. Metrikine (eds.), *Mechanics of Generalized Continua*, Advances in Mechanics and Mathematics 21,
DOI 10.1007/978-1-4419-5695-8_28, © Springer Science+Business Media, LLC 2010

The main characteristic of these models is that they necessarily introduce a material parameter with dimensions of length, thus introducing a material scale different from the geometrical scale of the problem. They can thus model problems where size effect is noticeable, either because the material and geometry scales are comparable or because localization of deformation occurs.

Independently of the details of the actual model used, the use of strain-gradient theories introduces the following problems, when compared to classical continuum theories:

1. Additional constitutive parameters must be determined. Even in the simplest case, there is always one additional parameter with dimensions of length.
2. Due to the higher order of the underlying partial differential equations, additional boundary conditions are needed. Their physical meaning, the way they can be applied in reality and their values for a given problem are not always clear.
3. The higher order of the governing equations necessitates a different, more complicated numerical treatment.

This work considers only the last point, it presents the complications this introduces into a finite element formulation, it considers one possible way of overcoming these difficulties and compares it to other methods that have been proposed.

28.2 Summary of Basic Equations

This section presents the main equations of Form II of gradient elasticity [8, 9], as they apply in Cartesian coordinates for small strains and static loading.

For a given displacement field u_i, a potential energy density $W = W(\varepsilon_{ij}, \kappa_{ijk})$ is defined as a function of the strains $\varepsilon_{ij} = \frac{1}{2}(u_{i,j} + u_{j,i})$ and the strain gradients $\kappa_{ijk} = \varepsilon_{jk,i}$. Computing the variation

$$\delta W = \frac{\partial W}{\partial \varepsilon_{ij}}\delta\varepsilon_{ij} + \frac{\partial W}{\partial \kappa_{ijk}}\delta\kappa_{ijk} = \tau_{ij}\delta\varepsilon_{ij} + \mu_{ijk}\delta\kappa_{ijk} \tag{28.1}$$

introduces the stresses τ_{ij} and double-stresses μ_{ijk}. Integrating the variation δW over the body volume V and using the divergence theorem leads, after extensive manipulation, to the virtual work equation

$$\int_V (\tau_{ij}\delta\varepsilon_{ij} + \mu_{ijk}\delta\kappa_{ijk})\,\mathrm{d}V$$
$$= \int_V F_k\delta u_k\,\mathrm{d}V + \int_S (P_k\delta u_k + R_k\delta(Du_k))\,\mathrm{d}S + \oint_C E_k\delta u_k\,\mathrm{d}C, \tag{28.2}$$

where F_k are body forces, P_k and R_k are tractions and double tractions on the surface S of the body, E_k are edge tractions along the edges C of the body and $D \equiv n_i\partial_i$ is the normal derivative operator.

In linear gradient elasticity, stresses and double stresses are linear combinations of strains and strain gradients. W is generally assumed to be the quadratic form

$$W = \frac{1}{2} c_{ijkl} \varepsilon_{ij} \varepsilon_{kl} + f_{ijklm} \kappa_{ijk} \varepsilon_{lm} + \frac{1}{2} a_{ijklmn} \kappa_{ijk} \kappa_{lmn}, \tag{28.3}$$

while in the linear isotropic case the above expression becomes

$$\begin{aligned} W = {} & \frac{1}{2} \tilde{\lambda} \varepsilon_{ii} \varepsilon_{jj} + \tilde{\mu} \varepsilon_{ij} \varepsilon_{ij} + \hat{a}_{(1)} \frac{\kappa_{iik} + \kappa_{iki}}{2} \kappa_{kjj} \\ & + \hat{a}_{(2)} \kappa_{ijj} \kappa_{ikk} + \hat{a}_{(3)} \kappa_{jjk} \kappa_{iki} + \hat{a}_{(4)} \kappa_{ijk} \kappa_{ijk} + \hat{a}_{(5)} \kappa_{ikj} \kappa_{kji}. \end{aligned} \tag{28.4}$$

28.3 Displacement-Only Finite Element Discretization

The usual displacement-only numerical discretization (see, e.g., [19]) can be used for gradient elasticity. The resulting formulae, when expressed in matrix notation, are very similar to the ones for the classical case, although the nature of their components introduces additional constraints.

The displacement field $\mathbf{u}(\mathbf{x})$ can be interpolated as

$$\mathbf{u}(\mathbf{x}) = \mathbf{N}(\mathbf{x}) \mathbf{u}^N, \tag{28.5}$$

where $\mathbf{N}$ is a matrix of shape functions and $\mathbf{u}^N$ is a column vector of degrees of freedom. Note that the components of $\mathbf{u}^N$ are not necessary displacements.

Stress-strain relations can be written in a compact and computationally efficient way by introducing the "generic strain" vector $\breve{\boldsymbol{\varepsilon}}$ and "generic stress" vector $\breve{\boldsymbol{\tau}}$ defined so that

$$\breve{\boldsymbol{\varepsilon}}^T \breve{\boldsymbol{\tau}} = \tau_{ij} \varepsilon_{ij} + \mu_{ijk} \kappa_{ijk}. \tag{28.6}$$

The components of $\breve{\boldsymbol{\varepsilon}}$ are linear combinations of the first and second order derivatives of the displacements $\mathbf{u}$, so using (28.5) the discretization of the generic strain is

$$\breve{\boldsymbol{\varepsilon}}(\mathbf{x}) = \mathbf{B}(\mathbf{x}) \mathbf{u}^N, \tag{28.7}$$

where $\mathbf{B}(\mathbf{x})$ is a matrix containing linear combinations of the first and second derivatives of the components of $\mathbf{N}(\mathbf{x})$.

Substituting (28.5), (28.6) and (28.7) into the virtual work equation (28.2) yields

$$\begin{aligned} \delta(\mathbf{u}^N)^T \int_V \mathbf{B}^T \breve{\boldsymbol{\tau}} \, \mathrm{d}V = {} & \delta(\mathbf{u}^N)^T \Bigg(\int_V \mathbf{N}^T \mathbf{F} \, \mathrm{d}V + \int_S (\mathbf{N}^T \mathbf{P} + D(\mathbf{N}^T) \mathbf{R}) \, \mathrm{d}S \\ & + \oint_C \mathbf{N}^T \mathbf{E} \, \mathrm{d}C \Bigg), \end{aligned} \tag{28.8}$$

where the dependence of $\mathbf{B}$ and $\mathbf{N}$ on $\mathbf{x}$ has been omitted for clarity. The above scalar equation must hold for every $\delta \mathbf{u}^N$, so it is equivalent to the vector equation

$$\int_V \mathbf{B}^T \breve{\tau}\, \mathrm{d}V = \int_V \mathbf{N}^T \mathbf{F}\, \mathrm{d}V + \int_S (\mathbf{N}^T \mathbf{P} + D(\mathbf{N}^T)\mathbf{R})\, \mathrm{d}S + \oint_C \mathbf{N}^T \mathbf{E}\, \mathrm{d}C. \quad (28.9)$$

Generic stresses are generally a nonlinear function of generic strains, so the above nonlinear equation can be solved using the Newton–Raphson method. For linear gradient elasticity the stress-strain relation is

$$\breve{\tau} = \mathbf{D}\breve{\varepsilon} = \mathbf{D}\mathbf{B}\mathbf{u}^N, \quad (28.10)$$

where $\mathbf{D}$ is a matrix of material parameters incorporating the components of tensors c_{ijkl}, f_{ijklm} and a_{ijklmn} appearing in (28.3). Therefore, (28.9) becomes

$$\mathbf{K}\mathbf{u}^N = \mathbf{f}, \quad (28.11)$$

where

$$\mathbf{K} = \int_V \mathbf{B}^T \mathbf{D} \mathbf{B}\, \mathrm{d}V \quad (28.12)$$

is the stiffness matrix and $\mathbf{f}$ is the vector of external actions. The solution of the linear system (28.11) yields the values of the degrees of freedom $\mathbf{u}^N$ and, using (28.5), the approximate solution for the displacement field.

28.4 Finite Elements with C^1 Continuity

The procedure described in the previous section differs from the one for classical elasticity in the following points:

1. The expression for the external actions $\mathbf{f}$ includes additional terms for surface double tractions and edge tractions.
2. The $\breve{\varepsilon}$ vector contains both strains and strain gradients and the $\breve{\tau}$ vector contains both stresses and double stresses.
3. The $\mathbf{B}$ matrix contains both first *and* second derivatives of the shape functions.

It is this last difference that introduces difficulties in the numerical treatment of the problem. Calculation of the integral of (28.12) requires that the first derivatives present in the $\mathbf{B}$ matrix must be continuous on the whole domain, i.e., the shape functions must ensure C^1 continuity of the interpolation. This is generally not a problem within each element, since the interpolation is usually polynomial, but is indeed a problem when interfaces between different elements are considered.

In one-dimensional problems, it is easy to satisfy the C^1 requirement, since elements have only single nodes in common. The discretization can employ Hermite elements, with cubic interpolation, where at each node both the value of the discretized field and its derivative are considered as degrees of freedom. An example of the use of such elements with strain-gradient theories can be found in [5].

In two-dimensional problems the C^1 requirement is much harder to satisfy, since the inter-element boundaries consist of both nodes and element edges, along which

C^1 continuity must be ensured. Appropriate elements have been however developed for problems of plate-bending and have been successfully adapted for strain-gradient problems.

One such element is the triangle presented in [2] and then in [6], which has been used for problems of gradient elastoplasticity [16, 17], elastic deformation [18] and fracture mechanics [10]. This element has straight edges, uses only corner nodes, and uses as degrees of freedom the displacement and all its derivatives of first and second order. The displacement is interpolated by a quintic polynomial, but along the edges the normal derivative is constrained to be cubic.

Development of isoparametric elements with C^1 continuity, on the other hand, is hindered by the necessity to map degrees of freedom which are spatial derivatives of the displacement from the parametric to the global (e.g., Cartesian) coordinate system. It is not therefore straightforward to extend the one-dimensional Hermite element to two dimensions. Such extension to two dimensions has nevertheless been presented by Petera and Pittman [12], who developed a C^1 isoparametric Hermite quadrilateral. This element employs as degrees of freedom at each node the displacement, its first derivatives and the mixed second derivative. Due to the details of the element construction, the first derivatives are in the global coordinate system while the mixed derivative is in the parametric system. Due to the isoparametric formulation, the derivatives of the global coordinates with respect to the parametric coordinates must be calculated in a pre-processing step. Additionally, the element can only be used in structured meshes, otherwise special interface-type elements would be needed to enable correct mapping of degrees of freedom to neighboring elements. Despite these restrictions, this element has been successfully used for problems of elastic deformation [18] and fracture mechanics [10].

The development of C^1 elements has generally not been extended to three dimensions. The only three-dimensional C^1 element, a Hermite hexahedron, has been recently developed by the authors [11], based on the respective quadrilateral.

28.5 Alternative Formulations

In Sect. 28.3, it was shown that discretizing only the displacement field leads to the requirement for C^1 continuity. A variety of techniques have been developed to avoid this requirement; a brief description of a number of such techniques and the resulting elements is given in this section.

Shu et al. [13] used a mixed formulation where, besides the displacements, the displacement gradients are discretized as a separate field, called the *relaxed* displacement gradient field. The relation between the two fields is enforced through the use of Lagrange multipliers, which constitute a third field to be discretized. Both quadrilateral and triangular elements are presented, with varying degrees of numerical performance. Quadratic interpolation is used to interpolate the displacements while either linear or quadratic interpolation is used to interpolate the relaxed displacement gradients. The Lagrange multipliers are either constant or are interpolated linearly. Full integration is used for the displacements, but reduced integration is used for integrating terms containing the Lagrange multipliers.

Amanatidou and Aravas [1] have presented a series of quadrilateral elements based on a mixed formulation for Form I or Form III of gradient elasticity. For Form I relaxed displacement gradients are discretized, while for Form III relaxed strains and relaxed rotations are discretized. Lagrange multipliers are interpolated linearly while all other fields are interpolated quadratically. Reduced integration is used for all quantities.

Matsushima et al. [7] presented a quadrilateral element, also based on a formulation using Lagrange multipliers. The displacements are interpolated quadratically while the relaxed displacement gradients are interpolated linearly. The Lagrange multipliers are constant within each element. Uniquely among the elements presented here, a large-strain formulation is presented for this element. The stiffness matrix is given in detail, showing that it is non-symmetric.

Askes and Aifantis [3] propose a meshless formulation, using the element-free Galerkin method, which provides directly the required C^1 continuity, as no inter-element boundaries exist.

Askes and Gutiérrez [4] propose a mixed formulation based on a reformulation of gradient elasticity as an *implicit* gradient elasticity. Only two fields are discretized: the displacements and the nonlocal strains, the latter defined as a function of the strains and their Laplacian. Counter to intuition, the quadrilateral element implementing this formulation uses a quadratic interpolation of the nonlocal strains, which are considered the primary variable, and only a linear interpolation of the displacements which are considered the constraint variables. Full integration with respect to the nonlocal strains is used.

Zervos [15] presents a series of elements (triangle, quadrilateral, tetrahedron and hexahedron) which discretize the more general theory of elasticity with microstructure [8]. Displacements and micro-deformations are interpolated using either linear or quadratic interpolations, and full integration is used. Through an appropriate choice of material parameters, the same elements can be used for solving gradient elasticity problems, using one of the material parameters as a penalty parameter that forces the micro-deformations to coincide with the gradients of the displacement. A comparison of the results obtained using the two dimensional elements based on this formulation with similar results obtained using C^1 elements is given in [18].

28.6 Comparison Between Different Formulations

As shown in Sect. 28.3, the use of C^1 elements is a direct consequence of choosing a simple, displacement-only discretization. Due to the limited number of such elements and their perceived complexity and computational cost, other elements such as the ones listed in Sect. 28.5 were developed to avoid the need for C^1 continuity, usually based on a mixed formulation where multiple fields are discretized. This section attempts a comparison of C^1 and mixed-formulation elements, considering various strengths and weaknesses of each approach.

One clear advantage of pursuing a mixed formulation is that, due to its relaxed requirements, it allows creating a much larger variety of elements, while there are only a few elements that satisfy the C^1 requirement. Of course, not all mixed-formulation elements perform equally well as seen, for example, in works that present different variations of a given element [13, 1].

Considering ease of implementation, almost all elements considered here can be implemented with relative ease within any finite element code that allows for arbitrary, non-displacement degrees of freedom. The exceptions are the C^1 Hermite quadrilateral and hexahedron, whose implementation requires the ability to compute and store for each node additional "coordinates" which are the derivatives of global coordinates with respect to parametric coordinates.

An inherent advantage of the C^1 approach is that the strain field is easily and accurately calculated at the nodes without additional effort. In mixed formulations there may be a discretization of the *relaxed* strain field, but it is not the exact strain field corresponding to the computed displacements. Especially for the C^1 triangle, the nodal values of the strain gradient are also computed directly, although the strain gradient is not continuous along element edges. It should be noted that the easy and accurate evaluation of the strains is very useful in problems of gradient elasticity where it can be used to directly calculate the stresses as well. An example of such calculations for fracture mechanics can be found in [10]. A related advantage of the C^1 formulation is the accurate enforcement of boundary conditions on the normal derivative of the displacement.

The usual criterion used, however, when comparing different formulations is their computational performance. A direct comparison of the performance is quite difficult, especially in terms of execution speed, as it requires implementing different elements within the same numerical code. A first indirect measurement of the element performance, both in terms of execution speed and memory requirements, can however be attempted by examining the number of degrees of freedom employed by each element, as done in Table 28.1 for quadrilaterals and hexahedra.

Table 28.1 shows that it is not only the total number of degrees of freedom that is important, but also the number of degrees of freedom actually used to interpolate the requested solution for the displacement field. It is seen for example that the

Table 28.1 Number of degrees of freedom for different quadrilaterals and hexahedra used with gradient models [11]. Numbers in parentheses are theoretical estimates, as the respective elements have not been presented in the literature

Element	2D DOFs		3D DOFs	
	Total	u field	Total	u field
C^1 Hermite element [12, 18, 11]	32	32	192	192
Penalty Quad8U4P/Bri20U8P [15]	32	16	132	60
Implicit element [4]	32	8	(144)	(24)
Element with Lagrange multipliers [7]	36	16	(141)	(60)
Element QU34L4 with Lagrange multipliers [13]	38	18	(162)	(81)
Form III, III9-70 [1]	70	18	(396)	(81)

C^1 Hermite quadrilateral is one of the "cheapest" elements, while the respective hexahedron is one of the most "expensive", in terms of degrees of freedom used. They both are, however, the "richer" in their category, since they provide a cubic interpolation for the displacements instead of the quadratic interpolation provided by the other elements.

It is important to note that this is a general advantage of the C^1 approach, shared for example by the C^1 triangle. Discretizing an additional field allows the use of C^0 elements, but introduces a significant overhead in degrees of freedom that do not enrich the interpolation being employed for the primary field. In other words, where a C^1 element exists, it is computationally cheaper than a C^0 element using a mixed formulation with the same order of interpolation for the displacements.

28.7 Conclusions

A variety of solutions have been proposed to address the problems arising in the numerical implementation of strain-gradient models. This work presents the way these problems arise and how the requirement for elements with C^1 continuity naturally follows the simple choice of a displacement-only discretization. Alternative techniques are presented as well, and an attempt is made to indicate the main advantages and disadvantages of different formulations.

Although the numerical behavior of each element should be tested in practice for different types of problems to be solved, the comparison presented here indicates that C^1 elements have, in theory at least, some interesting advantages. Importantly, it is shown that, where they exist, C^1 elements are computationally cheaper than other elements providing the same order of interpolation for the displacement field. This means that it is worthwhile to implement existing C^1 elements for strain-gradient problems, and also to look for new C^1 elements, especially in three dimensions.

Acknowledgements The research leading to these results has received funding from the European Research Council under the European Community's Seventh Framework Programme (FP7/2007-2013)/ERC grant agreement No. [228051].

References

1. Amanatidou, E., Aravas, N.: Mixed finite element formulations of strain-gradient elasticity problems. Comput. Methods Appl. Mech. Eng. **191**, 1723–1751 (2002). doi:10.1016/S0045-7825(01)00353-X
2. Argyris, J.H., Fried, I., Scharpf, D.W.: The TUBA family of plate elements for the matrix displacement method. Aeronaut. J. R. Aeronaut. Soc. **72**, 701–709 (1968)
3. Askes, H., Aifantis, E.C.: Numerical modelling of size effects with gradient elasticity—formulation, meshless discretization and examples. Int. J. Fract. **117**(4), 347–358 (2002). doi:10.1023/A:1022225526483

4. Askes, H., Gutiérrez, M.A.: Implicit gradient elasticity. Int. J. Numer. Methods Eng. **67**(3), 400–416 (2006). doi:10.1002/nme.1640
5. Chambon, R., Caillerie, D., El Hassan, N.: One-dimensional localisation studied with a second grade model. Eur. J. Mech. A, Solids **17**(4), 637–656 (1998)
6. Dasgupta, S., Sengupta, D.: A higher-order triangular plate bending element revisited. Int. J. Numer. Methods Eng. **30**, 419–430 (1990). doi:10.1002/nme.1620300303
7. Matsushima, T., Chambon, R., Caillerie, D.: Large strain finite element analysis of a local second gradient model: application to localization. Int. J. Numer. Methods Eng. **54**, 499–521 (2002). doi:10.1002/nme.433
8. Mindlin, R.D.: Micro-structure in linear elasticity. Arch. Ration. Mech. Anal. **16**(1), 51–78 (1964). doi:10.1007/BF00248490
9. Mindlin, R.D., Eshel, N.N.: On first strain-gradient theories in linear elasticity. Int. J. Solids Struct. **4**, 109–124 (1968). doi:10.1016/0020-7683(68)90036-X
10. Papanicolopulos, S.A., Zervos, A.: Numerical solution of crack problems in gradient elasticity. Eng. Comput. Mech. (2009, accepted)
11. Papanicolopulos, S.A., Zervos, A., Vardoulakis, I.: A three dimensional C^1 finite element for gradient elasticity. Int. J. Numer. Methods Eng. **77**(10), 1396–1415 (2009). doi:10.1002/nme.2449
12. Petera, J., Pittman, J.F.T.: Isoparametric Hermite elements. Int. J. Numer. Methods Eng. **37**(20), 3489–3519 (1994). doi:10.1002/nme.1620372006
13. Shu, J.Y., King, W.E., Fleck, N.A.: Finite elements for materials with strain gradient effects. Int. J. Numer. Methods Eng. **44**(3), 373–391 (1999). doi:10.1002/(SICI)1097-0207(19990130)44:3<373::AID-NME508>3.0.CO;2-7
14. Toupin, R.A.: Elastic materials with couple stresses. Arch. Ration. Mech. Anal. **11**(1), 385–414 (1962). doi:10.1007/BF00253945
15. Zervos, A.: Finite elements for elasticity with microstructure and gradient elasticity. Int. J. Numer. Methods Eng. **73**(4), 564–595 (2008). doi:10.1002/nme.2093
16. Zervos, A., Papanastasiou, P., Vardoulakis, I.: A finite element displacement formulation for gradient elastoplasticity. Int. J. Numer. Methods Eng. **50**(6), 1369–1388 (2001). doi:10.1002/1097-0207(20010228)50:6<1369::AID-NME72>3.0.CO;2-K
17. Zervos, A., Papanastasiou, P., Vardoulakis, I.: Modelling of localisation and scale effect in thick-walled cylinders with gradient elastoplasticity. Int. J. Solids Struct. **38**(30–31), 5081–5095 (2001)
18. Zervos, A., Papanicolopulos, S.A., Vardoulakis, I.: Two finite element discretizations for gradient elasticity. J. Eng. Mech. ASCE **135**(3), 203–213 (2009). doi:10.1061/(ASCE)0733-9399(2009)135:3(203)
19. Zienkiewicz, O.C., Taylor, R.L.: The Finite Element Method, 4th edn. McGraw-Hill, New York (1989)

Chapter 29
C^1 Discretizations for the Application to Gradient Elasticity

Paul Fischer, Julia Mergheim, and Paul Steinmann

Abstract For the numerical solution of gradient elasticity, the appearance of strain gradients in the weak form of the equilibrium equation leads to the need for C^1-continuous discretization methods. In the present work, the performances of a variety of C^1-continuous elements as well as the C^1 Natural Element Method are investigated for the application to nonlinear gradient elasticity. In terms of subparametric triangular elements the Argyris, Hsieh–Clough–Tocher and Powell–Sabin split elements are utilized. As an isoparametric quadrilateral element, the Bogner–Fox–Schmidt element is used. All these methods are applied to two different numerical examples and the convergence behavior with respect to the L^2, H^1 and H^2 error norms is examined.

29.1 Gradient Elasticity

In contrast to the classical (Boltzmann) continuum, the free energy density of a gradient continuum is not solely dependent on $\mathbf{F} := \nabla_{\mathbf{X}}\varphi$ but also on the curvature measure $\mathbf{G} := \nabla_{\mathbf{X}}\mathbf{F}$, i.e.,

$$W_0 := W_0(\mathbf{F}, \mathbf{G}),$$

leading to the following weak form of equilibrium equation [10]:

P. Fischer (✉) · J. Mergheim · P. Steinmann
University Erlangen-Nuremberg, Egerlandstrasse 5, 91058 Erlangen, Germany
e-mail: paul.fischer@ltm.uni-erlangen.de

J. Mergheim
e-mail: julia.mergheim@ltm.uni-erlangen.de

P. Steinmann
e-mail: paul.steinmann@ltm.uni-erlangen.de

G.A. Maugin, A.V. Metrikine (eds.), *Mechanics of Generalized Continua*, Advances in Mechanics and Mathematics 21,
DOI 10.1007/978-1-4419-5695-8_29, © Springer Science+Business Media, LLC 2010

$$\delta\Pi = \int_{\mathscr{B}_0} [\mathbf{P} : \delta\mathbf{F} + \mathbf{Q} : \cdot\delta\mathbf{G} - \mathbf{b}\cdot\delta\varphi]\,\mathrm{d}V - \int_{\partial\mathscr{B}_0^{\mathrm{P}}} \bar{\mathbf{t}}^{\mathrm{P}}\cdot\delta\varphi\,\mathrm{d}A + \int_{\partial\mathscr{B}_0^{\mathrm{Q}}} \bar{\mathbf{t}}^{\mathrm{Q}}\cdot(\nabla_{\mathbf{X}}\varphi)\cdot\mathbf{N}\,\mathrm{d}A = 0, \tag{29.1}$$

where $\mathbf{P} := \partial W_0/\partial\mathbf{F}$ and $\mathbf{Q} := \partial W_0/\partial\mathbf{G}$. The following boundary conditions are applied

$$\begin{aligned} \varphi &= \overline{\varphi} \quad \text{on } \partial\mathscr{B}_0^{\varphi}, & \mathbf{t}^{\mathrm{P}} &= \bar{\mathbf{t}}^{\mathrm{P}} \quad \text{on } \partial\mathscr{B}_0^{\mathrm{P}}, \\ (\nabla_{\mathbf{X}}\varphi)\cdot\mathbf{N} &= (\nabla_{\mathbf{X}}\overline{\varphi})\cdot\mathbf{N} \quad \text{on } \partial\mathscr{B}_0^{\mathrm{F}}, & \mathbf{t}^{\mathrm{Q}} &= \bar{\mathbf{t}}^{\mathrm{Q}} \quad \text{on } \partial\mathscr{B}_0^{\mathrm{Q}}. \end{aligned} \tag{29.2}$$

The two types of independent boundaries have the properties $\partial\mathscr{B}_0^{\mathrm{P}} \cap \partial\mathscr{B}_0^{\varphi} = \partial\mathscr{B}_0^{\mathrm{Q}} \cap \partial\mathscr{B}_0^{\mathrm{F}} = \emptyset$ and $\partial\mathscr{B}_0^{\mathrm{P}} \cup \partial\mathscr{B}_0^{\varphi} = \partial\mathscr{B}_0^{\mathrm{Q}} \cup \partial\mathscr{B}_0^{\mathrm{F}} = \partial\mathscr{B}_0$.

29.2 Numerical Methods

For the numerical solution of gradient elastic problems, usually alternative methods not requiring C^1-continuous elements are used. These can either be implicit methods [4, 3], mixed formulations [1] and micromorphic continuum formulations with Lagrange multipliers [15] or penalty parameters [9, 10]. These methods are employed due to the increase of complexity caused by the C^1 finite elements. However, the better performance of the C^1-continuous elements could legitimate the additional implementational effort, which is shown in [8, 18, 11]. Nevertheless, an extensive overview on the behavior of C^1-continuous methods is still missing.

29.2.1 Subparametric Triangular Elements

C^1-continuous finite elements are mostly used in subparametric meshes, where the geometry is approximated linearly. This has the major advantage that the mesh construction is as easy as for C^0-continuous elements. However, a poor geometry approximation can lead to significant numerical errors.

The triangular C^1 finite elements which are compared in this contribution are the Argyris and Bell elements [2], the Hsieh–Clough–Tocher (HCT) element [6] and the group of Powell–Sabin split (PS) elements [13].

The shape functions of the Argyris and Bell elements are completely polynomial on the triangle. The other elements are macroelements, and their shape functions are piecewise cubic on the three subelements of the HCT element and quadratic on the 6/12 subelements of the PS elements. We apply all elements in the original and reduced form. For the Powell–Sabin 12 split element (PS-12), the normal derivative along the edge has been left piecewise linear for the original form. The PS-6 element is slightly modified with respect to the usual case, i.e., instead of the circumcircle midpoint, the barycenter of the triangle is used for the subdivision of the

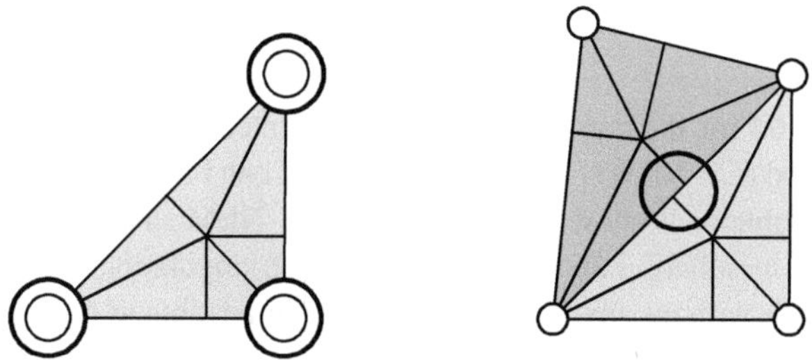

Fig. 29.1 (*Left*) Schematic representation of the PS-6 element. (*Right*) Since the lines dividing the macroelement edges have to be perpendicular to the element sides, the shape functions are piecewise polynomials on different subsections of the edge resulting in discontinuous displacements and strains

element. Therefore, the usually used heuristic criterion that the maximum angle of the element has to be smaller than $75°$ in order to avoid skinny subtriangles is not necessary. However, the displacement and the first derivative in tangential direction loose their continuity along the element edges, which is sketched in Fig. 29.1. Nevertheless, since the normal derivative is defined to be linear along the element sides, it is obviously continuous. Fortunately, the jumps in the stresses are not significant, i.e., see Fig. 29.3.

29.2.2 Isoparametric Quadrilateral Element

Since subparametric meshes with a linear geometry approximation for quadrilateral elements are strongly restricted, i.e., the mesh has to be composed of rhomboids, these elements should be used isoparametrically. In this contribution, the Bogner–Fox–Schmidt element [5] is used. It has been successfully applied to gradient elasticity in [18, 11]. The major advantage of the isoparametric quadrilateral element is its ability of approximating different kinds of geometries much better than the subparametric elements of the previous section. However, it has the disadvantage that the mesh is restricted to four elements at a node otherwise the C^1-continuity will be lost. One algorithm for the mesh construction is given in [12]. However, this method may not be resulting in an optimal mesh, see [8]. Therefore, the following fictitious energy is used which is minimized in order to obtain an optimal isoparametric mesh:

$$\bar{\Pi} = \frac{1}{2} \int_{\mathscr{B}_\xi} \bar{\mathbf{G}} : \cdot \bar{\mathbf{G}} \, \mathrm{d}V_{\boldsymbol{\xi}}, \tag{29.3}$$

where $\mathrm{d}V_{\boldsymbol{\xi}}$ is the volume element in the local coordinate system of each reference element and $\bar{\mathbf{G}} := \partial^2 \mathbf{X} / \partial \boldsymbol{\xi}^2$. As Dirichlet boundary conditions, all nodal coordinates are given.

29.2.3 Natural Element Method

The third method for the comparison of the presented examples is the C^1-Natural Element Method (C^1-NEM) [17]. That the method can be successfully used for gradient elastic problems is shown in [8, 14]. The C^1-NEM shape functions are based on the Voronoi tessellation and its dual Delaunay triangulation [16], i.e., their construction is done without any given mesh. Its main advantage over different meshfree methods is that essential boundary conditions can directly be applied. As extension to non-convex domains, the α-shape method is used [7]. The integration is performed with a 16 point Gauss-integration on the Delaunay triangulation.

In this particular case, the set of points will be arranged in a regular way, i.e., they are created by the nodal points of a regular bilinear mesh.

29.3 Numerical Examples

In the following section, the performance of the different numerical methods is tested on two experiments. One is a specimen with external pressure and the other a cracked specimen under uniaxial loading. We employ the same material model for both, i.e.,

$$W_0(\mathbf{F},\mathbf{G}) = W_0^H(\mathbf{F}) + W_0^G(\mathbf{G}),$$

whereby a compressible Neo–Hookean model is chosen for W_0^H

$$W_0^H = \frac{1}{2}\lambda \ln^2 J + \frac{1}{2}\mu\left[\mathbf{F}:\mathbf{F} - n^{\mathrm{dim}} - 2\ln J\right]$$

and a quadratic expression for W_0^G

$$W_0^G = \frac{1}{2}\mu l^2 \mathbf{G} : \cdot \mathbf{G}.$$

For the material parameters, a Young's modulus of 200 kN/mm^2 and a Poisson's ratio of 0.3 are used. For the internal length parameter, $l = L$ is applied. For both numerical examples, homogeneous Neumann boundary conditions for the double tractions are assumed, i.e., $\bar{\mathbf{t}}^{\mathrm{Q}} = \mathbf{0}$.

Since to the authors best knowledge, no analytical solution for these nonlinear gradient elastic problems is available, the error is computed with respect to another numerical solution with a significantly higher amount of degrees of freedom. Additionally, we want to point out that all the meshes are regular, i.e., the rates of convergence with respect to the number of degrees of freedom are approximately half the convergence rate with respect to the mesh diameter h.

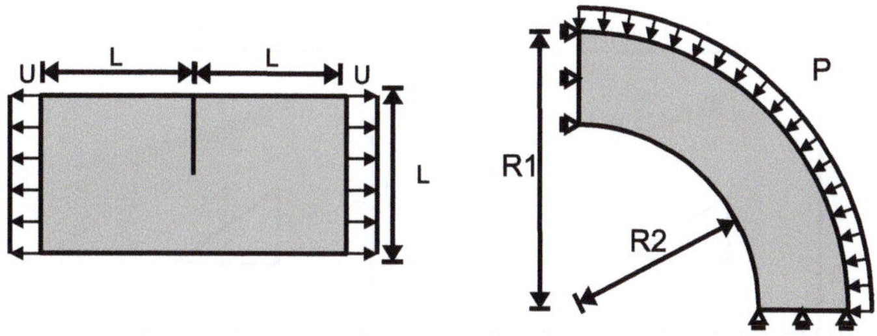

Fig. 29.2 The two tested specimen. (*Left*) Uniaxial loading of cracked specimen. (*Right*) Thick hollow cylinder

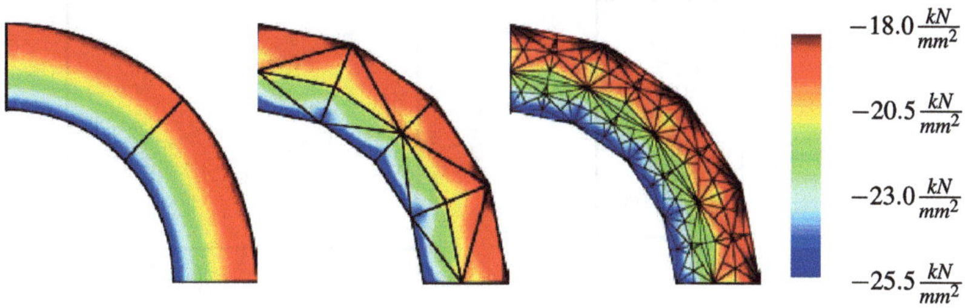

Fig. 29.3 First principal σ stress. (*Left*) two BFS elements. (*Center*) 15 given computational points for the C^1 NEM. (*Right*) 32 PS-6 elements

29.3.1 Thick Hollow Cylinder

In the first example, a thick hollow cylinder under external pressure is considered. The specimen is loaded with an external pressure $\bar{\mathfrak{t}}^{\mathrm{P}} = -18$ kN/mm^2 at the outside of the cylinder. The external radius $R = 3L$ and the internal radius $r = 2L$ are chosen. A detailed description of the geometry is presented in Fig. 29.2. For symmetry reasons, only a quarter of the geometry is computed. The symmetry conditions lead to the following set of Dirichlet boundary conditions along the thickness of the cylinder:

$$\begin{aligned} &u_n = 0, \qquad \partial_n u_t = 0, \qquad \partial_t u_n = 0, \\ &\partial_n^2 u_n = 0, \qquad \partial_t^2 u_n = 0, \quad \text{and} \quad \partial_n \partial_t u_t = 0. \end{aligned} \tag{29.4}$$

The higher order derivatives are necessary for the use of the Argyris and Bell element since they have higher order derivatives as nodal degrees of freedom. In Fig. 29.4, it can be observed that the error values for the subparametric methods are much higher than those of the isoparametric elements. The differences in the contour plots for the first principal Cauchy-stress can be observed in Fig. 29.3. However, the rates of convergence for the HCT and the BFS elements are similar to each other. Second best are the Argyris and Bell elements. All the other methods seem to be very similar. Except the NEM, they all have in common that the normal derivative across the element edges is linear, which seems to reduce the convergence behavior

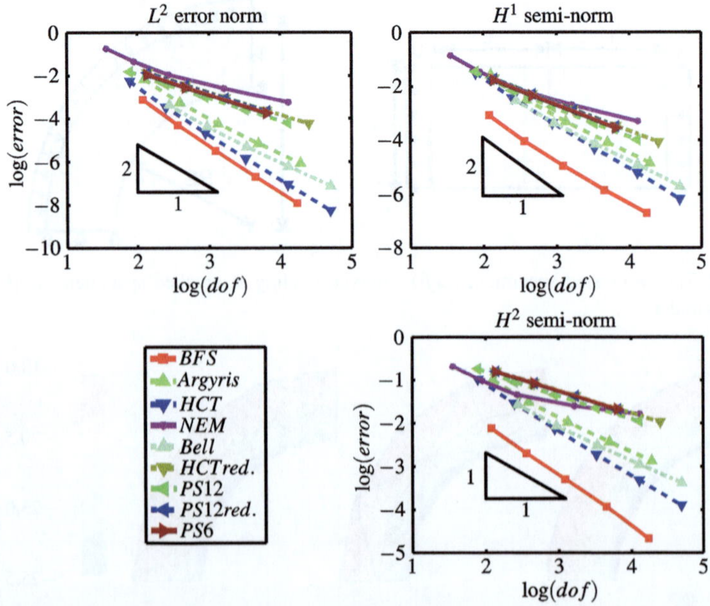

Fig. 29.4 Convergence behavior for the thick hollow cylinder

of the elements. In this particular example, the higher numerical costs for the NEM shows to be inefficient.

29.3.2 Uniaxial Loading of Cracked Specimen

The second numerical example is a cracked specimen under uniaxial loading. Due to the fact, that the singularity in the stresses is shifted to the double stress $\mathbf{Q}(\mathbf{G})$, our aim is in the investigation of the L^2, H^1 and H^2 errors while using mesh refinement. Again the symmetry of the specimen is utilized with the conditions (29.4). At the upper part of the specimen, the displacement in normal direction is prescribed.

In contrast to the previous example, all methods seem to behave similarly. The convergence rates are far from the rates of the previous examples, compare Table 29.1. This is caused by the crack tip singularity.

29.4 Discussion

In this contribution, the Argyris, Bell, Hsieh–Clough–Tocher, Powell–Sabin and the Bogner–Fox–Schmidt elements are analyzed for the simulation of nonlinear gradient elasticity. Additionally, the results for the C^1-NEM are investigated. For the

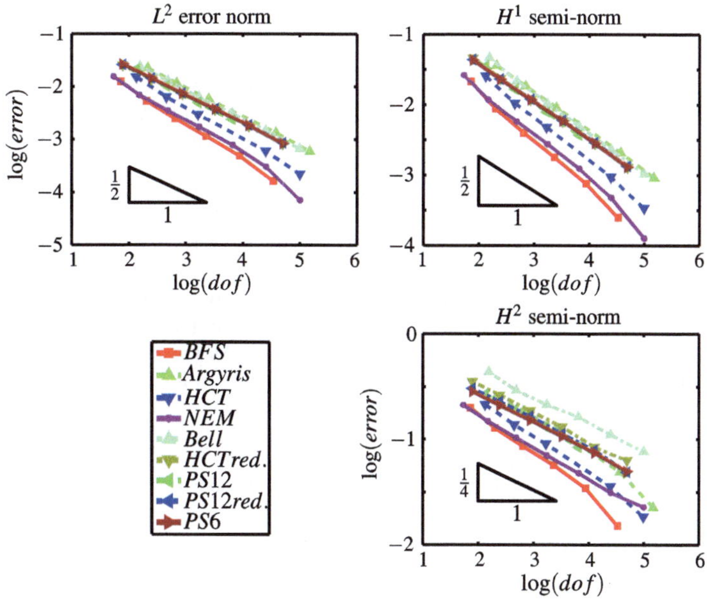

Fig. 29.5 Convergence behavior for the cracked specimen

Table 29.1 Convergence rates of the different elements. (*Left*) Thick hollow cylinder. (*Right*) Cracked specimen

Method	L^2	H^1	H^2
BFS	2.23	1.68	1.19
Argyris	1.79	1.50	0.86
HCT	2.14	1.66	1.04
NEM	0.97	0.95	0.43
Bell	1.65	1.40	0.80
HCT red.	1.05	1.05	0.52
PS12	1.08	1.15	0.55
PS12 red.	1.07	1.07	0.53
PS6	1.07	1.07	0.50

Method	L^2	H^1	H^2
BFS	0.71	0.72	0.45
Argyris	0.55	0.56	0.40
HCT	0.64	0.65	0.40
NEM	0.72	0.71	0.33
Bell	0.56	0.58	0.30
HCT red.	0.53	0.55	0.30
PS12	0.57	0.58	0.35
PS12 red.	0.53	0.54	0.32
PS6	0.54	0.54	0.31

thick hollow cylinder problem, an optimal geometry approximation seems to be the most important issue for good performance of the elements, showing up in the superior quantitative behavior of the BFS element in the H^2-seminorm, see Fig. 29.4. The best choice in terms of triangular elements seems to be the HCT element in its original form. Compared to the reduced form of the HCT element, the Powell–Sabin elements have shown a better performance with respect to the number of global degrees of freedom.

For the cracked specimen under uniaxial loading, due to the singularity all elements show very poor performance. Therefore, special crack tip elements have to be developed which are able to capture singularities in the gradient elasticity formulation.

References

1. Amanatidou, E., Aravas, N.: Mixed finite element formulations of strain-gradient elasticity problems. Comput. Methods Appl. Mech. Eng. **191**, 1723–1751 (2002)
2. Argyris, J.H., Fried, I., Scharpf, D.W.: The TUBA family of elements for the matrix displacement method. Aeronaut. J. **72**, 701–709 (1968)
3. Askes, H., Bennett, T., Aifantis, E.C.: A new formulation and C^0-implementation of dynamically consistent gradient elasticity. Int. J. Numer. Methods Eng. **72**, 111–126 (2007)
4. Askes, H., Gutiérrez, M.A.: Implicit gradient elasticity. Int. J. Numer. Methods Eng. **67**, 400–416 (2006)
5. Bogner, F.K., Fox, R.L.: The generation of inter-element-compatible stiffness and mass matrices by the use of interpolation formulas. In: Proceedings of the Conference held at Wright-Patterson Air Force Base, pp. 397–443. Wright-Patterson Air Force Base, Ohio (1965)
6. Clough, R.W., Tocher, J.L.: Finite element stiffness matrices for analysis of plate bending. In: Proceedings of the Conference held at Wright-Patterson Air Force Base, pp. 515–545. Wright-Patterson Air Force Base, Ohio (1965)
7. Cueto, E., Sukumar, N., Calvo, B., Cegonio, J., Doblare, M.: Overview and recent advantages in natural neighbor Galerkin methods. Arch. Comput. Methods Eng. **10**, 307–384 (2002)
8. Fischer, P., Mergheim, J., Steinmann, P.: On the C^1 continuous discretization of nonlinear gradient elasticity: a comparison of NEM and FEM based on Bernstein–Bézier patches. Int. J. Numer. Methods Eng. (2009). doi:10.1002/nme.2802
9. Hirschberger, C.B., Kuhl, E., Steinmann, P.: On deformational and configurational mechanics for micromorphic hyperelasticity—theory and computation. Comput. Methods Appl. Mech. Eng. **196**, 4027–4044 (2007)
10. Kirchner, N., Steinmann, P.: A unifying treatise on variational principles for gradient and micro-morphic continua. Philos. Mag. **85**, 3875–3895 (2005)
11. Papanicolopulos, S.A., Zervos, A., Vardoulakis, I.: A three-dimensional C^1 finite element for gradient elasticity. Int. J. Numer. Methods Eng. **77**, 1396–1415 (2009)
12. Petera, J., Pittman, J.F.T.: Isoparametric Hermite elements. Int. J. Numer. Methods Eng. **37**, 3489–3519 (1994)
13. Powell, M.J., Sabin, M.A.: Piecewise quadratic approximation on triangles. ACM Trans. Math. Softw. **3**, 316–325 (1977)
14. Rajagopal, A., Scherer, M., Steinmann, P., Sukumar, N.: Smooth conformal α-nem for gradient elasticity. Int. J. Struct. Changes Solids **1**, 83–109 (2009)
15. Shu, J.Y., King, W.E., Fleck, N.E.: Finite elements for materials with strain gradient effects. Int. J. Numer. Methods Eng. **44**, 373–391 (1999)
16. Sibson, R.A.: A vector identity for the Dirichlet tessellation. Math. Proc. Camb. Philos. Soc. **87**, 151–155 (1980)
17. Sukumar, N., Moran, B., Belytschko, T.: The natural element method in solid mechanics. Int. J. Numer. Methods Mech. **43**, 839–887 (1998)
18. Zervos, A., Papanicolopulos, S.A., Vardoulakis, I.: Two finite-element discretizations for gradient elasticity. J. Eng. Mech. **135**, 203–213 (2009)

Chapter 30
A Generalized Framework and a Multiplicative Formulation of Electro-Mechanical Coupling

Carlo Sansour, Sebastian Skatulla, and A. Arunachalakasi

Abstract Electro-active polymers (EAP) recently attracted much interest because, upon electrical loading, EAP exhibit a large amount of deformation while sustaining large forces. On the other hand, generalized continuum frameworks are relevant to electro-mechanical coupled problems as they naturally incorporate couples or higher order stresses and can describe scale effects. Here, we want to adopt a strain gradient approach based on the generalized continuum framework as formulated in (Sansour in J. Phys. IV, Proc. 8:341–348, 1998; Sansour and Skatulla in Geomech. Geoeng. 2:3–15, 2007) and extend it to encompass the electro-mechanically coupled behavior of EAP. A new aspect of the electro-mechanical formulation relates to the multiplicative decomposition of the deformation gradient, well known from plasticity, into a purely elastic part and a further part which relates to the electric field. The formulation is elegant, makes for clarity and is numerically efficient. A numerical example of coupled large deformations is presented as well.

C. Sansour (✉)
Division of Materials, Mechanics, and Structures, The University of Nottingham, University Park, Nottingham NG7 2RD, UK
e-mail: carlo.sansour@nottingham.ac.uk

S. Skatulla
CERECAM, Department of Civil Engineering, The University of Cape Town, Private Bag X3, Rondebosch 7701, South Africa
e-mail: sebastian.skatulla@uct.ac.za

A. Arunachalakasi
Department of Applied Mechanics, Indian Institute of Technology Madras, Chennai 600 036, India
e-mail: aarajan@iitm.ac.in

G.A. Maugin, A.V. Metrikine (eds.), *Mechanics of Generalized Continua*, Advances in Mechanics and Mathematics 21,
DOI 10.1007/978-1-4419-5695-8_30, © Springer Science+Business Media, LLC 2010

30.1 A Generalized Continuum Framework and Higher Gradients

The generalized continuum framework [6] is based on the mathematical concept of a fiber bundle, where in the simplest case the generalized space is constructed as the Cartesian product of a macro space $\mathscr{B} \subset \mathbb{E}(3)$ and a micro space $\mathscr{S}$ which we write as $\mathscr{G} := \mathscr{B} \times \mathscr{S}$. Appropriate projection maps are to be defined. The macro-space $\mathscr{B}$ is parameterized by the curvilinear co-ordinates ϑ^i, $i = 1, 2, 3$ and the micro-space $\mathscr{S}$ by the curvilinear coordinates ζ^α. Here, and in what follows, Greek indices take the values $1, \ldots, n$. The dimension of $\mathscr{S}$ denoted by n is arbitrary, but finite.

We assume that the placement vector $\tilde{\mathbf{x}}$ of a material point $P(\tilde{\mathbf{X}} \in \mathscr{G})$ can be formulated as the sum of its position in the macro-continuum $\mathbf{x} \in \mathscr{B}_t$ and in the micro-continuum $\boldsymbol{\xi} \in \mathscr{S}_t$ as follows

$$\tilde{\mathbf{x}} = \mathbf{x}(\vartheta^k, t) + \boldsymbol{\xi}(\vartheta^k, \zeta^\beta, t). \tag{30.1}$$

Accordingly, for $t = t_0$ the placement of a material point (30.1) takes

$$\tilde{\mathbf{X}} = \mathbf{X}(\vartheta^k) + \Xi(\vartheta^k, \zeta^\beta), \quad \text{with } \Xi \in \mathscr{S}. \tag{30.2}$$

The fundamental idea now is to assume for $\boldsymbol{\xi}(\vartheta^k, \zeta^\beta, t)$ explicit expressions as to its dependency on the coordinates ζ^β. While the framework allows for general expressions, for the time being we want to restrict ourselves to a linear function according to

$$\tilde{\mathbf{x}} = \mathbf{x}(\vartheta^k, t) + \zeta^\alpha \mathbf{a}_\alpha(\vartheta^k, t). \tag{30.3}$$

The vector functions $\mathbf{a}_\alpha(\vartheta^k, t)$ correspond to the micro-coordinates ζ^α, the number of which must be chosen according to a specific topology of the micro-space as well as certain physical properties of the material.

In order to avoid the incorporation of additional degrees of freedom, other than the displacement degrees of freedom, we restrict the dimensionality of the microspace to three and Greek indices take the values 1, 2, or 3. This allows us to define the directors of the micro-continuum $\mathbf{a}_\alpha$ from now on as follows

$$\mathbf{a}_\alpha = \frac{\partial \mathbf{x}}{\partial \vartheta^\alpha} = \mathbf{g}_\alpha(\vartheta^k, t), \tag{30.4}$$

with $\mathbf{g}_\alpha \in \mathscr{T}\mathscr{B}_t$, the tangent space of $\mathscr{B}_t$. Then, the generalized deformation field (30.3) takes the following form

$$\tilde{\mathbf{x}} = \mathbf{x}(\vartheta^k, t) + \zeta^\alpha \mathbf{g}_\alpha(\vartheta^k, t), \tag{30.5}$$

which essentially constitutes a higher gradient continuum. Note that even if the micro-continuum $\mathscr{S}$ is defined by the spatial derivatives of the macroscopic placement vector (30.4), it is important to realize that the dimension of the micro-space does not have to coincide with the dimension of the macro-space, but must not be larger than three. A variable orientation of the microspace with respect to the macrospace can be modeled by defining (30.5) as follows

$$\tilde{\mathbf{x}} = \mathbf{x}(\vartheta^k, t) + \zeta^\alpha Q_{i\alpha}\mathbf{g}_i, \tag{30.6}$$

where, in general, $Q_{i\alpha}$ denotes a constant rotation matrix of dimension (3×3).

Taking the derivatives of $\tilde{\mathbf{x}}$ with respect to the coordinates ϑ^i and ζ^α

$$\tilde{\mathbf{x}}_{,i} = \frac{\partial \tilde{\mathbf{x}}}{\partial \vartheta^i} = \mathbf{x}_{,i}(\vartheta^k, t) + \zeta^\alpha \mathbf{g}_{\alpha,i}(\vartheta^k, t), \qquad \tilde{\mathbf{x}}_{,\alpha} = \frac{\partial \tilde{\mathbf{x}}}{\partial \zeta^\alpha} = \mathbf{g}_\alpha(\vartheta^k, t), \tag{30.7}$$

the generalized deformation gradient tensor can be expressed as follows

$$\tilde{\mathbf{F}} = \left[\mathbf{x}_{,i}(\vartheta^k, t) + \zeta^\alpha \mathbf{g}_{\alpha,i}(\vartheta^k, t)\right] \otimes \tilde{\mathbf{G}}^i + \mathbf{g}_\alpha(\vartheta^k, t) \otimes \mathbf{I}^\alpha, \tag{30.8}$$

where the operator $\otimes$ denotes the dyadic product of two vectors and $\tilde{\mathbf{G}}^i$, $\mathbf{I}^\alpha$ are contra-variant base vectors in the generalized space. In order to formulate a generalized strain measure based on (30.8) we proceed in analogy to the definition of the classical right *Cauchy–Green* deformation tensor. Thus, its generalized equivalent is expressed as

$$\tilde{\mathbf{C}} = \tilde{\mathbf{F}}^T \tilde{\mathbf{F}}, \qquad \tilde{\mathbf{C}} = \mathbf{C} + \zeta^\alpha \mathbf{K}_\alpha, \tag{30.9}$$

where higher order terms in ζ^α are neglected. By denoting the scalar products of vectors by a dot, we arrive at

$$\begin{aligned} \mathbf{C} = {} & \mathbf{x}_{,k} \cdot \mathbf{x}_{,l} \tilde{\mathbf{G}}^k \otimes \tilde{\mathbf{G}}^l + \mathbf{x}_{,k} \cdot \mathbf{g}_\alpha \left(\tilde{\mathbf{G}}^k \otimes \mathbf{I}^\alpha + \mathbf{I}^\alpha \otimes \tilde{\mathbf{G}}^k\right) \\ & + \mathbf{g}_\alpha \cdot \mathbf{g}_\beta \mathbf{I}^\alpha \otimes \mathbf{I}^\beta, \end{aligned} \tag{30.10}$$

$$\begin{aligned} \mathbf{K}_\alpha = {} & (\mathbf{x}_{,k} \cdot \mathbf{g}_{\alpha,l} + \mathbf{g}_{\alpha,k} \cdot \mathbf{x}_{,l}) \tilde{\mathbf{G}}^k \otimes \tilde{\mathbf{G}}^l \\ & + \mathbf{g}_{\alpha,k} \cdot \mathbf{g}_\beta \left(\tilde{\mathbf{G}}^k \otimes \mathbf{I}^\beta + \mathbf{I}^\beta \otimes \tilde{\mathbf{G}}^k\right). \end{aligned} \tag{30.11}$$

Note that the first term of $\mathbf{C}$ represents the conventional right *Cauchy–Green* deformation tensor. Note also that the terms of the strain measures with the base vectors $\tilde{\mathbf{G}}^k \otimes \tilde{\mathbf{G}}^l$ are the most dominant ones. Unless otherwise stated we confine ourselves to them. The equilibrium equations of a continuum with the above strain measures are derived in [9].

30.2 Basic Relations of Electro-Elasticity

In this section, we present the fundamental electric fields and the field equations in electrostatics in both forms as defined at the actual configuration as well as at the reference one. For further details of electrostatics, the reader is referred to [3–5].

30.2.1 Electro-Mechanical Fields and Their Governing Equations

For an electro-mechanically coupled material behavior we consider, besides the mechanical fields previously introduced, the electric field and the electric displacement field (or the electric charge potential). In the current configuration, the electric field and the electric displacement are given by

$$\tilde{\mathbf{e}}_t = -\text{grad}\,\tilde{\phi}, \qquad \tilde{\mathbf{d}}_t = \varepsilon_0 \tilde{\mathbf{e}}_t + \tilde{\mathbf{p}}_t, \tag{30.12}$$

where $\tilde{\phi}$ denotes the electric potential defined in the generalized space, ε_0 denotes the *vacuum electric permittivity* and $\tilde{\mathbf{p}}_t$ the dielectric polarization or simply the polarization. The polarization represents a derived quantity as it depends on the electric field and the material and its state which includes density, temperature and strain [2].

With the restriction to electrostatics only two of *Maxwell's* equations are of interest. *Gauss's* law for electricity and *Faraday's* law of induction read:

$$\nabla_{\tilde{\mathbf{x}}} \cdot \tilde{\mathbf{d}}_t = 0 \quad \text{in } \mathscr{B}_t \times \mathscr{S}_t \quad \text{and} \quad \nabla_{\tilde{\mathbf{x}}} \times \tilde{\mathbf{e}}_t = \mathbf{0} \quad \text{in } \mathscr{B}_t \times \mathscr{S}_t, \tag{30.13}$$

respectively, where the del operator is defined as $\nabla_{\tilde{\mathbf{x}}} = (\tilde{\mathbf{g}}_i \cdot \tilde{\mathbf{g}}_i)^{-1/2}\, \partial/\partial\vartheta^i$ (no summation over i) and its counterpart in $\mathscr{B} \times \mathscr{S}$ is given by $\nabla_{\tilde{\mathbf{X}}} = (\tilde{\mathbf{G}}_i \cdot \tilde{\mathbf{G}}_i)^{-1/2}\, \partial/\partial\vartheta^i$. Here $\tilde{\mathbf{g}}_i$, $\tilde{\mathbf{G}}_i$ are base vectors in the actual and reference configurations in the generalized space, respectively. The governing equations are supplemented with corresponding Neumann and Dirichlet boundary conditions which read as follows:

$$\tilde{\mathbf{d}}_t \cdot \tilde{\mathbf{n}}_t = -\tilde{q}_t^s \quad \text{on } \partial\mathscr{B}_{t,N}^q \times \mathscr{S}_t, \qquad \tilde{\phi} = \tilde{h}_\phi \quad \text{on } \partial\mathscr{B}_{t,D}^\phi \times \mathscr{S}_t, \tag{30.14}$$

respectively, $\tilde{q}_t^s$ denotes the electric surface charge density in the deformed configuration, $\tilde{\mathbf{n}}$ the normal vector on $\partial\mathscr{B} \times \mathscr{S}$ with $\tilde{\mathbf{n}}_t$ as its equivalent in the deformed configuration and $\partial\mathscr{B}_D^\phi \times \mathscr{S} \subset \partial\mathscr{B} \times \mathscr{S}$ and $\partial\mathscr{B}_N^q \times \mathscr{S} = \partial\mathscr{B} \times \mathscr{S} \setminus \partial\mathscr{B}_D^\phi \times \mathscr{S}$. In this sense, the electric boundary conditions are represented as a microscopical average.

The above equations can be reformulated in a purely material setting. First, the following material quantities are defined by means of pull-back operations:

$$\tilde{\mathbf{e}} = \tilde{\mathbf{F}}^T \tilde{\mathbf{e}}_t, \qquad \tilde{\mathbf{d}} = \tilde{J}\tilde{\mathbf{F}}^{-1}\tilde{\mathbf{d}}_t, \qquad \tilde{\mathbf{p}} = \tilde{J}\tilde{\mathbf{F}}^{-1}\tilde{\mathbf{p}}_t, \tag{30.15}$$

where $\tilde{J} = \det \tilde{\mathbf{F}}$ denotes the Jacobian. Using (30.12) and (30.15), the material electric displacement vector and the electric field are given as

$$\tilde{\mathbf{d}} = \varepsilon_0 \tilde{J} \tilde{\mathbf{C}}^{-1} \tilde{\mathbf{e}} + \tilde{\mathbf{p}}, \quad \tilde{\mathbf{e}} = -\text{Grad}\,\tilde{\phi}, \tag{30.16}$$

where it is assumed that $\tilde{\mathbf{p}}_t$ is given by a constitutive equation.

Finally, the governing equations of the electric field and the electric displacement can be recast in a material setting as

$$\nabla_{\tilde{\mathbf{X}}} \times \tilde{\mathbf{e}} = \mathbf{0} \quad \text{in } \mathscr{B} \times \mathscr{S} \quad \text{and} \quad \nabla_{\tilde{\mathbf{X}}} \cdot \tilde{\mathbf{d}} = 0 \quad \text{in } \mathscr{B} \times \mathscr{S}, \tag{30.17}$$

respectively. The electric Neumann and Dirichlet boundary conditions (30.14) are expressed in the undeformed configuration as

$$\tilde{\mathbf{d}} \cdot \tilde{\mathbf{n}} = -\tilde{q}^s \quad \text{on } \partial \mathscr{B}_N^q \times \mathscr{S}, \qquad \tilde{\phi} = \tilde{h}_\phi \quad \text{on } \partial \mathscr{B}_D^\phi \times \mathscr{S}, \tag{30.18}$$

respectively, where q^s denotes the electric surface charge density in the reference configuration. For further details, the reader is referred to [9].

It can be shown that all field equations can be derived from a coupled free energy function suitably defined. Assuming that such a function exists and defined per unit volume, one has

$$\tilde{\mathbf{d}} = -\frac{\partial \Psi_{\text{coupled}}}{\partial \tilde{\mathbf{e}}}, \qquad \tilde{\boldsymbol{\sigma}} = 2\tilde{\mathbf{F}} \frac{\partial \Psi_{\text{coupled}}}{\partial \tilde{\mathbf{C}}} \tilde{\mathbf{F}}^T, \tag{30.19}$$

where $\tilde{\boldsymbol{\sigma}}$ is a Kirchhoff-type tensor which also captures Maxwell's stress tensor.

The question now arises of how to define $\Psi_{\text{coupled}}(\mathbf{C}, \mathbf{e})$. In previous work on the subject, simple formulations have been suggested such as, e.g., [10]

$$\Psi_{\text{coupled}}(\mathbf{C}, \mathbf{e}) = \Psi_{\text{mech}}(\mathbf{C}) + c_1 \mathbf{1} : \mathbf{e} \otimes \mathbf{e} + c_2 \mathbf{C} : (\mathbf{e} \otimes \mathbf{e}) - \frac{c_3}{2} \tilde{J} \tilde{\mathbf{C}}^{-1} : \tilde{\mathbf{e}} \otimes \tilde{\mathbf{e}}. \tag{30.20}$$

While the first term $\Psi_{\text{mech}}(\mathbf{C})$ reflects the usual formulation of the elastic energy, two further terms are added to account for the electro-mechanical coupling while the last term is an extra term which accounts for an energy contribution of the nearly rigid body. The material constants involved are to be determined by experiments. Very often, however, the nature of these coupling terms is not clear from the outset and one is motivated to resort to very simple forms of them, as, in fact, has been done in (30.20). We notice also that no linear terms in $\mathbf{e}$ are included in the first place and it is also not clear how to proceed should higher order terms (higher order nonlinearities) be deemed relevant. While it is true that one can resort to representation theorems of tensor-valued functions, such representations are by far not adequate as they result, in general, in a large number of material constants and entail complicated issues regarding convexity and existence of solutions.

30.2.2 Multiplicative Electro-Mechanical Coupling Theory

Now, instead of incorporating the electro-mechanical coupling by adding additional terms to the free energy function to account for the polarization of dielectric material, we alternatively aim to achieve this via a proper decomposition of the deformation gradient. We consider now the multiplicative decomposition of the deformation gradient

$$\tilde{\mathbf{F}} = \tilde{\mathbf{F}}_{\text{mech}} \tilde{\mathbf{F}}_{\text{elec}}, \tag{30.21}$$

which clearly constitutes a point of departure from conventional electro-mechanical coupling formulations. Such a decomposition is best known from plasticity. However, plasticity is a dissipative process and the constitutive laws are defined in rate form, in contrast to the present case. The electric part of the deformation gradient tensor is now defined as

$$\tilde{\mathbf{F}}_{\text{elec}} := \exp(\mathbf{D}), \tag{30.22}$$

where the second order tensor $\mathbf{D}$ is given via a separate electro-mechanical constitutive law. The choice of the exponential map is motivated by the fact that polymers are nearly incompressible materials. Accordingly, $\mathrm{J} = \det \tilde{\mathbf{F}} = 1$. Since the relation

$$\det \tilde{\mathbf{F}}_{\text{elec}} = \exp(\operatorname{tr} \mathbf{D}) \tag{30.23}$$

holds, the incompressibility condition $\det \tilde{\mathbf{F}}_{\text{elec}} = 1$, highly nonlinear in the components of $\tilde{\mathbf{F}}_{\text{elec}}$, becomes simply linear in terms of components of $\mathbf{D} : \operatorname{tr} \mathbf{D} = 0$.

For $\mathbf{D}$ an appropriate constitutive law is to be formulated. It is assumed that it depends only on $\tilde{\mathrm{e}}$, the electric field vector in the un-deformed configuration:

$$\mathbf{D} = \overset{4}{\mathbb{C}}_{\text{elec}} : \tilde{\mathrm{e}} \otimes \tilde{\mathrm{e}}, \tag{30.24}$$

where $\overset{4}{\mathbb{C}}_{\text{elec}}$ denotes a fourth order constitutive tensor and the : operator represents the double contraction of a fourth order tensor with a second order tensor defined as $D_{ij} = \mathbb{C}_{(\text{elec})ijkl}\tilde{e}_k\tilde{e}_l$. The component matrix $\mathbb{C}_{(\text{elec})ijkl}$ is considered to be symmetric with respect to indices k, l. For polymers, due to near incompressibility of the material, the resulting deformation in compression and expansion needs to accommodate the volume preservation. That is, it holds

$$\operatorname{tr} \mathbf{D} = 0 \quad \Rightarrow \quad D_{11} + D_{22} + D_{33} = 0. \tag{30.25}$$

Now, as the electro-mechanical constitutive law (30.24) is assumed to take into account only contraction and expansion due to the electric field but no shear, consequently, the component matrix of the electro-mechanical constitutive tensor is sparsely set as shown below:

$$\mathbb{C}_{(\text{elec})1111} = \mathbb{C}_{(\text{elec})2222} = \mathbb{C}_{(\text{elec})3333} = c_1$$

as well as

$$\begin{aligned} \mathbb{C}_{(\text{elec})1122} &= \mathbb{C}_{(\text{elec})1133} = \mathbb{C}_{(\text{elec})2211} = \mathbb{C}_{(\text{elec})2233} \\ &= \mathbb{C}_{(\text{elec})3311} = \mathbb{C}_{(\text{elec})3322} = c_2. \end{aligned}$$

The additional electro-mechanical material constants c_1 and c_2 are required to be determined by experiments. The remaining entries of the component matrix have zero values. Accordingly, with (30.25) we find that the two material constants are related by

$$c_2 = -0.5c_1. \tag{30.26}$$

With (30.21), the purely mechanical right *Cauchy–Green* deformation tensor $\tilde{\mathbf{C}}_{\text{mech}}$ can be expressed as

$$\tilde{\mathbf{C}}_{\text{mech}} = \tilde{\mathbf{F}}_{\text{elec}}^{-T} \tilde{\mathbf{C}} \tilde{\mathbf{F}}_{\text{elec}}^{-1} \tag{30.27}$$

where $\tilde{\mathbf{C}} = \tilde{\mathbf{F}}^T \tilde{\mathbf{F}}$ represents the conventional right *Cauchy–Green* deformation tensor. Subsequently, the electro-mechanical coupling is incorporated into the free energy function per unit volume $\Psi_{\text{coupled}}(\tilde{\mathbf{C}}, \tilde{\mathbf{e}})$ by simply formulating its mechanical part Ψ_{mech} as a function of the mechanical right *Cauchy–Green* deformation tensor $\tilde{\mathbf{C}}_{\text{mech}}$ and we have

$$\Psi_{\text{coupled}}(\tilde{\mathbf{C}}, \tilde{\mathbf{e}}) = \Psi_{\text{mech}}(\tilde{\mathbf{C}}_{\text{mech}}) - \frac{1}{2} c_3 \tilde{J} \tilde{\mathbf{C}}^{-1} : \tilde{\mathbf{e}} \otimes \tilde{\mathbf{e}}. \tag{30.28}$$

The last term refers to the polarization of a rigid body but transferred to the reference configuration. For further details, the reader is referred to [8].

30.3 Numerical Examples

We present an example of a coupled deformation. A fully 3-D computation is carried out via a mesh-free Moving Least Square (MLS)-approximation scheme. The MLS-approximation functions have to meet the continuity and consistency requirements which is $C^1(\Omega)$ continuity and consistency of order 2 due the incorporated strain gradients. The numerical integration over the micro-continuum $\mathscr{S}$ is carried out with the help of the *Gauss quadrature*, the order of which has to be second according to the used basis polynomial. The enforcement of the essential boundary conditions is carried out using the penalty method.

The mechanical part of the stored energy density function (30.28) is assumed to be of the nonlinear statistically based hyperelastic type [1]. It makes use of three constants, the shear modulus C_R, the bulk modulus κ and parameter N which addresses the limited extensibility of the macromolecular network structure of the polymer material.

The example features a cantilever beam which has the shape of angle consisting of two equally long pieces. One side of the beam is clamped and the electric potential is assumed as $\phi = 0$. The other free end is subjected to electric surface

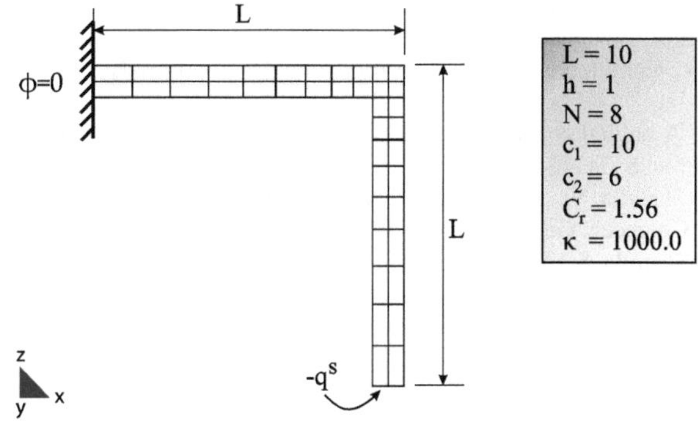

Fig. 30.1 Problem configuration

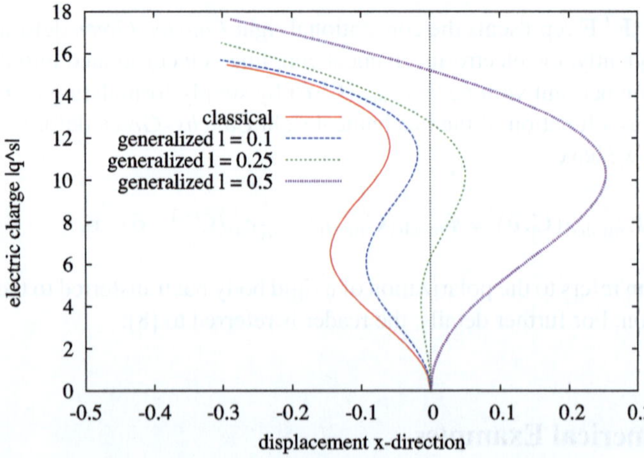

Fig. 30.2 Electric surface charge versus displacement at the tip of the beam in x-direction simulating the generalized approach based on a one-dimensional microspace using as micro-director $\mathbf{a} = \mathbf{x}_{,x}$ with different magnitudes for the internal length scale parameter l_x

charge loading depicted in Fig. 30.1. The problem is discretized by 189 particles, where each cross-section consists of 9 particles as shown by the mesh of the problem configuration. The generalized solutions illustrated in Fig. 30.2 are obtained with a one-dimensional micro-space choosing the micro-director as $\mathbf{a} = \mathbf{x}_{,1} = \mathbf{x}_{,x}$. Three scaling levels are considered setting the internal length scale parameter as $l_x = 0.1, 0.25$, and 0.5 denoted by the blue, green, and purple line, respectively.

References

1. Arruda, E.M., Boyce, M.C.: A three-dimensional constitutive model for the large stretch behavior of rubber elastic materials. J. Mech. Phys. Solids **41**(2), 389–412 (1993)
2. Griffiths, D.: Introduction to Electrodynamics. Prentice-Hall, New Jersey (1999)
3. Jackson, J.D.: Classical Electrodynamics, 3rd edn. Wiley, New York (1999)
4. Kovetz, A.: Electromagnetic Theory. Oxford Science Publications, New York (2000)
5. Maugin, G.A.: Continuum Mechanics of Electromagnetic Solids. North-Holland, Amsterdam (1988)
6. Sansour, C.: A unified concept of elastic-viscoplastic Cosserat and micromorphic continua. J. Phys. IV Proc. **8**, 341–348 (1998)
7. Sansour, C., Skatulla, S.: A higher gradient formulation and meshfree-based computation for elastic rock. Geomech. Geoeng. **2**, 3–15 (2007)
8. Sansour, C., Skatulla, S., Arunachalakasi, A.: A multiplicative formulation for electromechanical coupling at finite strains (2009, to appear)
9. Skatulla, S., Arunachalakasi, A., Sansour, C.: A nonlinear generalized continuum approach for electro-elasticity including scale effects. J. Mech. Phys. Solids **57**, 137–160 (2009)
10. Vu, D.K., Steinmann, P., Possart, G.: Numerical modelling of non-linear electroelasticity. Int. J. Numer. Methods Eng. **70**, 685–704 (2007)

Part VIII
Beyond the Cosserats: Original Approaches (Kinematics, Geometry, Fractals)

Chapter 31
Generalized Variational Principle for Dissipative Continuum Mechanics

German A. Maximov

Abstract The generalization of the Hamilton's and Onsager's variational principles for dissipative hydrodynamical systems is represented in terms of the mechanical and the heat displacement fields. A system of equations for these fields is derived from the extreme condition for action with a Lagrangian in the form of the difference between the kinetic and the free energies minus the time integral of the dissipation function. The generalized hydrodynamic equation system is then evaluated on the basis of the generalized variational principle. At low frequencies, this system corresponds to the traditional Navier–Stokes equation system, and in the high frequency limit it describes propagation of acoustical and heat modes with the finite propagation velocities.

31.1 Introduction

A system of hydrodynamic equations for a viscous, heat conducting fluid is usually derived on the basis of the mass, momentum and energy conservation laws [7]. Certain assumptions about forms of viscous stresses tensor and energy density flow vector are made to complete it. This system is considered presently as the one describing quite adequately a large set of hydrodynamical phenomena. However, there are some aspects which suggest that this system is only an approximation.

For example, if we consider propagation of small perturbations described by this system, then it is possible to separate formally the longitudinal, shear and heat and/or entropy waves. The coupling of the longitudinal and heat waves results in their splitting into independent acoustic-thermal and thermal-acoustic modes [1, 2]. For these modes, the limits of phase velocities tend to infinity at high frequencies so that the system is in formal contradiction with the requirements for a finite propaga-

G.A. Maximov (✉)
N.N. Andreyev Acoustical Institute, Shvernika str. 4, Moscow 117036, Russia
e-mail: maximov@dpt39.mephi.ru

G.A. Maugin, A.V. Metrikine (eds.), *Mechanics of Generalized Continua*, Advances in Mechanics and Mathematics 21,
DOI 10.1007/978-1-4419-5695-8_31, © Springer Science+Business Media, LLC 2010

tion velocity of any perturbation. For this reason, it is possible to suggest that such a hydrodynamic equation system is a mere low frequency approximation.

In particular, taking into account the viscosity relaxation phenomenon [8], it allows guaranteeing the limit for the propagation velocity of the shear mode and introducing the heat relaxation term [3, 14, 9] which, in turn, ensures the finite propagation velocities of the acoustic-thermal and thermal-acoustic modes.

However, the introduction of such relaxation processes requires serious efforts especially for more complicated cases, e.g., in the case of a multi-phase medium which possesses additional internal degrees of freedom.

Classical mechanics provides us with the Lagrange's variational principle which allows deriving easily the equations of motion for a mechanical system knowing the forms of kinetic and potential energies. The difference between these energies determines the Lagrange's function. This approach translates easily into continuum mechanics by introduction of the Lagrangian density for non-dissipative media. In this approach, the dissipation forces can be accounted for by the introduction of the dissipation function derivatives into the corresponding equations of motion in accordance with Onsager's principle of symmetry of kinetic coefficients [6]. There is an established opinion that for a dissipative system it is impossible to formulate the variational principle analogously to the least action principle of Hamilton [6]. At the same time, there are successful approaches in which the variational principles for heat conduction theory and for irreversible thermodynamics are applied to account explicitly for the dissipation processes [4, 5].

Therefore, there are good reasons to attempt to formulate the generalized Hamilton's variational principle for dissipative systems which argues against its established opposition [7]. It is shown [12, 13] that such variational principle can be formulated in terms of the displacements of the mechanical and thermal fields.

31.2 Hamilton's Variational Principle

The non-dissipative case of the Hamilton's variational principle can be formulated for a continuous medium in the form of extremum of the action functional $\delta S = 0$:

$$S = \int_{t_1}^{t_2} \mathrm{d}t \int_V \mathrm{d}\mathbf{r}\, L, \tag{31.1}$$

where, by analogy with mechanics, the Lagrangian density is represented as a difference of kinetic K and potential U energies:

$$L(\dot{\mathbf{u}}, \nabla\mathbf{u}) = K(\dot{\mathbf{u}}) - U(\nabla\mathbf{u}). \tag{31.2}$$

Here, as it is seen from (31.2), the Lagrangian is considered as a function of velocities of displacements $\dot{\mathbf{u}} = \mathrm{d}\mathbf{u}/\mathrm{d}t$ and deformations $\nabla\mathbf{u} = \operatorname{div}\mathbf{u}$.

Motion equations derived from the variational principle (31.1)–(31.2) have form

$$\frac{\mathrm{d}}{\mathrm{d}t}\frac{\partial L}{\partial \dot{\mathbf{u}}} + \nabla \frac{\partial L}{\partial \nabla \mathbf{u}} = 0. \tag{31.3}$$

In the simplest case, when kinetic and potential energies are determined by quadratic forms

$$2K\left(\dot{\mathbf{u}}^2\right) = \rho_0 \dot{\mathbf{u}}^2, \qquad 2U = \lambda \varepsilon_{ll}^2 + 2\mu \varepsilon_{ik}^2, \quad \varepsilon_{ik} = \frac{1}{2}\left(\frac{\partial u_i}{\partial x_k} + \frac{\partial u_k}{\partial x_i}\right), \tag{31.4}$$

the well known motion equation for elastic medium is derived as

$$\rho_0 \frac{\mathrm{d}}{\mathrm{d}t}\dot{\mathbf{u}} - \mu \Delta \mathbf{u} - (\lambda + \mu)\nabla(\nabla \mathbf{u}) = 0, \tag{31.5}$$

where the constants of the quadratic form (31.4) are the density ρ_0, and Lame's constants λ and μ.

31.3 Onsager's Variational Principle

Further, if we were to consider the quasi-equilibrium systems only, then the variational principle of the least energy dissipation was formulated for them by Onsager [15, 16]. This principle is based on the symmetry of kinetic coefficients and can be formulated as an extreme of a functional built as a difference between entropy increase rate $\dot{s}$ and dissipation function D, considered as functions of rates $\dot{\alpha}$ of thermodynamical relaxation processes:

$$\delta_{\dot{\alpha}}\left[\dot{s}(\alpha) - D(\dot{\alpha})\right] = 0. \tag{31.6}$$

The kinetic equation, derived from the variational principle (31.6) and describing relaxation of the thermodynamical system to equilibrium state, can be written in the form:

$$\frac{\mathrm{d}}{\mathrm{d}t}s(\alpha) = 2D(\dot{\alpha}), \tag{31.7}$$

which is satisfied by the symmetry principle for kinetic coefficients.

31.4 Variational Principle for Mechanical Systems with Dissipation

As it was mentioned, generalization of the motion equation (31.3) in the presence of dissipation is fulfilled by introducing the dissipation function differentiated with respect to velocities into the right part of (31.3), that is, in correspondence with Onsager's symmetry principle for the kinetic coefficients [6], we have

$$\frac{\mathrm{d}}{\mathrm{d}t}\frac{\partial L}{\partial \dot{\mathbf{u}}} + \nabla \frac{\partial L}{\partial \nabla \mathbf{u}} = -\frac{\partial D}{\partial \dot{\mathbf{u}}}. \tag{31.8}$$

It is possible to show that the motion equation in the form (31.8) can be obtained, if Hamilton's variational principle is considered with Lagrangian function in the form

$$L(\dot{\mathbf{u}}, \nabla \mathbf{u}) = K(\dot{\mathbf{u}}) - U(\nabla \mathbf{u}) - \int_0^t D(\dot{\mathbf{u}})\,\mathrm{d}t', \tag{31.9}$$

where the term in the form of a time integral of dissipation function is added, which is different from (31.2). The initial time in the integral (31.9) denoted as 0 corresponds to the time t_1 in the functional (31.1). In particular, if the dissipation function is considered as a quadratic form of deformation velocities

$$D(\nabla \dot{\mathbf{u}}) = \eta' \left(\frac{\partial \dot{u}_i}{\partial x_k} + \frac{\partial \dot{u}_k}{\partial x_i} \right)^2 + \varsigma' \left(\frac{\partial \dot{u}_l}{\partial x_l} \right)^2, \tag{31.10}$$

then the derived motion equation with an account of (31.4) corresponds to the linearized Navier–Stokes equation

$$\rho \frac{\mathrm{d}}{\mathrm{d}t} \dot{\mathbf{u}} - (\lambda + \mu)\Delta \mathbf{u} - \lambda \nabla(\nabla \mathbf{u}) = (\eta' + \varsigma')\Delta \dot{\mathbf{u}} + \varsigma' \nabla(\nabla \dot{\mathbf{u}}), \tag{31.11}$$

where the shear η and volume ς viscosities are determined by the following combination of the constants in (31.10): $\eta' - \varsigma'$ and $(2\varsigma' - \eta')/3$, respectively.

31.5 Generalized Variational Principle for Dissipative Hydrodynamics

The considered example with derivation of the motion equations for dissipative systems on the base of Hamilton's variational principle with Lagrangian (31.10) demonstrates that there exists a possibility to formulate a generalized variational principle for dissipative hydrodynamical systems, and this formulation can be obtained by the simple combination of Hamilton's variational principle (31.1)–(31.2) and Onsager's variational principle (31.6), if the latter is integrated with respect to time and multiplied by temperature. The Lagrangian density in this case can be written in the form

$$L = K - E + T\left[s - \int_0^t D\,\mathrm{d}t' \right] = K - F - T\int_0^t D\,\mathrm{d}t', \tag{31.12}$$

where E and F are the internal (potential for the dissipationless case) and the free energies. In this case, as it was shown in the report, the equation system for dissipative hydrodynamics corresponds to the motion equations for mechanical and thermal fields, derived from stationary condition for action built on the Lagrangian (31.12) with quadratic forms for all the terms. If, in accordance with Biot [2], we consider the field of heat displacements $\mathbf{u}_T$, together with mean mass displacement $\mathbf{u}$, then the temperature is determined as the divergence of the field $\mathbf{u}_T$:

$$T = T_0(1 - \theta \nabla \mathbf{u}_T). \tag{31.13}$$

The dimensionless constant θ is purposely introduced in the definition (31.13) for simplification of the dissipation function. Then in these terms the generalized Lagrangian can be written in the form:

$$L(\dot{\mathbf{u}}, \nabla \mathbf{u}, \nabla \mathbf{u}_T) = K(\dot{\mathbf{u}}) - F(\nabla \mathbf{u}, \nabla \mathbf{u}_T) - T_0 \int_0^t D(\dot{\mathbf{u}}, \dot{\mathbf{u}}_T)\, \mathrm{d}t', \tag{31.14}$$

where the kinetic energy is given by the quadratic form (31.4), the free energy is given by the usual thermo-elasticity quadratic form [8]:

$$2F(\nabla \mathbf{u}, T) = \mu \varepsilon_{ik}^2 + \lambda \varepsilon_{ll}^2 + \kappa \left(\frac{T - T_0}{\theta T_0}\right)^2 + 2\alpha \varepsilon_{ll} \left(\frac{T - T_0}{\theta T_0}\right) \tag{31.15}$$

(with substitution of expression (31.13)) in temperature terms), and the dissipation function is a square of the difference between mean mass and heat displacements

$$2D(\dot{\mathbf{u}}, \dot{\mathbf{u}}_T) = \beta(\dot{\mathbf{u}} - \dot{\mathbf{u}}_T)^2. \tag{31.16}$$

Then the motion equations for the mean displacement field and for temperature field derived on the base of generalized variational principle coincide practically with the linearized traditional hydrodynamical system:

$$\rho_0 \frac{\mathrm{d}}{\mathrm{d}t} \dot{\mathbf{u}} - \mu \Delta \mathbf{u} - (\lambda + \mu + \alpha) \nabla(\nabla \mathbf{u}) = \frac{\alpha + \kappa}{\theta T_0} \nabla T, \tag{31.17}$$

$$\beta(\dot{T} - T_0 \theta \nabla \dot{\mathbf{u}}) - \kappa \Delta T = \alpha T_0 \theta \Delta(\nabla \mathbf{u}). \tag{31.18}$$

31.6 Comparison with the System of Hydrodynamics Equations

The coefficients of the quadratic forms in (31.15) and (31.16) can be determined by comparison of the equation system (31.17)–(31.18) with the linearized hydrodynamics equation system in the variables $\mathbf{u}$, T:

$$\rho = \rho_0(1 - \nabla \mathbf{u}),$$

$$\rho_0 \frac{\mathrm{d}^2 \mathbf{u}}{\mathrm{d}t^2} - \rho_0 c_0^2 \Delta \mathbf{u} = -\rho_0 \tilde{\alpha} \nabla T + \eta \Delta \dot{\mathbf{u}} + \left(\zeta + \frac{\eta}{3}\right) \nabla(\nabla \dot{\mathbf{u}}), \tag{31.19}$$

$$\rho_0 C_V \frac{\mathrm{d}T}{\mathrm{d}t} + \rho_0 T_0 \tilde{\alpha} \nabla \dot{\mathbf{u}} - \kappa \Delta T = 0. \tag{31.20}$$

In the absence of viscosity $\eta = 0$, $\zeta = 0$, which was not taken into account in the dissipation function (31.16), the structure of (31.17) and (31.18) practically coincides with the second (31.19) and the third (31.20) equations of the hydrodynamics system. The only difference is the additional term in the right part of (31.18) in comparison with (31.20). We briefly note here that the reason for the introduction of this term is related with the generalized form of Fourier law for heat energy flow, and it is discussed in detail in [11]. A direct comparison of the coefficients of (31.17), (31.18) and (31.19), (31.20) with the condition rot $\dot{\mathbf{u}} = 0$ gives us a relationship between them. One only needs to take into account the different dimension of (31.18) and (31.20), and hence the presence of common dimension multiplier when comparing the coefficients for these equations. In the explicit form, the parameters of the quadratic forms are expressed through the known parameters as

$$\beta\frac{\rho_0 c_0^2}{\chi}(\gamma^2-1), \qquad \theta = -\frac{\gamma-1}{\alpha T_0}, \qquad \tilde{\alpha} = \rho_0 c_0^2(\gamma-1),$$
$$\lambda + 2\mu = \rho_0 c_0^2 \gamma, \qquad \tilde{\kappa} = \rho_0 c_0^2(\gamma^2-1),$$

where γ is the specific heat capacity ratio $\gamma = C_P/C_V$, and $\chi = \kappa/\rho_0 C_V$ is the heat conductivity coefficient. It is remarkable that the coefficient of the dissipation function happens to be inversely proportional to the heat conductivity coefficient.

31.7 Mandelshtam–Leontovich Approach for Description of Viscosity Relaxation

Let us consider now how to derive equations accounting for viscosity on the basis of a generalized variational principle. In complete analogy with Mandelshtam–Leontovich approach [1], let us consider together with the basic thermodynamical parameters such as specific volume and temperature (in our description the fields of mean u and heat u_T displacements) some additional (internal) parameters $\{\xi_i\}$, by which the state of the system is characterized in a vicinity of the thermodynamical equilibrium. If there is only one scalar internal parameter, the deviation of which from its thermodynamically equilibrium value is denoted by ξ, then the quadratic form of the free energy expansion can be written as

$$2F(\nabla \mathbf{u}, T, \xi) = \mu\varepsilon_{ik}^2 + \lambda\varepsilon_{ll}^2 + \kappa\left[\frac{T-T_0}{T_0}\right]^2,$$
$$+\, 2\alpha\varepsilon_{ll}\left[\frac{T-T_0}{T_0}\right] a\xi^2 + 2b\xi\varepsilon_{ll} + 2c\xi\left[\frac{T-T_0}{T_0}\right], \qquad (31.21)$$

analogously, the dissipation function will have the form

$$2D(\dot{\mathbf{u}}, \dot{\mathbf{u}}_T, \dot{\xi}) = \beta(\dot{\mathbf{u}} - \dot{\mathbf{u}}_T)^2 + \gamma\dot{\xi}^2. \qquad (31.22)$$

Then, taking the variation by the fields $\mathbf{u}$ and $\mathbf{u}_T$ together with the variation by the field of the internal parameter ξ, we obtain the following system of motion equations

$$\begin{aligned}
&\rho_0 \frac{\mathrm{d}}{\mathrm{d}t}\mathbf{u} - \mu\Delta\mathbf{u} - (\lambda+\mu)\nabla(\nabla\mathbf{u}) - \alpha\nabla(\nabla\mathbf{u}_T) - b\nabla\xi = \beta(\mathbf{u}_T - \mathbf{u}),\\
&\beta(\dot{\mathbf{u}}_T - \dot{\mathbf{u}}) - \kappa\nabla(\nabla\mathbf{u}_T) = \alpha\nabla(\nabla\mathbf{u}) + c\nabla\xi,\\
&\gamma\frac{\mathrm{d}\xi}{\mathrm{d}t} + a\xi + b(\nabla\mathbf{u}) + c(\nabla\mathbf{u}_T) = 0,
\end{aligned} \tag{31.23}$$

where the last equation represents by itself the linear kinetic equation for the internal parameter ξ, which is simply postulated in the Mandelshtam–Leontovich approach [10]. Since in relation to the parameter ξ this equation is an ordinary differential equation of the first order, it can be integrated, and its solution can be represented in the form

$$\xi = -\frac{1}{\gamma}\int_0^t \exp\left\{-\frac{\alpha}{\gamma}(t-t')\right\}\left(b\nabla\mathbf{u} - c\frac{T-T_0}{T_0}\right)\mathrm{d}t'. \tag{31.24}$$

It is seen that this solution describes the relaxation of the internal parameter to the current values of the basic thermodynamical fields. Substituting (31.24) into rested equations of the system (31.23), it is possible to obtain the motion equations which account for the relaxation of the internal parameter:

$$\begin{aligned}
&\rho_0\frac{\mathrm{d}}{\mathrm{d}t}\dot{\mathbf{u}} - \mu\Delta\mathbf{u} + (\mu+\lambda+\tilde{\alpha})\nabla(\nabla\mathbf{u}) + (\alpha+\tilde{\kappa})\nabla T\\
&\quad = \frac{b+c}{a}\int_0^t \exp\left\{-\frac{\alpha}{\gamma}(t-t')\right\}\nabla(b\nabla\mathbf{u} - c\dot{T})\,\mathrm{d}t',
\end{aligned} \tag{31.25}$$

$$\begin{aligned}
&\beta(\dot{T} - \nabla\dot{\mathbf{u}}) - \left(\kappa - \frac{c^2}{a}\right)\Delta T\\
&\quad = \left(\alpha - \frac{bc}{a}\right)\Delta\nabla\mathbf{u} + \frac{c}{a}\int_0^t \exp\left\{-\frac{\alpha}{\gamma}(t-t')\right\}\Delta(b\nabla\mathbf{u} - c\dot{T})\,\mathrm{d}t',
\end{aligned} \tag{31.26}$$

where the following notations are used

$$\tilde{\alpha} = \alpha - \frac{b}{a}(b+c), \qquad \tilde{\kappa} = \kappa - \frac{c}{a}(b+c).$$

Now, it is easy to see that at the long times in comparison with the relaxation time $\tau = \gamma/a$ (or in the limit $\tau \to 0$) the integrand can be taken out in the vicinity of the top integration limit, and the resulting equation in this limiting case can be written in the form

$$\begin{aligned}
&\rho_0\frac{\mathrm{d}}{\mathrm{d}t}\dot{\mathbf{u}} - \mu\Delta\mathbf{u} + (\mu+\lambda+\tilde{\alpha})\nabla(\nabla\mathbf{u}) + (\alpha+\tilde{\kappa})\nabla T\\
&\quad = \gamma(b+c)\nabla(b\nabla\mathbf{u} - c\dot{T}),
\end{aligned} \tag{31.27}$$

$$\beta(\dot{T} - \nabla\dot{\mathbf{u}}) - \left(\kappa - \frac{c^2}{a}\right)\Delta T = \left(\alpha - \frac{bc}{a}\right)\Delta\nabla\mathbf{u} + \gamma c\Delta(b\nabla\mathbf{u} - c\dot{T}). \tag{31.28}$$

It is easy to see that at $c = 0$ and $\mu = 0$ (31.27) coincides in structure with (31.19), so we will have also the relation $\gamma b^2 = \varsigma + (4/3)\eta$.

Thus, the account of internal parameters allows us to introduce naturally the viscous terms in the motion equation, that is usually the principal difficulty in other variational approaches. Moreover, the presence of viscosity relaxation appears to be natural. The introduction of this relaxation for viscosity is required, for example, for finite speed of perturbation propagation in dissipative hydrodynamics.

It should also be mentioned that additional terms in the quadratic forms (31.21)–(31.22) dealt with the internal parameter ξ, assuming its scalar nature. However, such parameters can possess vector or tensor properties. In the latter case, relaxation of viscosity will be attributed to the shear viscosity together with the bulk viscosity.

31.8 Conclusion

The following principal results are obtained in the framework of the given article: The generalization of the Hamilton's and Osager's variational principles for dissipative hydrodynamical systems is represented in terms of the mechanical and heat displacement fields. A system of equations for these fields is derived from the extreme condition for action with a Lagrangian in the form of the difference between the kinetic and free energies minus the time integral of the dissipation function. The generalized hydrodynamic equation system is then evaluated on the basis of the generalized variational principle. At low frequencies, this system corresponds to the traditional Navier–Stokes equation system

It is shown how to introduce a viscosity into the fluid motion equation on the basis of the generalized variational principle. An internal parameter is used for the description of quasi-equilibrium state in analogy to Mandelshtam–Leontovich approach. The derived motion equation describes viscosity relaxation and generalizes the well known Navier–Stokes equation.

Acknowledgements The work was supported by ISTC grant 3691 and by RFBR grant 09-02-00927-a.

References

1. Biot, M.A.: Theory of propagation of elastic waves in fluid-saturated porous solid. I. Low-frequency range. J. Acoust. Soc. Am. **28**, 168–178 (1956)
2. Biot, M.: Variational principles in theory of heat transfer. Mosc. Energy (1975)
3. Deresiewicz, H.: Plane wave in a thermoelastic solids. J. Acoust. Soc. Am. **29**, 204–209 (1957)
4. Glensdorf, P., Prigogine, I.: Thermodynamic Theory of Structure, Stability and Fluctuations. New York (1971)
5. Gyarmati, I.: Non-Equilibrium Thermodynamics. Field Theory and Variational Principles. Springer, Berlin (1970)
6. Landau, L.D., Lifshitz, E.M.: Theoretical Physics. Statistical Physics, vol. 5. Nauka, Moscow (1964)

7. Landau, L.D., Lifshitz, E.M.: Theoretical Physics. Hydrodynamics, vol. 6. Nauka, Moscow (1986)
8. Landau, L.D., Lifshitz, E.M.: Theoretical Physics. Theory of Elasticity, vol. 7. Nauka, Moscow (1972)
9. Lykov, A.V.: Theory of Heat Conduction. Vysshaya Shkola, Moscow (1967)
10. Mandelshtam, L.I., Leontovich, M.A.: To the sound absorption theory in liquids. J. Exp. Theor. Phys. **7**(3), 438–444 (1937). In Russian
11. Martynov, G.A.: Hydrodanamic theory of sound wave propagation. Theor. Math. Phys. **129**, 1428–1438 (2001)
12. Maximov, G.A.: On the variational principle for dissipative hydrodynamics. Preprint 006-2006, Moscow, MEPhI (2006)
13. Maximov, G.A.: Generalized variational principle for dissipative hydrodynamics and its application to the Biot's equations for multicomponent, multiphase media with temperature gradient. In: Weis, B.N. (ed.) New Research in Acoustics, pp. 21–61. Nova Science, New York (2008). ISNB:978-1-60456-403-7
14. Nettleton, R.E.: Relaxation theory of thermal conduction in liquids. Phys. Fluids **3**, 216–223 (1960)
15. Onsager, L.: Reciprocal relations in irreversible process I. Phys. Rev. **37**, 405–426 (1931)
16. Onsager, L.: Reciprocal relations in irreversible process II. Phys. Rev. **38**, 2265–2279 (1931)

Chapter 32
Cosserat Continua Described by Mesoscopic Theory

Wolfgang Muschik and Christina Papenfuss

Abstract Beyond the usual 5-field theory (the basic fields are the mass density, velocity, internal energy), additional variables are needed for the unique description of complex media. Beside the conventional method of introducing additional fields by their balances, another procedure, the mesoscopic theory, is here discussed and applied to Cosserat continua.

32.1 Introduction

Continuum mechanics is based on the balance equations of mass, momentum, angular momentum or spin, total or kinetic energy, and internal energy. Additionally, one has to consider the balance of entropy for taking into account the second law. In non-relativistic physics, all these balances are defined with respect to time and position (x, t). Beyond the quantities whose balance equations are mentioned above, complex materials need more variables for their unique description. Examples for these additional quantities are internal variables, order and damage parameters, Cosserat triads, directors and alignment and conformation tensors.

In principle, there are two possibilities to include these additional quantities into the continuum theoretical description: One can introduce additional fields and their balance equations defined with respect to (x, t) as in the case of Cosserat continua, or the additional quantities, the so-called mesoscopic variables, are introduced as variables extending space–time to the so-called mesoscopic space. This description introducing the mesoscopic space is called the mesoscopic theory. According to its

W. Muschik (✉)
Institut für Theoretische Physik, TU Berlin, Hardenbergstr. 36, 10623 Berlin, Germany
e-mail: muschik@physik.tu-berlin.de

C. Papenfuss
e-mail: c.papenfuss@gmx.de

G.A. Maugin, A.V. Metrikine (eds.), *Mechanics of Generalized Continua*, 307
Advances in Mechanics and Mathematics 21,
DOI 10.1007/978-1-4419-5695-8_32, © Springer Science+Business Media, LLC 2010

construction, the dimension of the mesoscopic space is equal to the number of the mesoscopic variables plus four.

The balances of mass, momentum, etc. can be easily written down on the mesoscopic space because only the number of dimensions changed in comparison with the usual balances on space-time. The different values of the mesoscopic variables in a given volume element are described by a mesoscopic distribution function (MDF), representing the statistical tool of the mesoscopic description. The moments of the MDF, generated by integration over the mesoscopic variables, are additional macroscopic fields which are beyond the fields appearing in the balance equations. The Cosserat continuum is generated by integrating the mesoscopic balances over the mesoscopic variables.

32.2 The Mesoscopic Concept

In principle, there are two possibilities to include additional quantities into the continuum theoretical description [21]. One can introduce additional fields and their balance equation defined with respect to $(\mathbf{x}, t) \in R^3 \times R^1$, or the additional quantities are introduced as variables extending $R^3 \times R^1$ to the so-called *mesoscopic space* on which now the balances of mass, momentum, etc. are defined. The first possibility for describing complex materials by introducing additional fields has a long history in continuum mechanics. Starting out with the first contribution of the Cosserat brothers [6, 7], the development of mechanics of generalized continua is lasting even today [12, 13, 16, 5]. The second possibility to introduce the mesoscopic space is called the *mesoscopic concept* which stems historically from the theory of liquid crystals [9, 11, 18, 15, 25, 14, 17] by taking the orientation distribution function of the molecules into consideration [8, 2, 4].

As discussed above, the mesoscopic concept introduces the mesoscopic space

$$(\mathbf{m}, \mathbf{x}, t) \in \mathscr{M} \times R^3 \times R^1 \tag{32.1}$$

on which the balances are defined. Here $\mathbf{m} \in \mathscr{M}$ is a set of mesoscopic variables which is an element of a suitable manifold $\mathscr{M}$ on which an integration can be defined.

Beyond the use of additional variables $\mathbf{m}$, the mesoscopic concept introduces a statistical element, the so-called *mesoscopic distribution function* (MDF) $f(\mathbf{m}, \mathbf{x}, t)$ generated by the different values of the mesoscopic variables of the molecules in a volume element

$$f(\mathbf{m}, \mathbf{x}, t) \equiv f(\cdot), \quad (\cdot) \equiv (\mathbf{m}, \mathbf{x}, t) \in \mathscr{M} \times R^3 \times R^1. \tag{32.2}$$

The MDF is defined on the mesoscopic space $\mathscr{M} \times R^3 \times R^1$ describing the distribution of $\mathbf{m}$ in a volume element around $\mathbf{x}$ at time t, and therefore it is always normalized

$$\int f(\mathbf{m}, \mathbf{x}, t)\, \mathrm{d}\mathscr{M} = 1. \tag{32.3}$$

32.3 Local Mesoscopic Balances

Starting out with a usual local balance equation on $R^3 \times R^1$

$$\frac{\partial}{\partial t}\mathbf{X}(\mathbf{x},t) + \nabla_x \cdot \big[\mathbf{v}(\mathbf{x},t)\mathbf{X}(\mathbf{x},t)\big] = \boldsymbol{\Sigma}(\mathbf{x},t), \tag{32.4}$$

we introduce the mesoscopic part by replacing $R^3 \times R^1$ with $\mathscr{M} \times R^3 \times R^1$ and by adding the differentiated part belonging to $\mathscr{M}$. Thus, we obtain the shape of a local mesoscopic balance

$$\frac{\partial}{\partial t}\mathbf{X}(\cdot) + \nabla_x \cdot \big[\mathbf{v}(\cdot)\mathbf{X}(\cdot) - \mathbf{S}(\cdot)\big] + \nabla_m \cdot \big[\mathbf{u}(\cdot)\mathbf{X}(\cdot) - \mathbf{R}(\cdot)\big] = \boldsymbol{\Sigma}(\cdot). \tag{32.5}$$

The special balances are obtained by a special physical identification of $\mathbf{X}(\cdot)$, $\mathbf{S}(\cdot)$, $\mathbf{R}(\cdot)$ and $\boldsymbol{\Sigma}(\cdot)$. Beside the mesoscopic velocity $\mathbf{v}(\cdot)$, the *mesoscopic change velocity* $\mathbf{u}(\cdot)$ occurs, which is defined as follows

$$(\mathbf{m},\mathbf{x},t) \longrightarrow \big(\mathbf{m}+\mathbf{u}(\cdot)\Delta t, \mathbf{x}+\mathbf{v}(\cdot)\Delta t, t+\Delta t\big). \tag{32.6}$$

In particular, we obtain the following local mesoscopic balance equations:

1. *Mass*

$$\frac{\partial}{\partial t}\varrho(\cdot) + \nabla_x \cdot \{\mathbf{v}(\cdot)\varrho(\cdot)\} + \nabla_m \cdot \{\mathbf{u}(\cdot)\varrho(\cdot)\} = 0, \tag{32.7}$$

2. *Momentum*

$$\begin{aligned}\frac{\partial}{\partial t}[\varrho(\cdot)\mathbf{v}(\cdot)] &+ \nabla_x \cdot \big[\mathbf{v}(\cdot)\varrho(\cdot)\mathbf{v}(\cdot) - \mathbf{T}^\top(\cdot)\big]\\ &+ \nabla_m \cdot \big[\mathbf{u}(\cdot)\varrho(\cdot)\mathbf{v}(\cdot) - \mathscr{T}^\top(\cdot)\big] = \varrho(\cdot)\mathbf{k}(\cdot),\end{aligned} \tag{32.8}$$

3. *Angular Momentum and Spin*

$$\mathbf{M}(\cdot) := \mathbf{x} \times \mathbf{v}(\cdot) + \mathbf{s}(\cdot), \quad \mathbf{s}(\cdot) := \text{mesoscopic specific spin}, \tag{32.9}$$

$$\begin{aligned}\frac{\partial}{\partial t}\big[\varrho(\cdot)\mathbf{s}(\cdot)\big] &+ \nabla_x \cdot \big[\mathbf{v}(\cdot)\varrho(\cdot)\mathbf{s}(\cdot) - \mathbf{W}(\cdot)^T\big]\\ &+ \nabla_m \cdot \big[\mathbf{u}(\cdot)\varrho(\cdot)\mathbf{s}(\cdot) - \mathscr{W}(\cdot)^T\big] = \boldsymbol{\varepsilon} : \mathbf{T}(\cdot) + \varrho(\cdot)\mathbf{g}(\cdot).\end{aligned} \tag{32.10}$$

The meaning of the used quantities is as follows: $\mathbf{T}$ is the mesoscopic Cauchy stress tensor, $\mathscr{T}$ its analogue on the mesoscopic space. $\mathbf{k}$ is the mesoscopic force density and $\mathbf{g}$ the mesoscopic angular momentum exerted by the external forces on $\mathbf{m}$. $\mathbf{W}$ is the mesoscopic couple stress and $\mathscr{W}$ its analogue acting on the mesoscopic variables $\mathbf{m}$.

The mesoscopic balances can be written down very easily because only the number of dimensions differ with respect to the macroscopic balances. Consequently, a setting of new balances is not necessary: the quantities beyond the 5-field theory are included in the mesoscopic space as variables. Of course, not all variables beyond 5-field theory are mesoscopic ones. If there is no mesoscopic distribution function for a special quantity, it is not a mesoscopic variable and cannot be treated with the mesoscopic theory.

32.4 Cosserat Media

Finally, we are not interested in the mesoscopic fields, but we have to create macroscopic ones from them. To do so, it is obvious that we have to integrate over the mesoscopic variables to obtain the macroscopic balances of Cosserat media

$$\int \cdots \mathrm{d}\mathscr{M} \longrightarrow \text{balances of Cosserat media.} \tag{32.11}$$

Taking into account that the integral over the divergence with respect to the mesoscopic variables vanish [2], i.e.,

$$\int \nabla_m \cdot [\cdots] \, \mathrm{d}\mathscr{M} = 0, \tag{32.12}$$

we obtain by integrating the mesoscopic mass balance (32.7) the following mesoscopic definitions of the macroscopic mass density and velocity

$$\varrho(\mathbf{x},t) := \int \varrho(\cdot) \, \mathrm{d}\mathscr{M}, \qquad \mathbf{v}(\mathbf{x},t) := \int f(\cdot)\mathbf{v}(\cdot) \, \mathrm{d}\mathscr{M}. \tag{32.13}$$

In the same way by integrating balances, we obtain the mesoscopic definitions of the macroscopic spin, stress tensor, couple stress, and external momentum

$$\mathbf{s}(\mathbf{x},t) := \int f(\cdot)\mathbf{s}(\cdot) \, \mathrm{d}\mathscr{M}, \tag{32.14}$$

$$\mathbf{T}(\mathbf{x},t) := \int [\mathbf{T}(\cdot) - \mathbf{v}(\cdot)\varrho(\cdot)\mathbf{v}(\cdot)] \, \mathrm{d}\mathscr{M} + \mathbf{v}(\mathbf{x},t)\varrho(\mathbf{x},t)\mathbf{v}(\mathbf{x},t), \tag{32.15}$$

$$\mathbf{W}(\mathbf{x},t) := \int [\mathbf{W}(\cdot) - \mathbf{s}(\cdot)\varrho(\cdot)\mathbf{v}(\cdot)] \, \mathrm{d}\mathscr{M} + \mathbf{s}(\mathbf{x},t)\varrho(\mathbf{x},t)\mathbf{v}(\mathbf{x},t), \tag{32.16}$$

$$\varrho(\mathbf{x},t)\mathbf{g}(\mathbf{x},t) := \int [\varrho(\cdot)\mathbf{g}(\cdot) + \boldsymbol{\varepsilon} : \mathbf{T}(\cdot) - \nabla_m \cdot \mathscr{W}^{\top}(\cdot)] \, \mathrm{d}\mathscr{M} - \boldsymbol{\varepsilon} : \mathbf{T}(\mathbf{x},t). \tag{32.17}$$

Interesting is that the mesoscopic stress tensors and the mesoscopic external moments are not additive forming the macroscopic quantities.

The macroscopic spin balance which is the main equation of Cosserat continua results from integration of (32.10) [1]

$$\frac{\partial}{\partial t}[\varrho(\mathbf{x},t)\mathbf{s}(\mathbf{x},t)] + \nabla_x \cdot [\mathbf{v}(\mathbf{x},t)\varrho(\mathbf{x},t)\mathbf{s}(\mathbf{x},t) - \mathbf{W}^T(\mathbf{x},t)] \tag{32.18}$$

$$= \boldsymbol{\varepsilon} : \mathbf{T}(\mathbf{x},t) + \varrho(\mathbf{x},t)\mathbf{g}(\mathbf{x},t). \tag{32.19}$$

With this, the mesoscopic background of the Cosserat continua is elucidated.

32.5 Entropy Balance

Because of the second law which can be formulated only macroscopically, the entropy balance is only interesting in its macroscopic form. It is written

$$\frac{\partial}{\partial t}\big[\varrho(\mathbf{x},t)\eta(\mathbf{x},t)\big] + \nabla_x \cdot \big[\varrho(\mathbf{x},t)\eta(\mathbf{x},t)\mathbf{v}(\mathbf{x},t) + \phi(\mathbf{x},t)\big]$$
$$= \zeta(\mathbf{x},t) + \sigma(\mathbf{x},t). \tag{32.20}$$

Here, the fields have the following meaning: $\eta(\mathbf{x},t)$ = is specific entropy, $\phi(\mathbf{x},t)$ = entropy flux density, $\zeta(\mathbf{x},t)$ = entropy supply, $\sigma(\mathbf{x},t)$ = entropy production density. The second law is now expressed by the *dissipation inequality*

$$\sigma(\mathbf{x},t) \geq 0 \tag{32.21}$$

which has to be taken into account for writing down constitutive equations [19, 26].

32.6 Distribution Function Balance

Taking $(32.13)_1$ and (32.3) into account, we obtain

$$\varrho(\cdot) = \varrho(\mathbf{x},t)f(\cdot). \tag{32.22}$$

The macroscopic mass balance results from integrating (32.7) over the mesoscopic variables by taking (32.12) into account

$$\frac{\partial}{\partial t}\varrho(\mathbf{x},t) + \nabla_x \cdot \big\{\mathbf{v}(\mathbf{x},t)\varrho(\mathbf{x},t)\big\} = 0. \tag{32.23}$$

The macroscopic mass balance induces a differential equation for the mesoscopic distribution function [20]

$$\frac{\partial}{\partial t}f(\cdot) + \nabla_x \cdot \big[\mathbf{v}(\cdot)f(\cdot)\big] + \nabla_m \cdot \big[\mathbf{u}(\cdot)f(\cdot)\big]$$
$$+ f(\cdot)\left[\frac{\partial}{\partial t} + \mathbf{v}(\cdot)\cdot\nabla_x\right]\ln\varrho(\mathbf{x},t) = 0. \tag{32.24}$$

This differential equation for the MDF is a derived one, and consequently, it does not represent an approximation or an ad-hoc equation in the frame of the mesoscopic theory. From the mathematical point of view, (32.24) is an integro-differential equation due to the appearance of the macroscopic mass density which is an integral over the mesoscopic mass density according to $(32.13)_1$. Form a physical point of view, (32.24) is a mean field equation because the mesoscopic distribution function is also determined by the macroscopic mass density.

32.7 Beyond Cosserat, the Order Parameters

Cosserat continua are characterized by a non-trivial macroscopic spin balance (32.10) with non-vanishing couple stress which can be derived mesoscopically. But beyond Cosserat, the mesoscopic theory creates macroscopic balances by the moments of the MDF. These moments generate the so-called family of the order parameters

$$1 = \int f(\cdot)\,\mathrm{d}\mathscr{M}, \tag{32.25}$$

$$\mathbf{A}(\mathbf{x},t) := \int f(\cdot)\mathbf{m}\,\mathrm{d}\mathscr{M}, \tag{32.26}$$

$$\mathbf{a}(\mathbf{x},t) := \int f(\cdot)\,\overline{\mathbf{m}\mathbf{m}}\,\mathrm{d}\mathscr{M}, \tag{32.27}$$

$$\underline{\mathbf{a}}_N(\mathbf{x},t) := \int f(\cdot)\,\overline{\overbrace{\mathbf{m}\cdots\mathbf{m}}^{N \text{ times}}}\,\mathrm{d}\mathscr{M}. \tag{32.28}$$

Here $\overline{\mathbf{A}\cdots\mathbf{B}}$ denotes a symmetric and traceless tensor. As an example, $\mathbf{a}(\mathbf{x},t)$ is the alignment tensor of liquid crystal theory which is beyond the theory of Cosserat continua.

32.8 Remarks on Constitutive Theory

All balance equations contain constitutive equations whose choice makes the system of balances mathematically complete for solution and which determine the material for which the considered system of balance equations is valid. The domain of the constitutive equations is called the *state space* or the *constitutive space* which characterizes the material and which has to be chosen. The mesoscopic constitutive theory is up to now poorly developed. The reason for that is that there are different kinds of state spaces whose final choice is difficult. There are

1. Purely mesoscopic state spaces without any mean field influence

$$\mathscr{Z} = \left(\mathbf{m}, \varrho(\cdot),\ \overline{\nabla\mathbf{v}(\cdot)},\ \left[\mathbf{u}(\cdot)\right]^{\mathrm{obj}}, \varepsilon(\cdot)\right); \tag{32.29}$$

2. Purely macroscopic state spaces whose macroscopic variables are defined by a mesoscopic background [3]

$$\mathscr{Z} = \left(\varrho(\mathbf{x},t)\ \ \overline{\nabla\mathbf{v}(\mathbf{x},t)}, \varepsilon(\mathbf{x},t), \mathbf{a}(\mathbf{x},t), \underline{\mathbf{a}}_4(\mathbf{x},t), \nabla\mathbf{a}(\mathbf{x},t)\right), \tag{32.30}$$

and finally we have

3. Mixed state spaces [10]

$$\mathscr{Z} = \left(\mathbf{m}, \varrho(\cdot),\ \overline{\nabla\mathbf{v}(\cdot)}, \nabla_n \ln\varrho(\cdot),\ \overline{\nabla\mathbf{v}(\mathbf{x},t)}, \mathbf{a}(\mathbf{x},t)\right). \tag{32.31}$$

Often it is difficult to find the suitable state space describing a special material. That is one of the reasons why the mesoscopic constitutive theory is poorly developed.

32.9 Summary

Cosserat continua have a mesoscopic background because the macroscopic spin balance including the couple stress can be derived by using a mesoscopic description. The former is characterized by introducing the mesoscopic space, the product of space–time and the mesoscopic variables, and by introducing the mesoscopic distribution function defined on the mesoscopic space describing the distribution of the values of the mesoscopic variables around a point in space–time. No additional balances for the mesoscopic variables are necessary because they are introduced as variables of the mesoscopic space and not as additional fields which need also additional, first of all, unknown balances. Instead of them, the mesoscopic description needs a model for the transient behavior of the mesoscopic variables, the mesoscopic change velocity. Having solved the mesoscopic balances by introducing constitutive equations (a difficult task), the macroscopic quantities, the only that can be measured, are derived by integrating over the mesoscopic variables. Beyond Cosserat, additional macroscopic fields, the moments of the mesoscopic distribution function, the so-called order parameters can be derived which includes the additional information stemming from the mesoscopic background.

Successful application of the mesoscopic theory is always possible, if mesoscopic variables can be defined. Up to now, liquid crystals [22], micro-cracks [27, 24] and ferrofluids [23] are mesoscopically described.

References

1. Blenk, S., Muschik, W.: Orientational balances for nematic liquid crystals. J. Non-Equilib. Thermodyn. **16**, 67–87 (1991)
2. Blenk, S., Ehrentraut, H., Muschik, W.: Statistical foundation of macroscopic balances for liquid crystals in alignment tensor formulation. Physica A **174**, 119–138 (1991)
3. Blenk, S., Ehrentraut, H., Muschik, W.: Macroscopic constitutive equations for liquid crystals induced by their mesoscopic orientation distribution. Int. J. Eng. Sci. **30**, 1127–1143 (1992)
4. Blenk, S., Ehrentraut, H., Muschik, W.: A continuum theory for liquid crystals describing different degrees of orientational order. Liquid Cryst. **14**, 1221–1226 (1993)
5. Capriz, G.: Continua with Microstructure. Springer, Berlin (1989)
6. Cosserat, E., Cosserat, F.: Sur la mécanique générale. Acad. Sci. Paris **145**, 1139–1142 (1907)
7. Cosserat, E., Cosserat, F.: Théorie des Corps Déformables. Hermann, Paris (1909)
8. Condiff, D.W., Brenner, H.: Transport mechanics in systems of orientable particles. Phys. Fluids **12**, 539 (1969)
9. De Gennes, P.G., Prost, J.: The Physics of Liquid Crystals. Clarendon, Oxford (1993). Sect. 2.1
10. Ehrentraut, H., Muschik, W., Papenfuss, C.: Mesoscopically derived orientation dynamics of liquid crystals. J. Non-Equilib. Thermodyn. **22**, 285–298 (1997)

11. Ericksen, J.L.: Anisotropic fluids. Arch. Ration. Mech. Anal. **4**, 231 (1960)
12. Eringen, A.C., Suhubi, E.S.: Nonlinear theory of simple microelastic solids I. Int. J. Eng. Sci. **2**, 189 (1964)
13. Eringen, A.C., Suhubi, E.S.: Nonlinear theory of simple microelastic solids II. Int. J. Eng. Sci. **2**, 389 (1964)
14. Eringen, A.C.: Continuum theory of cholesteric liquid crystals. In: Johnson, J.F., Porter, R.S. (eds.) Liquid Crystals and Ordered Fluids, vol. 3, pp. 443–474. Plenum, New York (1974)
15. Eringen, A.C., Lee, J.D.: Relation of two continuum theories of liquid crystals. In: Johnson, J.F., Porter, R.S. (eds.) Liquid Crystals and Ordered Fluids, vol. 2, pp. 315–330. Plenum, New York (1974)
16. Eringen, A.C., Kafadar, C.B.: Polar field theories. In: Eringen, A.C. (ed.) Continuum Physics, vol. IV, pp. 1–73. Academic Press, New York (1976)
17. Hess, S.: Irreversible thermodynamics of nonequilibrium phenomena in molecular liquids and in liquid crystals. Z. Naturforsch. **30a**, 728 (1975)
18. Leslie, F.J.: Some constitutive equations for liquid crystals. Arch. Ration. Mech. Anal. **28**, 265 (1965)
19. Muschik, W.: Formulations of the second law—Recent developments. J. Phys. Chem. Solids **49**, 709–720 (1988)
20. Muschik, W., Ehrentraut, H., Papenfuss, C.: Mesoscopic continuum mechanics. In: Maugin, G.A. (ed.) Geometry, Continua and Microstructure. Collection Travaux en Cours, vol. 60, pp. 49–60. Herrman, Paris (1999)
21. Muschik, W., Ehrentraut, H., Papenfuss, C.: Concepts of mesoscopic continuum physics. J. Non-Equilib. Thermodyn. **25**, 179 (2000)
22. Muschik, W., Papenfuss, C., Ehrentraut, H.: Sketch of the mesoscopic description of nematic liquid crystals. J. Non-Newton. Fluid Mech. **119**, 91 (2004)
23. Papenfuss, C., Ciancio, V., Rogolino, P.: Application of the mesoscopic theory to dipolar media. Tech. Mech. **22**, 132–140 (2002)
24. Papenfuss, C., Van, P., Muschik, W.: Mesoscopic theory of microcracks. Arch. Mech. **55**, 481–499 (2003)
25. Stephen, M.J., Straley, J.P.: Physics of liquid crystals. Rev. Mod. Phys. **46**, 617 (1974)
26. Triani, V., Papenfuss, C., Cimmelli, V., Muschik, W.: Exploitation of the second law: Coleman-Noll and Liu procedure in comparison. J. Non-Equilib. Thermodyn. **33**, 47–60 (2008)
27. Van, P., Papenfuss, C., Muschik, W.: Mesoscopic dynamics of microcracks. Phys. Rev. E **62**, 6206 (2000)

Chapter 33
Fractal Solids, Product Measures and Continuum Mechanics

Jun Li and Martin Ostoja-Starzewski

Abstract This paper builds on the recently begun extension of continuum thermomechanics to fractal porous media which are specified by a mass (or spatial) fractal dimension D, a surface fractal dimension d, and a resolution length scale R. The focus is on pre-fractal media (i.e., those with lower and upper cut-offs) through a theory based on a dimensional regularization, in which D is also the order of fractional integrals employed to state global balance laws. In effect, the governing equations may be cast in forms involving conventional (integer-order) integrals, while the local forms are expressed through partial differential equations with derivatives of integer order but containing coefficients involving D, d and R. The formulation allows a generalization of the principles of virtual work, and virtual stresses, which, in turn, allow us to extend the extremum and variational theorems of elasticity and plasticity, as well as handle flows in fractal porous media. In all the cases, the derived relations depend explicitly on D, d and R, and, upon setting $D = 3$ and $d = 2$, reduce to conventional forms of governing equations for continuous media with Euclidean geometries. While the original formulation was based on a Riesz measure—and thus more suited to isotropic media—the new model is based on a product measure making it capable of grasping local material anisotropy. The product measure allows one to grasp the anisotropy of fractal dimensions on a mesoscale and the ensuing lack of symmetry of the Cauchy stress, as noted in condensed matter physics. On this basis, a framework of micropolar mechanics of fractal media is developed.

J. Li (✉) · M. Ostoja-Starzewski
University of Illinois at Urbana-Champaign, Urbana, IL 61801, USA
e-mail: junli3@illinois.edu

M. Ostoja-Starzewski
e-mail: martinos@illinois.edu

G.A. Maugin, A.V. Metrikine (eds.), *Mechanics of Generalized Continua*, Advances in Mechanics and Mathematics 21,
DOI 10.1007/978-1-4419-5695-8_33, © Springer Science+Business Media, LLC 2010

33.1 Background

Mandelbrot's seminal work on fractals [11] was first followed, outside mathematics itself, by the condensed matter physicists who focused on the effects of fractal geometries on bulk material responses, e.g., [5]. That work concentrated on explaining physical phenomena and properties of materials whose fractal (non-Euclidean) geometry plays a key role, but a field theory—as an analogue of continuum physics/mechanics—has been sorely lacking. Some progress in that respect has recently been made by mathematicians [8, 19] looking at classical problems, like Laplace's or heat equation, on fractal (albeit self-similar and non-random) sets. Also, various specialized models have also been developed for particular problems like wave scattering at fractals [1], fracture mechanics [3], or geomechanics [4]. A new step in the direction of continuum physics and mechanics, relying on dimensional regularization, was taken by Tarasov [20–22]. He developed continuum-type equations of conservation of mass, momenta, and energy for fractal porous media, and on that basis studied several fluid mechanics and wave motion problems. In principle, one can then map a mechanics problem of a fractal (which is described by its mass (D) and surface (d) fractal dimensions plus the spatial resolution (R)) onto a problem in the Euclidean space in which this fractal is embedded, while having to deal with coefficients explicitly involving D, d and R. As it turns out, D is also the order of fractional integrals employed to state global balance laws. This approach's great promise stems from the fact that much of the framework of continuum mechanics/physics may be generalized and partial differential equations (with derivatives of integer order) may still be employed [13, 14].

Whereas the original formulation of Tarasov was based on the Riesz measure—and thus more suited to isotropic media—the model proposed here is based on a product measure introduced very recently [18, 9]. That measure grasps the anisotropy of fractal geometry (i.e., different fractal dimensions in different directions) on mesoscale, which, in turn, leads to asymmetry of the Cauchy stress. This leads to a framework of micropolar mechanics of fractal materials, formulated here for the case of small strains and rotations. While all the derived relations depend explicitly on D, d and R, upon setting $D = 3$ and $d = 2$, they reduce to conventional forms of governing equations for continuous media with Euclidean geometries. Prior research has already involved an extension to continuum thermomechanics and fracture mechanics, a generalization of extremum and variational principles, and turbulent flows in fractal porous media [15–17].

33.2 Product Measures and Basic Integral Theorems

We begin with a fractal material whose mass m obeys a power law

$$m(R) = kR^D, \quad D < 3. \tag{33.1}$$

Here R is the length scale of measurement (or resolution), D is the fractal dimension of mass, and k is a proportionally constant. Certainly, (33.1) can be applied to a pre-fractal, i.e., a fractal-type, physical object with lower and upper cut-offs. Next, we use a fractional integral to represent mass in a 3D region

$$m(W) = \int_W \rho(\mathbf{r})\,\mathrm{d}V_D = \int_W \rho(\mathbf{r}) c_3(D,R)\,\mathrm{d}V_3, \tag{33.2}$$

where the first and second equalities, respectively, involve fractional (Riesz-type) integrals and conventional integrals, while the coefficient $c_3(D,R)$ provides a transformation between the two [20–22]. In order to deal with generally anisotropic rather than isotropic media, we replace (33.1) by a more general power law relation with respect to each coordinate

$$m(R) \sim x_1^{\alpha_1} x_2^{\alpha_2} x_3^{\alpha_3},$$

whereby the mass distribution is specified via a product measure

$$m(x_1,x_2,x_3) = \iiint_W \rho(x_1,x_2,x_3)\,\mathrm{d}\mu(x_1)\,\mathrm{d}\mu(x_2)\,\mathrm{d}\mu(x_3). \tag{33.3}$$

Here the length measurement in each coordinate is provided by

$$\mathrm{d}\mu(x_k) = c_1^{(k)}(\alpha_k, x_k)\,\mathrm{d}x_k, \quad k = 1,2,3. \tag{33.4}$$

Then, the total fractal dimension D of mass m is $\alpha_1 + \alpha_2 + \alpha_3$, while

$$c_3 = c_1^{(1)} c_1^{(2)} c_1^{(3)} = \prod_{i=1}^{3} c_1^{(i)}. \tag{33.5}$$

For the surface coefficient (c_2) we typically consider a cubic volume element, whose each surface element is specified by the normal vector (along axes; Fig. 33.1). Therefore, $c_2^{(k)}$ associated with the surface element $S_d^{(k)}$ is

$$c_2^{(k)} = c_1^{(i)} c_1^{(j)} = c_3 / c_1^{(k)}, \quad i \neq j \text{ and } i,j \neq k. \tag{33.6}$$

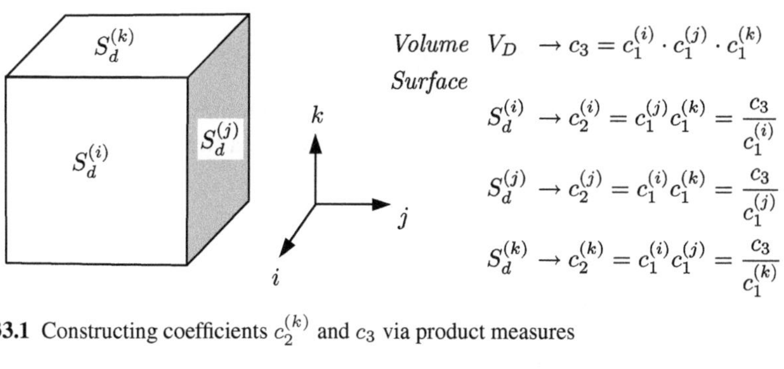

Fig. 33.1 Constructing coefficients $c_2^{(k)}$ and c_3 via product measures

We adopt a modified Riemann–Liouville fractional integral of Jumarie [6, 7], so that

$$c_1^{(k)} = \alpha_k (l_k - x_k)^{\alpha_k - 1}, \quad k = 1, 2, 3, \tag{33.7}$$

where l_k is the total length (integral interval) along x_k. Note that, for a point mass $\rho(x_1, x_2, x_3) = m_0 \delta(x_1)\delta(x_2)\delta(x_3)$, when $D \to 3$ (α_1, α_2, $\alpha_3 \to 1$), then $m(W) = \alpha_1 \alpha_2 \alpha_3 l^{D-3} m_0 \to m_0$, and so the conventional concept of a point mass is recovered [23]. However, the Riesz fractional integral has a non-smooth transition of mass with respect to its fractal dimension ($m(W) = 0$ except when $D = 3$, $m(W) = m_0 0^{D-3} = m_0$).

One can now obtain a fractional Gauss theorem via dimensional regularization, formulated within the framework of product measures discussed above,

$$\begin{aligned} \int_{\partial W} f_k n_k \, \mathrm{d}S_d &= \int_W \left(f_k c_2^{(k)}\right)_{,k} c_3^{-1} \, \mathrm{d}V_D \\ &= \int_W f_{k,k} \, c_2^{(k)} c_3^{-1} \, \mathrm{d}V_D = \int_W \frac{f_{k,k}}{c_1^{(k)}} \, \mathrm{d}V_D. \end{aligned} \tag{33.8}$$

It is important to observe three properties of the (fractal gradient) operator $\nabla_k^D = (\cdot)_{,k} / c_1^{(k)}$ under product measures:

(1) It is the "inverse" operator of fractional integrals.
(2) The rule of "term-by-term" differentiation is satisfied, i.e.,

$$\nabla_k^D (AB) = \nabla_k^D (A) B + A \nabla_k^D (B).$$

(3) Its operation on any constant is zero, which indeed is a desired property not possessed by the usual fractional derivative of Riemann–Liouville [12].

While we note that Properties (2) and (3) do not hold in Tarasov's Riesz measure formulations, the fractional generalization of Reynold's transport theorem is

$$\frac{\mathrm{d}}{\mathrm{d}t} \int_W P \, \mathrm{d}V_D = \int_W \left[\frac{\partial}{\partial t} P + (P v_k)_{,k} \right] \mathrm{d}V_D. \tag{33.9}$$

It follows that the fractal material time derivative is the same as the conventional one

$$\left(\frac{\mathrm{d}}{\mathrm{d}t}\right)_D P = \frac{\mathrm{d}}{\mathrm{d}t} P = \frac{\partial}{\partial t} P + P_{,k} v_k. \tag{33.10}$$

From a homogenization standpoint this allows an interpretation of the fractal (intrinsically discontinuous) medium as a continuum with a 'fractal metric' embedded in the equivalent homogenized continuum model, that is,

$$\mathrm{d}l_D = c_1 \, \mathrm{d}x, \qquad \mathrm{d}S_d = c_2 \, \mathrm{d}S_2, \qquad \mathrm{d}V_D = c_3 \, \mathrm{d}V_3. \tag{33.11}$$

Here $\mathrm{d}l_D$, $\mathrm{d}S_d$, $\mathrm{d}V_D$ represent the line, surface, and volume elements in the fractal body, while $\mathrm{d}x$, $\mathrm{d}S_2$, $\mathrm{d}V_3$, respectively, denote those in the homogenized model, see

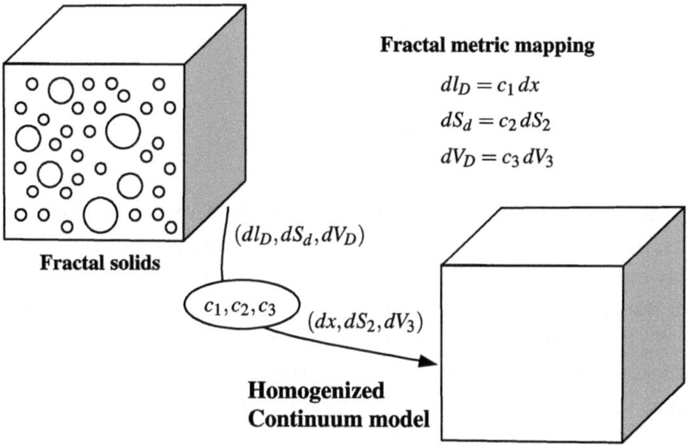

Fig. 33.2 An illustration of the homogenization process from geometry configurations

Fig. 33.2. The coefficients c_1, c_2, c_3 provide relations between both pictures. Standard image analysis techniques (such as the "box method" or the "sausage method") allow a quantitative calibration of these coefficients for every direction and every cross-sectional plane.

33.3 Micropolar Continuum Mechanics of Fractal Media

Now we proceed to develop a framework of continuum mechanics in fractal media. Analogous to the classical continuum mechanics, first we specify the surface force $\mathbf{T}^S$ in terms of the Cauchy stress tensor $\boldsymbol{\sigma}$ via fractional integrals

$$T_k^S = \int_{\partial W} \sigma_{lk} n_l \,\mathrm{d}S_d. \tag{33.12}$$

The conservation of linear and angular momenta in fractal media can be written as

$$\frac{\mathrm{d}}{\mathrm{d}t}\int_W \rho v_k \,\mathrm{d}V_D = \int_W X_k \,\mathrm{d}V_D + \int_{\partial W} \sigma_{lk} n_l \,\mathrm{d}S_d, \tag{33.13}$$

and

$$\frac{\mathrm{d}}{\mathrm{d}t}\int_W \rho e_{ijk} x_j v_k \,\mathrm{d}V_D = \int_W e_{ijk} x_j X_k \,\mathrm{d}V_D + \int_{\partial W} e_{ijk} x_j \sigma_{lk} n_l \,\mathrm{d}S_d, \tag{33.14}$$

where v_k denotes the velocity and X_k is the body force density; e_{ijk} is the permutation tensor. From the fractional Gauss theorem (33.8) and Reynold's transport theorem (33.9), we obtain the balance equations of linear and angular momenta in local form:

$$\rho\left(\frac{\mathrm{d}}{\mathrm{d}t}\right)_D v_k = X_k + \nabla_l^D \sigma_{lk} \tag{33.15}$$

and

$$e_{ijk}\frac{\sigma_{jk}}{c_1^{(j)}} = 0. \tag{33.16}$$

In general, $c_1^{(j)} \neq c_1^{(k)}$, meaning that the medium exhibits anisotropic fractal dimensions, thus making the Cauchy stress asymmetric, and suggesting a Cosserat (micropolar mechanics) formulation. This can be physically understood by noting that fractal media display a heterogeneous fine structure at arbitrarily small scales, also note [5, 10]. Motivated by this result, we introduce a couple-stress tensor $\boldsymbol{\mu}$ and a rotation vector $\boldsymbol{\varphi}$ augmenting, respectively, the Cauchy stress tensor $\boldsymbol{\tau}$ (thus denoted so as to distinguish it from the symmetric $\boldsymbol{\sigma}$) and the deformation vector $\mathbf{u}$. The surface force and surface couple can be specified by fractional integrals of $\boldsymbol{\tau}$ and $\boldsymbol{\mu}$, respectively, as

$$T_k^S = \int_{\partial W} \tau_{lk} n_l \,\mathrm{d}S_d, \qquad M_k^S = \int_{\partial W} \mu_{lk} n_l \,\mathrm{d}S_d. \tag{33.17}$$

Now, proceeding in a fashion similar as before, we arrive at the balance equations of linear and angular momenta

$$\rho\left(\frac{\mathrm{d}}{\mathrm{d}t}\right)_D v_i = X_i + \nabla_j^D \tau_{ji}, \tag{33.18}$$

$$I_{ij}\left(\frac{\mathrm{d}}{\mathrm{d}t}\right)_D w_j = Y_i + \nabla_j^D \mu_{ji} + e_{ijk}\frac{\tau_{jk}}{c_1^{(j)}}. \tag{33.19}$$

In the above, I_{ij} is the rotational inertia tensor determined by

$$I_{ii} = \rho\int_W \left[|\mathbf{x}|^2 - x_i^2\right] \mathrm{d}V_D, \qquad I_{ij} = \rho\int_W x_i x_j \,\mathrm{d}V_D, \tag{33.20}$$

which follows from considering a helicoidal vector field of velocity [23]. Also, X_i is the external body force density, Y_i is the body force couple, while $v_i (\equiv \dot{u}_i)$ and $w_i (\equiv \dot{\varphi}_i)$ are deformation and rotation velocities, respectively.

Let us now consider the conservation of energy. It has the following form

$$\frac{\mathrm{d}}{\mathrm{d}t}\int_W (u+k)\,\mathrm{d}V_D = \int_W (X_i v_i + Y_i w_i)\,\mathrm{d}V_D + \int_{\partial W} (t_i v_i + m_i w_i)\,\mathrm{d}S_d, \tag{33.21}$$

where $k = (1/2)(\rho v_i v_i + I_{ij} w_i w_j)$ is the kinetic energy density and u denotes the internal energy density. (Note here that, just like in conventional continuum mechanics, the balance equations of linear momentum (33.18) and angular momentum (33.19) can be consistently derived from the invariance of energy (33.21) with respect to rigid body translations ($v_i \to v_i + b_i$) and rotations ($v_i \to v_i + e_{ijk} x_j \omega_k$, $w_i \to w_i + \omega_i$), respectively.) Next, we want to obtain the expression for the rate of change of internal energy, and so we start with

$$\int_w \left[\left(\frac{\mathrm{d}}{\mathrm{d}t}\right)_D u + \rho v_i \left(\frac{\mathrm{d}}{\mathrm{d}t}\right)_D v_i + I_{ij} w_i \left(\frac{\mathrm{d}}{\mathrm{d}t}\right)_D w_j \right] \mathrm{d}V_D$$
$$= \int_W \left[(X_i v_i + Y_i w_i) + \nabla_j^D (\tau_{ji} v_i + \mu_{ji} w_i) \right] \mathrm{d}V_D, \qquad (33.22)$$

which yields the local form

$$\left(\frac{\mathrm{d}}{\mathrm{d}t}\right)_D u + \rho v_i \left(\frac{\mathrm{d}}{\mathrm{d}t}\right)_D v_i + I_{ij} w_i \left(\frac{\mathrm{d}}{\mathrm{d}t}\right)_D w_j$$
$$= (X_i v_i + Y_i w_i) + \nabla_j^D (\tau_{ji} v_i + \mu_{ji} w_i). \qquad (33.23)$$

In view of (33.18) and (33.19), and noting the "term by term" rule of ∇_j^D, we find

$$\left(\frac{\mathrm{d}}{\mathrm{d}t}\right)_D u = \tau_{ji} \left(\nabla_j^D v_i - e_{kji} \frac{w_k}{c_1{}^{(j)}} \right) + \mu_{ji} \nabla_j^D w_i. \qquad (33.24)$$

It is now convenient to define the strain tensor γ_{ji} and the curvature tensor κ_{ji} in fractal media as

$$\gamma_{ji} = \nabla_j^D u_i - e_{kji} \frac{\varphi_k}{c_1{}^{(j)}}, \qquad \kappa_{ji} = \nabla_j^D \varphi_i. \qquad (33.25)$$

For small motions $(\mathrm{d}/\mathrm{d}t)_D u = \dot{u}$, thus we can write the energy balance (33.24) as

$$\dot{u} = \tau_{ji} \dot{\gamma}_{ji} + \mu_{ji} \dot{\kappa}_{ji}. \qquad (33.26)$$

Assuming u to be a state function of γ_{ji} and κ_{ji} only, this leads to

$$\tau_{ji} = \frac{\partial u}{\partial \gamma_{ji}}, \qquad \mu_{ji} = \frac{\partial u}{\partial \kappa_{ji}}, \qquad (33.27)$$

which shows that, in the fractal setting, (τ_{ji}, γ_{ji}) and (μ_{ji}, κ_{ji}) are still conjugate pairs.

The specification of constitutive equations involves more physical arguments, and we choose to keep the stress–strain relations while modifying their definitions to the fractal setting. This is consistent with [2], where scale effects of material strength and stress are discussed from the standpoint of fractal geometry and confirmed by experiments of both brittle and plastic materials. Thus, we have

$$\tau_{ij} = C_{ijkl}^{(1)} \gamma_{kl} + C_{ijkl}^{(3)} \kappa_{kl}, \qquad \mu_{ij} = C_{ijkl}^{(3)} \gamma_{kl} + C_{ijkl}^{(2)} \kappa_{kl}. \qquad (33.28)$$

Equations (33.18), (33.19), (33.25), and (33.28) constitute a complete set of equations describing the initial-boundary value problems in fractal media.

33.4 Closure

Our approach builds on, but modifies, Tarasov's approach in that it admits an arbitrary anisotropic structure. This involves, in the first place, a specification of geometry of continua via 'fractal metric' coefficients, which then allows a construction of continuum mechanics of fractal solids. The anisotropy of fractal geometry on mesoscale leads to asymmetry of the Cauchy stress and to admission of the couple stress, i.e., to a fractal micropolar continuum. The proposed methodology broadens the applicability of continuum mechanics/physics to studies of material responses. The highly complex, fractal-type media which have so far been the domain of condensed matter physics, geophysics and biophysics, etc. (multiscale polycrystals, cracked materials, polymer clusters, gels, rock systems, percolating networks, nervous systems, pulmonary systems, ...) will become open to studies conventionally reserved for smooth materials. This will allow solutions of initial-boundary value problems of very complex, multiscale materials of both elastic and inelastic type.

References

1. Berry, M.V.: Diffractals. J. Phys. A, Math. Gen. **12**(6), 781–797 (1979)
2. Carpinteri, A., Pugno, N.: Are scaling laws on strength of solids related to mechanics or to geometry? Nature Mater. **4**, 421–423 (2005)
3. Carpinteri, A., Chiaia, B., Cornetti, P.: A disordered microstructure material model based on fractal geometry and fractional calculus. Z. Angew. Math. Mech. **84**, 128–135 (2004)
4. Dyskin, A.V.: Continuum fractal mechanics of the Earth's crust. Pure Appl. Geophys. **161**, 1979–1989 (2004)
5. Feder, J.: Fractals (Physics of Solids and Liquids). Springer, Berlin (2007)
6. Jumarie, G.: On the representation of fractional Brownian motion as an integral with respect to $(\mathrm{d}t)^a$. Appl. Math. Lett. **18**, 739–748 (2005)
7. Jumarie, G.: Table of some basic fractional calculus formulae derived from a modified Riemann–Liouville derivative for non-differentiable functions. Appl. Math. Lett. **22**(3), 378–385 (2008)
8. Kigami, J.: Analysis on Fractals. Cambridge University Press, Cambridge (2001)
9. Li, J., Ostoja-Starzewski, M.: Fractal solids, product measures and fractional wave equations. Proc. R. Soc. A **465**, 2521–2536 (2009)
10. Limat, L.: Micropolar elastic percolation: the superelastic problem. Phys. Rev. B **38**(10), 7219–7222 (1988)
11. Mandelbrot, B.B.: The Fractal Geometry of Nature. Freeman, New York (1982)
12. Oldham, K.B., Spanier, J.: The Fractional Calculus. Academic Press, San Diego (1974)
13. Ostoja-Starzewski, M.: Towards thermomechanics of fractal media. Z. Angew. Math. Phys. **58**, 1085–1096 (2007)
14. Ostoja-Starzewski, M.: Towards thermoelasticity of fractal media. J. Therm. Stress. **30**, 889–896 (2007)
15. Ostoja-Starzewski, M.: On turbulence in fractal porous media. Z. Angew. Math. Phys. **59**(6), 1111–1117 (2008)
16. Ostoja-Starzewski, M.: Continuum mechanics models of fractal porous media: integral relations and extremum principles. J. Mech. Mater. Struct. **4**(5), 912 (2009)
17. Ostoja-Starzewski, M.: Extremum and variational principles for elastic and inelastic media with fractal geometries. Acta Mech. **205**, 161–170 (2009)

18. Ostoja-Starzewski, M., Li, J.: Fractal materials, beams, and fracture mechanics. Z. Angew. Math. Phys. **60**, 1194–1205 (2009)
19. Strichartz, R.S.: Differential Equations on Fractals: A Tutorial. Princeton University Press, Princeton (2006)
20. Tarasov, V.E.: Continuous medium model for fractal media. Phys. Lett. A **336**, 167–174 (2005)
21. Tarasov, V.E.: Fractional hydrodynamic equations for fractal media. Ann. Phys. **318**(2), 286–307 (2005)
22. Tarasov, V.E.: Wave equation for fractal solid string. Mod. Phys. Lett. B **19**, 721–728 (2005)
23. Temam, R., Miranville, A.: Mathematical Modeling in Continuum Mechanics. Cambridge University Press, Cambridge (2005)

Chapter 34
Magnetoelasticity of Thin Shells and Plates Based on the Asymmetrical Theory of Elasticity

Smuel H. Sargsyan and Lusine S. Sargsyan

Abstract In the present, paper we aim at constructing three variants of general mathematical models of micropolar elastic electro-conducting and non-ferromagnetic shells and plates: one with independent fields of transition and rotation, one with constraint rotation and one with "small shift rigidity". The construction is based on the asymptotic method.

34.1 Introduction

In the course of constructing the theory of elasticity with the account of force and momental stresses, we sometimes face the problem of defining the stress-deformed state for materials which are endowed with electromagnetic properties [6, 2, 4, 5, 1]. From this point, the construction of general mathematical models of micropolar elastic electro-conducting and non-ferromagnetic shells and plates becomes actual.

In [7], we can follow the application of the asymptotic method while constructing general applied two-dimensional theory of magnetoelasticity of thin shells according to the classical theory of elasticity. The construction of theories of thin bars, shells and plates based on the above-mentioned principle is demonstrated in [9, 8, 10, 11].

In the present paper, our task consists in constructing the fundamental system of 2D equations of magnetoelasticity of electro-conducting and non-ferromagnetic thin plates and shells based on the three-dimensional asymmetrical theory of magnetoelasticity. In order to define the electromagnetic field in the region surrounding the shell or plate, it is convenient to represent the 3D space as a mathematical cut along the shell's or plate's mid-surface. The electro-conducting currents, averaged along the shell's or plates thickness, are supposed to flow along this cut. Thus, we

S.H. Sargsyan · L.S. Sargsyan (✉)
Gyumri State Pedagogical Institute, Sevak 4, Gyumri, Armenia
e-mail: slusin@yahoo.com

G.A. Maugin, A.V. Metrikine (eds.), *Mechanics of Generalized Continua*, Advances in Mechanics and Mathematics 21,
DOI 10.1007/978-1-4419-5695-8_34, © Springer Science+Business Media, LLC 2010

obtain 2D conditions on the boundary contour of the shell's or plate's mid-surface both for the mechanical and for corresponding electrodynamic values. As a result, with the account of dimensionless physical parameters, the above mentioned three variants of mathematical models are constructed.

34.2 Formulation of the Problem

Let us consider an isotropic shell with constant thickness $2h$ as a micropolar elastic electro-conducting (non-ferromagnetic) homogeneous body in a static three-orthogonal system of coordinates [7, 3]. The shell is supposed to be located in an external stationary homogeneous magnetic field described by the vector: $\mathbf{H}^0 = \{H_{01}, H_{02}, H_{03}\}$. We start from the main equations of the linearized theory of magnetoelasticity for 3D micropolar medium [2, 4, 5]. Taking into account volume forces of electromagnetic origin, the equations of motion become

$$\nabla_m \sigma^{mn} + f^n = \rho \frac{\partial^2 V^n}{\partial t^2}, \qquad \nabla_m \mu^{mn} + e^{nmk} \sigma_{mk} = J \frac{\partial^2 \omega^n}{\partial t^2}. \tag{34.1}$$

The generalized Hooke's law is

$$\begin{aligned} \sigma_{mn} &= (\mu + \alpha)\gamma_{mn} + (\mu - \alpha)\gamma_{nm} + \lambda \gamma_{kk} \delta_{nm}, \\ \mu_{mn} &= (\gamma + \varepsilon)\kappa_{mn} + (\gamma - \varepsilon)\kappa_{nm} + \beta \kappa_{kk} \delta_{nm}, \end{aligned} \tag{34.2}$$

with geometrical correlations

$$\gamma_{mn} = \nabla_m V_n - e_{kmn} \omega^k, \qquad \kappa_{mn} = \nabla_m \omega_n. \tag{34.3}$$

Quasi-stationary equations of electrodynamics with finite electro-conductivity in the region of moving medium consist of

$$\operatorname{rot} \mathbf{h} = \frac{4\pi}{c} \mathbf{j}, \qquad \operatorname{rot} \mathbf{E} = -\frac{1}{c} \frac{\partial \mathbf{h}}{\partial t}, \qquad \operatorname{div} \mathbf{h} = 0, \qquad \operatorname{div} \mathbf{E} = 4\pi \rho_e, \tag{34.4}$$

while in the external region (in the vacuum) they are

$$\begin{aligned} &\operatorname{rot} \mathbf{h}^{(e)} = 0, \qquad \operatorname{rot} \mathbf{E}^{(e)} = -\frac{1}{c} \frac{\partial \mathbf{h}^{(e)}}{\partial t}, \\ &\operatorname{div} \mathbf{E}^{(e)} = 0, \qquad \operatorname{div} \mathbf{h}^{(e)} = 0. \end{aligned} \tag{34.5}$$

Here, $m, n = 1, 2, 3$; σ^{nm}, μ^{nm} stand for contra-variant components of force and momental stress tensors; γ_{mn}, κ_{mn} are covariant components of deformation and bending-torsion tensors; V_n are covariant components of the displacement vector ($\mathbf{V}$); ω_n stands for covariant components of the vector of independent rotation ($\boldsymbol{\omega}$)

$$\mathbf{F} = \{f^1, f^2, f^3\} = \frac{1}{c} \mathbf{j} \times \mathbf{H}^0. \tag{34.6}$$

$\mathbf{F}$ denotes the vector of volume forces of electro-magnetic origin

$$\mathbf{j} = \sigma\left(\mathbf{E} + \frac{1}{c}\frac{\partial \mathbf{V}}{\partial t} \times \mathbf{H}^0\right). \tag{34.7}$$

$\mathbf{j}$ stands for the disturbed electrical current in the body; $\mathbf{E}$, $\mathbf{E}^{(e)}$ are the vectors of the induced electrical current in the body and vacuum, respectively; $\mathbf{h}$, $\mathbf{h}^{(e)}$ are vectors of the induced magnetic field in the body vacuum, respectively; ρ_e is the electrical charge in the body; λ, μ, α, β, γ, ε are the elastic constants of the shell's micropolar material; ρ is the thickness of the material; J is the measure for inertia during the rotation of the material; σ is the coefficient of electro-conduction; c is the electro dynamic constant which is equal to the speed of light in the vacuum; δ_{nm} stand for the Kroneker's symbols; e^{nmk} are contra-variant components of Levi–Civita tensor.

Mechanical boundary conditions on the shell's facial surface appear as

$$\sigma_{k3}|_{\alpha_3=\pm h} = \mp q_k^{\pm} \quad (k = 1, 2, 3). \tag{34.8}$$

It is assumed that the boundary conditions on the shell's boundary surface Σ correspond to either the first, second or mixed types of the theory of micropolar elasticity.

Electrodynamic boundary conditions both on the facial and boundary surfaces will be represented by the following correlations [6, 2]:

$$\begin{aligned} n^k[E_k]_- &= 4\pi\hat{\rho}_e, & e^{nmk}n_m[h_k]_- &= 0, \\ n^k[h_k]_- &= 0, & e^{nmk}n_m[E_k]_- &= 0, \end{aligned} \tag{34.9}$$

where $[\cdot]$ stands for the jump of the unknown at the boundary surface between the body and vacuum, and $\hat{\rho}_e$ is the surface charge. At infinity, we require that the electromagnetic vectors in vacuum decrease as [7]

$$\left|\vec{E}^{(e)}\right| = O(1/r), \qquad \left|\mathbf{h}^{(e)}\right| = O(1/r), \quad r \to \infty,\ 0 \le t < \infty, \tag{34.10}$$

where r is the distance from the origin of coordinates to the reference point.

The initial conditions of the defined problem characterize the positions and speeds of the body points, and the changes in electromagnetic field at $t = 0$ initial time moment.

34.3 The Asymptotic Method

Let us consider a problem of reducing the 3D initial condition–boundary value problem (34.1)–(34.10) of the asymmetric theory of magnetoelasticity for thin shells to 2D by the asymptotic method with boundary layer, including the problem of satisfying the boundary and initial conditions. Let us also mention that this problem is tightly connected with the construction of the internal iterative process, which is a 2D problem, as shown below.

For this purpose, in 3D equations of asymmetric theory of magnetoelasticity we introduce special stress tensors [7, 11, 3]:

$$\tau_{ii} = \left(1+\frac{\alpha_3}{R_j}\right)\sigma_{ii}, \qquad \tau_{ij} = \left(1+\frac{\alpha_3}{R_j}\right)\sigma_{ij}, \qquad \tau_{i3} = \left(1+\frac{\alpha_3}{R_j}\right)\sigma_{i3}$$
$$(i \leftrightarrow 3),$$
$$v_{ii} = \left(1+\frac{\alpha_3}{R_j}\right)\mu_{ii}, \qquad v_{ij} = \left(1+\frac{\alpha_3}{R_j}\right)\mu_{ij}, \qquad v_{i3} = \left(1+\frac{\alpha_3}{R_j}\right)\mu_{i3}$$
$$(i \leftrightarrow 3),$$
$$\tau_{33} = \left(1+\frac{\alpha_3}{R_1}\right)\left(1+\frac{\alpha_3}{R_2}\right)\sigma_{33}, \qquad v_{33} = \left(1+\frac{\alpha_3}{R_1}\right)\left(1+\frac{\alpha_3}{R_2}\right)\mu_{33},$$

and switch to dimensionless quantities and replace independent variables (coordinate α_n and time t)

$$\alpha_i = \frac{R}{\lambda^p}\xi_i, \qquad \alpha_3 = \frac{R}{\lambda^l}\zeta, \qquad t = \lambda^\omega \frac{h}{c_0}\tau, \qquad c_0 = \sqrt{\frac{E}{\rho}},$$
$$\lambda^{l(2k+1)}\bar{J} = \frac{J}{\rho h^2}, \qquad \bar{\tau}_{ij} = \frac{\tau_{ij}}{E}, \qquad \bar{\nu}_{ij} = \frac{\nu_{ij}}{RE},$$
$$\bar{V}_k = \frac{V_k}{h}, \qquad H^*_{0k} = \frac{H_{0k}}{\sqrt{E}}, \qquad \bar{H}_{0k} = \lambda^{k_1} H^*_{0k},$$
$$\bar{h}_k = \frac{h_k}{\sqrt{E}}, \qquad \bar{E}_k = \frac{c}{c_0}\frac{E_k}{\sqrt{E}}, \qquad \bar{\rho}_e = \frac{c}{c_0}\frac{h\rho_e}{\sqrt{E}}, \tag{34.11}$$
$$\bar{j}_k = \frac{c}{c_0}\frac{j_k}{\sigma\sqrt{E}}, \qquad R_m = \frac{c_0}{c}\frac{\sigma h}{c}, \qquad \bar{R}_i = \frac{R_i}{R},$$
$$\bar{q}^{\pm}_k = \frac{q^{\pm}_i}{E}, \qquad \bar{m}^{\pm}_{:} = \frac{m^{\pm}_{:}}{RE}, \qquad t_0 = \lambda^{-l(\omega-1)}\frac{R}{c_0}.$$

Here R_m is a dimensionless parameter of electro-conduction of shell material; ω characterizes the changeability of stress-deformable state (SDS) in time; the value p/l characterizes the changeability of SDS along the coordinates; p, l are whole numbers, $l > p \geq 0$ and λ is a large constant dimensionless geometrical parameter defined by $h = R\lambda^{-l}$ formula.

In defining the SDS of a shell, the values of physical constants of shell micropolar material are of great importance. From this point of view, we also introduce the following dimensionless parameters:

$$\frac{\alpha}{E}, \qquad \frac{\beta}{R^2E}, \qquad \frac{\gamma}{R^2E}, \qquad \frac{\varepsilon}{R^2E}, \tag{34.12}$$

where R is a characteristic radius of shell midplane curvature; c_0 is some characteristic speed (for example, the speed of longitudinal disturbance in long elastic bars in the classical theory of elasticity).

In what follows, we assume that $R_m \sim 1$. Following to asymptotic method of construction of the internal problem, we aim at an approximate reduction of 3D equations of magnetoelasticity with independent variables ξ_1, ξ_2, ζ, and time τ into 2D with independent variables ξ_1, ξ_2 and time τ.

34.4 Magnetoelasticity Theory of Micropolar Thin Shells with Independent Fields of Displacements and Rotations

We suppose that dimensionless physical parameters (34.12) are of order of unity:

$$\frac{\alpha}{E} \sim 1, \qquad \frac{\beta}{R^2E} \sim 1, \qquad \frac{\gamma}{R^2E} \sim 1, \qquad \frac{\varepsilon}{R^2E} \sim 1. \tag{34.13}$$

The numbers ω, k and k_1 are chosen as to obtain reconcilable equations in asymptotic approximations s and include the inertial terms in the initial system of equations. It leads to

$$\omega = 1 + \frac{p}{l}, \qquad k = \frac{1}{2} - \frac{p}{l}, \qquad 2k_1 = -l - p. \tag{34.14}$$

For the values of the internal problem in the 3D region of the shell body, we obtain the following asymptotic representations of $O(\lambda^{p-l})$ order:

$$\begin{aligned}
V_i &= h\lambda^{l-p}\big(V_i^0 + \lambda^{-l+2p-c}\zeta V_i^1\big),\\
V_3 &= h\lambda^{l-2p+c}\big(V_3^0 + \lambda^{-l+c}\zeta V_3^1\big),\\
\omega_i &= \lambda^{-p+c}\big(\omega_i^0 + \lambda^{-l+2p-c}\zeta\omega_i^1\big),\\
\omega_3 &= \lambda^0\big(\omega_3^0 + \lambda^{-l+c}\zeta\omega_3^1\big),\\
\tau_{ii} &= E\lambda^0\big(\tau_{ii}^0 + \lambda^{-l+c}\zeta\tau_{ii}^1\big),\\
\tau_{ij} &= E\lambda^0\big(\tau_{ij}^0 + \lambda^{-l+c}\zeta\tau_{ij}^1\big),\\
\tau_{3i} &= E\lambda^{-p+c}\big(\tau_{3i}^0 + \lambda^{-l+2p-c}\zeta\tau_{3i}^1 + \lambda^{-2l+2p}\tau_{3i}^2\big),\\
\tau_{i3} &= E\lambda^{-p+c}\big(\tau_{i3}^0 + \lambda^{-l+2p-c}\zeta\tau_{i3}^1\big),\\
v_{ii} &= RE\lambda^{-2p+c}\big(v_{ii}^0 + \lambda^{-l+2p-c}\zeta v_{ii}^1\big),\\
v_{ij} &= RE\lambda^{-2p+c}\big(v_{ij}^0 + \lambda^{-l+2p-c}\zeta v_{ij}^1\big),\\
v_{3i} &= RE\lambda^{-p}\big(v_{3i}^0 + \lambda^{-l+c}\zeta v_{3i}^1\big),\\
v_{i3} &= RE\lambda^{-p}\big(v_{i3}^0 + \lambda^{-l+c}\zeta v_{i3}^1\big),\\
\tau_{33} &= E\lambda^{-l+c}\big(\tau_{33}^0 + \zeta\tau_{33}^1 + \lambda^{-l+2p-c}\zeta^2\tau_{33}^2\big),\\
v_{33} &= RE\lambda^{-l}\big(v_{33}^0 + \zeta v_{33}^1 + \lambda^{-l+c}\zeta^2 v_{33}^2\big),
\end{aligned} \tag{34.15}$$

where $c = 2p - l$ for $l \le 2p$, $c = 2p$ for $l \ge 4p$, $c = l - 2p$ otherwise.

The fact that the rotation field of the shells' points is independent from the displacement field is important in the constructed iterative process.

We use the concept of the averaged force and momental stresses [11] and displacements and independent rotations of the shell's midplane surface:

$$u_i = V_i|_{\zeta=0}, \qquad w = -V_3|_{\zeta=0}, \qquad \Omega_i = \omega_i|_{\zeta=0}, \qquad \Omega_3 = -\omega_3|_{\zeta=0}. \tag{34.16}$$

As the main result, on the basis of the constructed internal problem, we obtain the principal system of equations of the general applied 2D theory of magnetoelasticity of micropolar thin shells with independent fields of displacement and rotation in the asymptotic approximation of $O(\lambda^{p-l})$ order:

The motion equations

$$\begin{cases} \frac{1}{A_i}\frac{\partial T_{ii}}{\partial \alpha_i} + \frac{1}{A_iA_j}\frac{\partial A_j}{\partial \alpha_i}(T_{ii} - T_{jj}) + \frac{1}{A_j}\frac{\partial S_{ji}}{\partial \alpha_j} + \frac{1}{A_iA_j}\frac{\partial A_i}{\partial \alpha_j}(S_{ji} + S_{ij}) \\ \quad + (-1)^j \frac{1}{c} j_{j0} H_{03} = 2\rho h \frac{\partial^2 u_i}{\partial t^2} + (q_i^+ + q_i^-), \end{cases}$$

$$\begin{cases} \frac{1}{A_i}\frac{\partial L_{ii}}{\partial \alpha_i} + \frac{1}{A_iA_j}\frac{\partial A_j}{\partial \alpha_i}(L_{ii} - L_{jj}) + \frac{1}{A_j}\frac{\partial L_{ji}}{\partial \alpha_j} + \frac{1}{A_iA_j}\frac{\partial A_i}{\partial \alpha_j}(L_{ji} + L_{ij}) \\ \quad + (-1)^j (N_{3j} - N_{j3}) = 2Jh \frac{\partial^2 \Omega_i}{\partial t^2} + (m_i^+ + m_i^-), \end{cases}$$

$$\begin{cases} \frac{T_{11}}{R_1} + \frac{T_{22}}{R_2} + \frac{1}{A_1A_2}\left[\frac{\partial(A_2N_{13})}{\partial \alpha_1} + \frac{\partial(A_1N_{23})}{\partial \alpha_2}\right] - \frac{1}{c} j_{10}H_{02} + \frac{1}{c} j_{20}H_{01} \\ \quad = 2\rho h \frac{\partial^2 w}{\partial t^2} - (q_3^+ + q_3^-), \end{cases} \tag{34.17}$$

$$\begin{cases} \frac{L_{11}}{R_1} + \frac{L_{22}}{R_2} + \frac{1}{A_1A_2}\left[\frac{\partial(A_2L_{13})}{\partial \alpha_1} + \frac{\partial(A_1L_{23})}{\partial \alpha_2}\right] - (S_{12} - S_{21}) \\ \quad = 2Jh \frac{\partial^2 \Omega_3}{\partial t^2} - (m_3^+ + m_3^-). \end{cases}$$

The elasticity relations

$$\begin{aligned} T_{ii} &= \frac{2Eh}{1-v^2}[\Gamma_{ii} + v\Gamma_{jj}] - \frac{v}{1-v}h(q_3^+ - q_3^-), \\ S_{ij} &= 2h\big[(\mu+\alpha)\Gamma_{ij} + (\mu-\alpha)\Gamma_{ji}\big], \\ L_{ii} &= 2h\left[\frac{4\gamma(\beta+\gamma)}{\beta+2\gamma}\chi_{ii} + \frac{2\gamma\beta}{\beta+2\gamma}\chi_{jj}\right] - h\frac{\beta}{\beta+2\gamma}m, \\ L_{ij} &= 2h\big[(\gamma+\varepsilon)\chi_{ij} + (\gamma-\varepsilon)\chi_{ji}\big], \\ N_{i3} &= -2h\frac{4\alpha\mu}{\alpha+\mu}\Gamma_{i3} + \frac{\mu-\alpha}{\mu+\alpha}h(q_i^+ - q_i^-), \\ L_{i3} &= -2h\frac{4\gamma\varepsilon}{\gamma+\varepsilon}\chi_{i3} + \frac{\gamma-\varepsilon}{\gamma+\varepsilon}h(m_i^+ - m_i^-). \end{aligned} \tag{34.18}$$

The geometrical relations

$$
\begin{aligned}
\Gamma_{i3} &= -\frac{1}{A_i}\frac{\partial w}{\partial \alpha_i} - \frac{u_i}{R_i} + (-1)^j \Omega_j, \qquad \chi_{i3} = -\frac{1}{A_i}\frac{\partial \Omega_3}{\partial \alpha_i} - \frac{\Omega_i}{R_i}, \\
\Gamma_{ii} &= \frac{1}{A_i}\frac{\partial u_i}{\partial \alpha_i} + \frac{1}{A_i A_j}\frac{\partial A_i}{\partial \alpha_j}u_j - \frac{w}{R_i}, \\
\Gamma_{ij} &= \frac{1}{A_i}\frac{\partial u_j}{\partial \alpha_i} - \frac{1}{A_i A_j}\frac{\partial A_i}{\partial \alpha_j}u_i + (-1)^j \Omega_3, \\
\chi_{ii} &= \frac{1}{A_i}\frac{\partial \Omega_i}{\partial \alpha_i} + \frac{1}{A_i A_j}\frac{\partial A_i}{\partial \alpha_j}\Omega_j - \frac{\Omega_3}{R_i}, \\
\chi_{ij} &= \frac{1}{A_i}\frac{\partial \Omega_j}{\partial \alpha_i} - \frac{1}{A_i A_j}\frac{\partial A_i}{\partial \alpha_j}\Omega_i.
\end{aligned}
\tag{34.19}
$$

The integro-differential equations of the conductivity surface current

$$
\begin{aligned}
j_{10}(P,t) &= -\frac{2\sigma h}{c^2}\frac{\partial}{\partial t}\int_\Omega \left[\frac{j_{10}(Q,t)\mathbf{e}_1(Q)\mathbf{e}_1(P)}{R_{PQ}} + \frac{j_{20}(Q,t)\mathbf{e}_2(Q)\mathbf{e}_1(P)}{R_{PQ}}\right] \mathrm{d}\Omega \\
&\quad + \frac{2\sigma h}{c}\left(H_{03}\frac{\partial u_2(P,t)}{\partial t} - H_{02}\frac{\partial w(P,t)}{\partial t}\right), \quad P \in \Omega, \\
j_{20}(P,t) &= -\frac{2\sigma h}{c^2}\frac{\partial}{\partial t}\int_\Omega \left[\frac{j_{10}(Q,t)\mathbf{e}_2(P)\mathbf{e}_1(Q)}{R_{PQ}} + \frac{j_{20}(Q,t)\mathbf{e}_2(Q)\mathbf{e}_2(P)}{R_{PQ}}\right] \mathrm{d}\Omega \\
&\quad + \frac{2\sigma h}{c}\left(H_{01}\frac{\partial w(P,t)}{\partial t} - H_{03}\frac{\partial u_1(P,t)}{\partial t}\right), \quad P \in \Omega.
\end{aligned}
\tag{34.20}
$$

Here T_{ii}, S_{ij}, N_{i3} are the averaged forces, L_{ii}, L_{ij}, L_{i3} are the averaged moments from momental stresses; Γ_{ii}, Γ_{ij}, Γ_{i3} are the components of deformation tensor; χ_{ii}, χ_{ij}, χ_{i3} are the components of bending-torsion tensor in shell's midplane; $\mathbf{e}_1$ are sorts of coordinate lines and α_2 on the midplane; R_{PQ} is the fundamental solution of the system (34.5) for the entire 3D vacuum space (R^3) [7] for 3D distance between a point $Q \in \Omega$ and an arbitrary reference point $P \in R^3$.

The mechanical boundary conditions on the midplane Γ are expressed as [11]:

$$
\begin{aligned}
T_{11}|_\Gamma &= \int_{-h}^{+h} p_1^* \,\mathrm{d}\alpha_3, \qquad S_{12}|_\Gamma = \int_{-h}^{+h} p_2^* \,\mathrm{d}\alpha_3, \qquad N_{13}|_\Gamma = -\int_{-h}^{+h} p_3^* \,\mathrm{d}\alpha_3, \\
L_{11}|_\Gamma &= \int_{-h}^{+h} m_1^* \,\mathrm{d}\alpha_3, \qquad L_{12}|_\Gamma = \int_{-h}^{+h} m_2^* \,\mathrm{d}\alpha_3, \\
L_{13}|_\Gamma &= -\int_{-h}^{+h} m_3^* \,\mathrm{d}\alpha_3.
\end{aligned}
\tag{34.21}
$$

The electrodynamic boundary conditions on the same contour look as follows [7]:

$$
j_1 = 0, \qquad j_2 = 0. \tag{34.22}
$$

By adding to the main equations (34.17)–(34.20) and boundary conditions (34.21)–(34.22) the corresponding initial conditions, we construct the general mathematical model of micropolar elastic electro-conducting non-ferromagnetic shells with independent fields of displacements and rotations.

34.5 Theory of Magnetoelasticity of Micropolar Thin Shells with Constraint Rotation

Let us consider now a different set of values of dimensionless physical parameters (34.12):

$$\begin{aligned}&\frac{\alpha}{E}\sim 1, \qquad \frac{\beta}{R^2E}=\lambda^{-2l}\beta_*, \qquad \frac{\gamma}{R^2E}=\lambda^{-2l}\gamma_*,\\&\frac{\varepsilon}{R^2E}=\lambda^{-2l}\varepsilon_*, \qquad \beta_*,\gamma_*,\varepsilon_*\sim 1.\end{aligned} \tag{34.23}$$

In this case, the numbers ω, k and k_1 are determined by

$$\begin{aligned}&\omega=1+\frac{c}{l}, \qquad k=-\frac{1}{2}-\frac{c}{l}+\frac{p}{l},\\&2k_1=-l-p+2c, \qquad c=2p-l, \quad l\le 2p,\end{aligned} \tag{34.24}$$

and 0 otherwise. Qualitatively, the rotations and displacements of the shell's midplane points are now coupled. If in the given case we introduce the averaged force and momental characteristics and use the notations (34.16), as the main result we obtain a system of the main equations of the applied 2D theory of magnetoelasticity of micropolar elastic thin shells with constraint rotation:

The motion equations

$$\begin{cases}\dfrac{1}{A_i}\dfrac{\partial T_{ii}}{\partial\alpha_i}+\dfrac{1}{A_iA_j}\dfrac{\partial A_j}{\partial\alpha_i}(T_{ii}-T_{jj})+\dfrac{1}{A_j}\dfrac{\partial S_{ji}}{\partial\alpha_j}+\dfrac{1}{A_iA_j}\dfrac{\partial A_i}{\partial\alpha_j}(S_{ji}+S_{ij})\\ \quad+(-1)^j\dfrac{1}{c}j_{j0}H_{03}=2\rho h\dfrac{\partial^2u_i}{\partial t^2}+(q_i^++q_i^-),\end{cases}$$

$$\begin{cases}\dfrac{T_{11}}{R_1}+\dfrac{T_{22}}{R_2}+\dfrac{1}{A_1A_2}\left[\dfrac{\partial(A_2N_{13})}{\partial\alpha_1}+\dfrac{\partial(A_1N_{23})}{\partial\alpha_2}\right]-\dfrac{1}{c}j_{10}H_{02}+\dfrac{1}{c}j_{20}H_{01}\\ \quad=2\rho h\dfrac{\partial^2w}{\partial t^2}-(q_3^++q_3^-),\end{cases}$$

$$\begin{cases}\dfrac{1}{A_i}\dfrac{\partial}{\partial\alpha_i}(G_{ii}-(-1)^jL_{ij})\\ \quad+\dfrac{1}{A_iA_j}\dfrac{\partial A_j}{\partial\alpha_i}[(G_{ii}-(-1)^jL_{ij})-(G_{jj}+(-1)^jL_{ji})]\\ -\dfrac{1}{A_j}\dfrac{\partial}{\partial\alpha_j}(H_{ji}+(-1)^jL_{jj})\\ \quad-\dfrac{1}{A_iA_j}\dfrac{\partial A_i}{\partial\alpha_j}[(H_{ji}+(-1)^jL_{jj})+(H_{ij}-(-1)^jL_{ii})]-N_{i3}\\ \quad=(-1)^j\left[2Jh\dfrac{\partial^2\Omega_j}{\partial t^2}+(m_j^++m_j^-)\right]-h(q_i^+-q_i^-).\end{cases} \tag{34.25}$$

The elasticity relations

$$
\begin{aligned}
T_{ii} &= \frac{2Eh}{1-v^2}[\Gamma_{ii} + v\Gamma_{jj}],\\
S_{ij} &= \frac{2Eh}{1+v}[\Gamma_{12} + \Gamma_{21}] + (-1)^j \frac{1}{2}[m_3^+ + m_3^-],\\
G_{ii} &= -\frac{2Eh^3}{3(1-v^2)}[K_{ii} + vK_{jj}],\\
H_{ij} &= \frac{Eh^3}{3(1+v)}[K_{12} + K_{21}] + (-1)^j \frac{1}{2}[m_3^+ + m_3^- + L_{33}],\\
L_{ii} &= 2h\left[\frac{4\gamma(\beta+\gamma)}{\beta+2\gamma}\chi_{ii} + \frac{2\gamma\beta}{\beta+2\gamma}\chi_{jj}\right] + \frac{\beta}{\beta+2\gamma}L_{33},\\
L_{ij} &= 2h\big[(\gamma+\varepsilon)\chi_{ij} + (\gamma-\varepsilon)\chi_{ji}\big].
\end{aligned}
\tag{34.26}
$$

The geometrical relations

$$
\begin{aligned}
\Gamma_{ii} &= \frac{1}{A_i}\frac{\partial u_i}{\partial \alpha_i} + \frac{1}{A_iA_j}\frac{\partial A_i}{\partial \alpha_j}u_j - \frac{w}{R_i},\\
\Gamma_{ij} &= \frac{1}{A_i}\frac{\partial u_j}{\partial \alpha_i} - \frac{1}{A_iA_j}\frac{\partial A_i}{\partial \alpha_j}u_i,\\
K_{ii} &= \frac{1}{A_i}\frac{\partial \beta_i}{\partial \alpha_i} + \frac{1}{A_iA_j}\frac{\partial A_i}{\partial \alpha_j}\beta_j,\\
K_{ij} &= \frac{1}{A_i}\frac{\partial \beta_j}{\partial \alpha_i} - \frac{1}{A_iA_j}\frac{\partial A_i}{\partial \alpha_j}\beta_i,\\
\chi_{ii} &= \frac{1}{A_i}\frac{\partial \Omega_i}{\partial \alpha_i} + \frac{1}{A_iA_j}\frac{\partial A_i}{\partial \alpha_j}\Omega_j - \frac{\Omega_3}{R_i},\\
\chi_{ij} &= \frac{1}{A_i}\frac{\partial \Omega_j}{\partial \alpha_i} - \frac{1}{A_iA_j}\frac{\partial A_i}{\partial \alpha_j}\Omega_i,\\
\beta_i &= \frac{1}{A_i}\frac{\partial w}{\partial \alpha_i} + \frac{u_i}{R_i},\\
\Omega_i &= -(-1)^j\frac{1}{2}(\psi_j - \beta_j), \qquad \Omega_3 = \frac{1}{2}(\Gamma_{21} - \Gamma_{12}),
\end{aligned}
\tag{34.27}
$$

where

$$
L_{33} = \left(\frac{th(hk_1)}{hk_1} - 1\right)4\gamma h(\chi_{11} + \chi_{22}) - \frac{th(hk_1)}{hk_1}(m_3^+ - m_3^-).
$$

The system of integro-differential equations (34.20) for the electrical currents j_{10}, j_{20} averaged along the shell's midplane should be added to (34.25)–(34.27).

The mechanical boundary conditions on the midplane contour Γ become [11]:

$$T_{11}|_\Gamma = \int_{-h}^{+h} p_1^* \, d\alpha_3, \qquad S_{12}|_\Gamma = \int_{-h}^{+h} p_2^* \, d\alpha_3,$$

$$(L_{12} - G_{11})|_\Gamma = \int_{-h}^{h} (\alpha_3 p_1^* + m_2^*) \, d\alpha_3, \tag{34.28}$$

$$\left[-N_{13} + \frac{1}{A_2}\frac{\partial}{\partial \alpha_2}(H_{12} - L_{11}) \right]\Big|_\Gamma = \int_{-h}^{+h} \left[p_3^* + \frac{1}{A_2}\frac{\partial}{\partial \alpha_2}(\alpha_3 p_2^* - m_1^*) \right] d\alpha_3.$$

The electrodynamic boundary conditions on Γ will be the same as (34.22).

34.6 Theory of Magnetoelastic Micropolar Thin Shells with a Small Shift Rigidity

Let us consider the third set of values for the physical dimensionless parameters (34.12):

$$\frac{\alpha}{E} = \lambda^{-2l+2p}\alpha_*, \qquad \frac{\beta}{R^2E} = \beta_*, \qquad \frac{\gamma}{R^2E} = \gamma_*,$$
$$\frac{\varepsilon}{R^2E} = \varepsilon_*, \qquad \alpha_*, \beta_*, \gamma_*, \varepsilon_* \sim 1. \tag{34.29}$$

In this case, the numbers ω, k and k_1 are determined by

$$\omega = 1 + \frac{c}{2l}, \qquad k = \frac{1}{2} + \frac{p}{l} - \frac{c}{2l}, \qquad 2k_1 = -l - p + c, \tag{34.30}$$

where $c = 2p - l$ if $l \le 2p$, $c = 2p$ if $l \ge 4p$ and $c = l - 2p$ otherwise.

Note that, on the basis of (34.29) and (34.30), in the obtained 2D equations of the internal problem of magnetoelasticity of micropolar shells on the basis of asymptotic accuracy $O(\lambda^{p-l})$, the variables of "pure momental" origin come apart and form a separate system of equations.

Let us formulate these groups of 2D equations. The equations of "pure momental" part of the magnetoelasticity problem of micropolar thin shells are:

The motion equations

$$\frac{1}{A_i}\frac{\partial L_{ii}}{\partial \alpha_i} + \frac{1}{A_iA_j}\frac{\partial A_j}{\partial \alpha_i}(L_{ii} - L_{jj}) + \frac{1}{A_j}\frac{\partial L_{ji}}{\partial \alpha_j} + \frac{1}{A_iA_j}\frac{\partial A_i}{\partial \alpha_j}(L_{ji} + L_{ij})$$
$$= 2Jh\frac{\partial^2 \Omega_i}{\partial t^2} + (m_i^+ + m_i^-),$$
$$\frac{L_{11}}{R_1} + \frac{L_{22}}{R_2} + \frac{1}{A_1A_2}\left[\frac{\partial(A_2L_{13})}{\partial \alpha_1} + \frac{\partial(A_1L_{23})}{\partial \alpha_2}\right] \tag{34.31}$$
$$= 2Jh\frac{\partial^2 \Omega_3}{\partial t^2} - (m_3^+ + m_3^-).$$

The elasticity relations

$$
\begin{aligned}
L_{ii} &= 2h\left[\frac{4\gamma(\beta+\gamma)}{\beta+2\gamma}\chi_{ii} + \frac{2\gamma\beta}{\beta+2\gamma}\chi_{jj}\right] + h\frac{\beta}{\beta+2\gamma}(m_3^+ - m_3^-),\\
L_{ij} &= 2h\left[(\gamma+\varepsilon)\chi_{ij} + (\gamma-\varepsilon)\chi_{ji}\right],\\
L_{i3} &= -2h\frac{4\gamma\varepsilon}{\gamma+\varepsilon}\chi_{i3} + \frac{\gamma-\varepsilon}{\gamma+\varepsilon}L_{3i}, L_{3i} = h(m_i^+ - m_i^-).
\end{aligned}
\tag{34.32}
$$

The geometrical relations

$$
\begin{aligned}
\chi_{ii} &= \frac{1}{A_i}\frac{\partial\Omega_i}{\partial\alpha_i} + \frac{1}{A_iA_j}\frac{\partial A_i}{\partial\alpha_j}\Omega_j - \frac{\Omega_3}{R_i},\\
\chi_{ij} &= \frac{1}{A_i}\frac{\partial\Omega_j}{\partial\alpha_i} - \frac{1}{A_iA_j}\frac{\partial A_i}{\partial\alpha_j}\Omega_i, \qquad \chi_{i3} = -\frac{1}{A_i}\frac{\partial\Omega_3}{\partial\alpha_i} - \frac{\Omega_i}{R_i}.
\end{aligned}
\tag{34.33}
$$

The equations of the "pure force" part of the magnetoelasticity problem of micropolar thin shells are:

$$
\begin{aligned}
&\frac{1}{A_i}\frac{\partial T_{ii}}{\partial\alpha_i} + \frac{1}{A_iA_j}\frac{\partial A_j}{\partial\alpha_i}(T_{ii}-T_{jj}) + \frac{1}{A_j}\frac{\partial S_{ji}}{\partial\alpha_j} + \frac{1}{A_iA_j}\frac{\partial A_i}{\partial\alpha_j}(S_{ji}+S_{ij})\\
&\quad + (-1)^j\frac{1}{c}j_{j0}H_{03} = 2\rho h\frac{\partial^2 u_i}{\partial t^2} + (q_i^+ + q_i^-),\\
&\frac{T_{11}}{R_1} + \frac{T_{22}}{R_2} + \frac{1}{A_1A_2}\left[\frac{\partial(A_2N_{13})}{\partial\alpha_1} + \frac{\partial(A_1N_{23})}{\partial\alpha_2}\right] - \frac{1}{c}j_{10}H_{02} + \frac{1}{c}j_{20}H_{01}\\
&\quad = 2\rho h\frac{\partial^2 w}{\partial t^2} - (q_3^+ + q_3^-),\\
&N_{3i} = \frac{1}{A_i}\frac{\partial G_{ii}}{\partial\alpha_i} + \frac{1}{A_iA_j}\frac{\partial A_j}{\partial\alpha_i}(G_{ii}-G_{jj}) - \frac{1}{A_j}\frac{\partial H_{ji}}{\partial\alpha_j}\\
&\quad - \frac{1}{A_iA_j}\frac{\partial A_i}{\partial\alpha_j}(H_{ji}+H_{ij}) + \frac{2}{3}\rho h^3\frac{\partial^2\psi_i}{\partial t^2} + h(q_i^+ - q_i^-),
\end{aligned}
\tag{34.34}
$$

$$
\begin{aligned}
T_{ii} &= \frac{2Eh}{1-v^2}\left[\Gamma_{ii} + v\Gamma_{jj}\right] - \frac{v}{1-v}h(q_3^+ - q_3^-),\\
S_{ij} &= 2\mu h(\Gamma_{12}+\Gamma_{21}) - (-1)^j 4\alpha h\Omega_3,\\
N_{i3} &= N_{3i} - 2h\cdot 4\alpha\Gamma_{i3},\\
G_{ii} &= -\frac{2Eh^3}{3(1-v^2)}\left[K_{ii} + vK_{jj}\right],\\
H_{ij} &= \frac{2h^3}{3}\mu(K_{12}+K_{21}),
\end{aligned}
\tag{34.35}
$$

$$\begin{aligned}
\Gamma_{i3} &= -\frac{1}{A_i}\frac{\partial w}{\partial \alpha_i} - \frac{u_i}{R_i} + (-1)^j \Omega_j, \\
\Gamma_{ii} &= \frac{1}{A_i}\frac{\partial u_i}{\partial \alpha_i} + \frac{1}{A_i A_j}\frac{\partial A_i}{\partial \alpha_j} u_j - \frac{w}{R_i}, \\
\Gamma_{ij} &= \frac{1}{A_i}\frac{\partial u_j}{\partial \alpha_i} - \frac{1}{A_i A_j}\frac{\partial A_i}{\partial \alpha_j} u_i, \\
K_{ii} &= \frac{1}{A_i}\frac{\partial \beta_i}{\partial \alpha_i} + \frac{1}{A_i A_j}\frac{\partial A_i}{\partial \alpha_j}\beta_j, \\
K_{ij} &= \frac{1}{A_i}\frac{\partial \beta_j}{\partial \alpha_i} - \frac{1}{A_i A_j}\frac{\partial A_i}{\partial \alpha_j}\beta_i,
\end{aligned} \tag{34.36}$$

Equations (34.20) should be added to (34.31)–(34.36) for the electrical currents averaged along the shell's midplane.

The mechanical boundary conditions on Γ are respectively expressed as

$$\begin{aligned}
L_{1i}|_\Gamma &= \int_{-h}^{h} m_i^* \, d\alpha_3, \\
L_{13}|_\Gamma &= -\int_{-h}^{h} m_3^* \, d\alpha_3 - \frac{\gamma-\varepsilon}{\gamma+\varepsilon}\frac{1}{A_2}\frac{\partial}{\partial \alpha_2}\int_{-h}^{h} \alpha_3 m_2^* \, d\alpha_3
\end{aligned} \tag{34.37}$$

and

$$\begin{aligned}
&T_{11}|_\Gamma = \int_{-h}^{h} p_1^* \, d\alpha_3, \qquad S_{12}|_\Gamma = \int_{-h}^{h} p_2^* \, d\alpha_3, \\
&G_{11}|_\Gamma = -\int_{-h}^{h} \alpha_3 p_1^* \, d\alpha_3, \\
&\left(-N_{13} + \frac{1}{A_2}\frac{\partial H_{12}}{\partial \alpha_2}\right)\bigg|_\Gamma = \int_{-h}^{h} p_3^* \, d\alpha_3 + \frac{1}{A_2}\frac{\partial}{\partial \alpha_2}\int_{-h}^{h} \alpha_3 p_2^* \, d\alpha_3.
\end{aligned} \tag{34.38}$$

The same as above boundary conditions remain valid on Γ. It is necessary to mention that from the constructed models we can obtain the main equations, boundary and initial conditions for micropolar elastic electro-conductive plates with free rotation, with constraint rotation and "with small shift rigidity".

References

1. Bagdasaryan, G.E., Asanyan, D.D.: The main equations and relations of asymmetrical theory of magnetoelasticity of ferromagnetic body. In: Problems of Mechanics of Deformable Bodies, Devoted to the 80th Anniversary of S.A. Ambartsumyan, pp. 37–47. Publ. NSA of Armenia, Yerevan (2002). In Russian
2. Eringen, A.C.: Microcontinuum Field Theories. I. Foundation and Solids. Springer, New York (1999) 319 p.
3. Goldenveiser, A.L.: Theory of Elastic Thin Shells. Nauka, Moscow (1976) p. 512. In Russian

4. Kaliski, S.: Thermo-magneto-microelasticity. Bull. Acad. Pol. Sci. **XVI**(1), 7–13 (1968)
5. Kaliski, S., Nowacki, W.: Wave-type equation of thermo-magneto-microelasticity. Bull. Acad. Pol. Sci. **XVII**(4), 155–159 (1970)
6. Maugin, G.A.: Continuum Mechanics of Electromagnetic Solids. North-Holland, Amsterdam (1988)
7. Sargsyan, S.H.: General Two-Dimensional Theory of Magnetoelasticity of Thin Shells. Publ. NSA of Armenia, Yerevan (1992) 232 p. In Russian
8. Sargsyan, S.H.: Dynamic problem of thin plates on the basis of asymmetric theory of elasticity. In: Proc. of XXXIV Summer School "Advanced Problems in Mechanics". Repino, Saint-Petersburg, Russia, 25 June–1 July, 2006, pp. 447–458 (2006)
9. Sargsyan, S.H.: Applied one-dimensional theory of bars on the basis of asymmetrical theory of elasticity. Phys. Mezomech. **11**(5), 41–54 (2008). In Russian
10. Sargsyan, S.H.: Boundary-value problems of asymmetric theory of elasticity for plates. J. Appl. Math. Mech. **72**, 77–86 (2008)
11. Sargsyan, S.H.: General theory of elastic thin shells on the basis of asymmetrical theory of elasticity. Rep. NSA Armenia **108**(4), 309–319 (2008). In Russian

MIX
Papier aus verantwortungsvollen Quellen
Paper from responsible sources
FSC® C105338

If you have any concerns about our products,
you can contact us on
ProductSafety@springernature.com

In case Publisher is established outside the EU,
the EU authorized representative is:
Springer Nature Customer Service Center GmbH
Europaplatz 3, 69115 Heidelberg, Germany

Printed by Libri Plureos GmbH
in Hamburg, Germany